Invasive Species in a
Globalized World

Invasive Species in a Globalized World

Ecological, Social, and Legal Perspectives on Policy

EDITED BY REUBEN P. KELLER, MARC W. CADOTTE, AND GLENN SANDIFORD

THE UNIVERSITY OF CHICAGO PRESS   CHICAGO AND LONDON

REUBEN P. KELLER is assistant professor of environmental science at Loyola University Chicago. MARC W. CADOTTE is the TD Professor of Urban Forest Conservation and Biology at the University of Toronto Scarborough. GLENN SANDIFORD is an adjunct instructor in the Department of Urban and Regional Planning at the University of Illinois in Urbana-Champaign.

The University of Chicago Press, Chicago 60637
The University of Chicago Press, Ltd., London
© 2015 by The University of Chicago
All rights reserved. Published 2015.
Printed in the United States of America

24 23 22 21 20 19 18 17 16 15   1 2 3 4 5

ISBN-13: 978-0-226-16604-9 (cloth)
ISBN-13: 978-0-226-16618-6 (paper)
ISBN-13: 978-0-226-16621-6 (e-book)
DOI: 10.7208/chicago/9780226166216.001.0001

Library of Congress Cataloging-in-Publication Data

Invasive species in a globalized world : ecological, social, and legal perspectives on policy / edited by Reuben P. Keller, Marc W. Cadotte, and Glenn Sandiford.
pages cm
Includes bibliographical references and index.
ISBN 978-0-226-16604-9 (cloth : alk. paper)—ISBN 978-0-226-16618-6 (pbk. : alk. paper)—ISBN 978-0-226-16621-6 (e-book) 1. Biological invasions. 2. Introduced organisms. I. Keller, Reuben P. II. Cadotte, Marc William, 1975– III. Sandiford, Glenn.
QH353.I593 2015
577'.18—dc23

2014001566

♾ This paper meets the requirements of ANSI/NISO Z39.48-1992 (Permanence of Paper).

Contents

A gallery of photographs follows page 184.

CHAPTER ONE

Working across Disciplines to Understand and Manage Invasive Species

Reuben P. Keller, Marc Cadotte, and Glenn Sandiford

Snakeheads in Maryland

In the summer of 2002, an Asian fish that the US Secretary of the Interior said was "like something from a bad horror movie" briefly became headline news all over America (US Fish and Wildlife Service 2002). The "fish from hell" (Fig. 1–1) generated hundreds of news stories, ranging from analysis on PBS's "NewsHour with Jim Lehrer" to satire by Jon Stewart on Comedy Central. David Letterman featured it on one of his Top Ten lists (Dolin 2002). The snakehead invasion, and the way that it was reported by journalists, so captured the public imagination that it has been used as the inspiration for no fewer than three horror movies.

The catalyst for this media frenzy was an 18-inch fish caught by an angler in a suburban pond twenty miles outside Washington, DC. The strange-looking fish had a mouthful of sharp teeth, and dorsal and anal fins running along most of its elongated body. It was soon identified as a northern snakehead (*Channa argus*). Native to rivers in China, the northern snakehead is a lie-in-wait predator that can surpass three feet in length. It is one of 28 snakehead species, many of them prized in their native ranges in Asia for sport and food (Fuller et al. 2013).

Several northern snakeheads had recently been caught in US waters. Officials presumed them to be aquarial specimens released illegally into the wild, as proved to be the case in Maryland. Northern snakeheads were

FIG. 1-1. Front page of the New York *Daily News*, July 13, 2002. Used with permission. See also color plate.

often sold live at ethnic markets. A few US experts who knew the species was a top-level predator without natural enemies worried that if it became established in open waters it could decimate native fish populations. Snakeheads tolerate a wide range of water temperatures, and can survive under ice. Adapted to the seasonal drying of shallow waters in their native habitat, snakeheads have a primitive lung that helps them survive for up to four days out of water, and longer when burrowed in mud (Snakehead Scientific Advisory Panel 2002).

However, most US state and federal biologists in 2002 were unfamiliar with northern snakehead, to the point that few could even identify the fish. This ignorance permeated early news stories about the discovery in Maryland, most notably in statements that northern snakeheads could travel overland to new bodies of water by wriggling across the ground (Thomson 2002). Some snakehead species are capable of this behavior, but not the northern snakehead. Nonetheless, the specter of a "Frankenfish" invasion in Maryland sparked coast-to-coast media coverage of the situation there. State and federal agencies sent "SWAT teams of biologists" with traps and electroshockers to the pond, which lay within one hundred yards of a major river. Their discovery of dozens of juvenile northern snakeheads transformed the fish into an international media superstar (Dolin 2002).

Coverage of the story ranged from serious to silly and brought unprecedented attention to the issue of invasive species. The episode highlighted the alarming ease with which potentially problematic plants and animals from abroad can enter US ecosystems, and the paucity of scientific information about their traits and potential impacts. Rhetoric about snakeheads featured alarmism, exaggeration, and vilification, and illustrated the difficulty of assessing and communicating risk. An eradication program, recommended by a panel of scientific experts, was stalled for weeks as the need for immediate action to protect a regional river fishery butted against property rights of the pond's owners.

Finally, four months after the initial discovery, state biologists killed all fish life in the pond with the chemical rotenone. Among the carcasses were 1,200 northern snakeheads, including six adults. Their demise ended the summer of the snakehead—but it was merely one chapter in a still unfolding story (Fields 2005). Despite federal and state bans, northern snakeheads have now been documented in eleven US states and one Canadian province (Fuller et al. 2013). Most cases involve single specimens, but reproducing populations have been discovered in California, Florida, and Maryland, where an explosive invasion of the Potomac River prompted an annual snakehead fishing tournament to help reduce the unwanted fish's abundance (Thomson 2011; Fears 2012). A population is also established in the White River drainage of Arkansas, from which it may spread into the Mississippi River. Elsewhere in the world, the northern snakehead is becoming invasive in Europe, as well as in Asian ecosystems outside its natural range.

Studying and Managing the Processes of Invasion

Northern snakehead is just one example from the many species that have been transported across the globe, have become established beyond their native range, and have caused negative impacts. As human populations have spread, and as societies have become more connected, the globalization of trade and travel has moved many thousands of species across the natural barriers (e.g., mountain ranges, oceans, deserts) that kept communities of native species ecologically separated for millions of years. Some of these species, such as the northern snakehead, find favorable conditions in the new regions, become established, and thrive. These species can have positive and/or negative outcomes, including impacts to the environment, economies, and human health. Preventing the arrival of new harmful nonnative species, and effectively managing those that are already established, has become a major policy goal at international, regional, national, and local levels.

The processes that move species, the factors that control species establishment and spread, and the ramifications for human societies, provide rich fodder for academic study. Ecologists have been the most active in this respect. Over recent decades they have built up a large literature about the mechanisms that move species and make them likely to become established (or not), the impacts of those species that do establish, and how impacts can be reduced. Although this ecological research has aided and in some cases stimulated policy, data show that limited success has been achieved at slowing the spread of invasive species. The number of known nonnative species in many ecosystems, for example, continues to grow at an exponential rate (e.g., Keller et al. 2009; Ricciardi 2006). Policy recommendations from the ecological community have, for the most part, been ignored by politicians.

A major reason for the failure of ecological results alone to prompt policy changes is that invasive species present inherently complex problems that are not confined to a single discipline. The two northern snakeheads that were introduced to Crofton Pond arrived through the trade in live food organisms and were kept in a private aquarium before being released. After they were discovered there were ethical issues to be considered (i.e., is it acceptable to kill all fish in the pond with a general fish poison?), along with property-rights issues, to determine whether and how authorities were allowed to manage the pond. Managing introductions of this and other species, then, could require preventing trade in a valuable commodity, regulating the species that can be kept as pets, determining what are acceptable side effects of eradication efforts, and many legal issues. In comparison, the problem of determining the ecological threat posed by northern snakehead can look quite simple!

Luckily, researchers in other disciplines have also become interested in nonnative species. Historians study the motivations for human transport of species, and how new species in turn affect human societies (e.g., Coates 2007). Legal scholars examine how regulations to control the spread of species can be enacted, and how and why they succeed or fail (e.g., Miller and Fabian 2004). Economists look at how trade and other human activities lead to species movement, and the economic consequences of those species (e.g., Perrings et al. 2001). Examples of other disciplines could be given, but the point here is that many people from many disciplines are trying to understand nonnative species.

Benefits from Greater Understanding among Disciplines

Although experts from many disciplines are now studying invasive species, interaction across these disciplines has so far been minimal. For example, many ecologists are interested in how new species arrive but are unfamiliar with the work of economists and legal scholars that describe how patterns in trade, and its regulation, affect the transport of cargo that may contain invasive species. The lack of interaction may reflect the relative infancy of nonnative species as a topic of study; the field of "invasion biology" is generally considered to have been established during the 1950s and it wasn't until decades later that it began to grow. Other disciplines have come to invasive species more recently. Additionally, intellectual cross-pollination has certainly been hindered by different paradigms and languages specific to each discipline. Regardless of the cause, the lack of interaction has hampered a deeper understanding of invasion processes and how they can be managed. Full understanding of the issues will require a synthetic approach, with contributions from a range of viewpoints and disciplines.

The necessity for interactions among disciplines is illustrated by the continuing spread of northern snakehead and other fishes across the United

States. In an ideal world, policy for these invasions would begin with a clear picture of the ecological risks posed by each species, and whether there are methods available to reduce population sizes and spread rates. Lawyers and legal scholars would contribute an understanding of how currently available legislation should be applied, and whether the creation of new legislation is warranted. Economists could calculate the effects of different management approaches on trade, and on local economies. The work of social scientists would also be necessary to determine society's level of desire to control the species, and society's tolerance for different management approaches, which may involve harmful chemicals. With all of this information, policy makers would be in a strong position to make wise decisions.

Although there will never be enough nonnative species experts to manage all species in the integrated way just described, continuing work from the disciplines described, and especially interaction among disciplines, has shown great promise for the creation of better management approaches. As the number of studies grows, it is inevitable that general rules will emerge to guide future policy. Interactions among economists and ecologists, for example, have begun to show when it is and is not economically rational to control a nonnative species, and how this depends on the basic biology of the species (e.g., Leung et al. 2002; Leung et al. 2005; Keller and Drake 2009; Springborn et al. 2011). Likewise, legal scholars have worked with biologists and economists to recommend new legislative approaches and institutions for managing nonnative species (Shine et al. 2010). Although these interactions are positive, they are too rare to have produced the integrated knowledge required to effectively respond to most nonnative species. The following two examples illustrate the complexities involved in managing nonnative species transport and impacts, and the difficulties in developing and implementing policy. In particular, these examples demonstrate the range of disciplines required to address invasive species threats, and the pressures put upon managers and policy makers who are responsible for mitigating those threats.

Ballast Water Regulations: A Long Time Coming

Most cargo moved around the world travels on ships, and the world merchant shipping fleet now includes more than 50,000 vessels, with these being registered in over 150 nations (International Chamber of Shipping

2012). Movement of these ships through the global shipping network now connects every port on earth by a small number of voyages (Keller et al. 2011). These close linkages are ecologically important because the safe operation of ships requires them to regularly take on and discharge ballast water. This water can contain organisms representing many species, and the movement of ballast water across the global shipping network means that all ports are effectively ecologically connected. Not surprisingly, ballast water has been the source of many of the world's most damaging marine and freshwater invasions (Ricciardi 2006; Molnar et al. 2008).

The movement of nonnative aquatic species in ballast water has long been acknowledged, as has the damage from those species that become established beyond their native range (Carlton and Geller 1993). In 2004 the international community, through the International Maritime Organization (IMO), agreed on ballast water discharge standards that are designed to reduce invasions. These have not yet come into force, however, because not enough countries have ratified them. At the national level there are few policies in place, although the United States is moving toward adopting ballast water standards that are likely to mirror those in the IMO regulations. Because of this lack of implemented international and national policy the problems of ballast water invasions remain unsolved and invasions continue.

The lack of regulations that address this well-known vector of invasive species illustrates several general problems for the creation of invasive species policy. First, effective regulations need to be based on scientific evidence that they will reduce the harm from invasive species to an acceptable level. This is difficult to calculate in any system, and in the case of ballast it requires reducing the number of organisms transported in ballast tanks to a level that gives an acceptably low risk of new invasions. Although this sounds simple, the relationship between the number of individual organisms released by ships and the likelihood of new invasions is unknown (Gollasch et al. 2007). In the case of ballast water, and for many other vectors that move nonnative species, it is difficult for scientists to make clear recommendations for what appropriate standards would be.

Second, treating ballast water will likely require most large ships to install expensive on-board water treatment plants. Ship owners are reluctant to invest in this technology, especially when there is doubt about what standards will be implemented and whether current technologies can achieve those standards. Reducing the risks from other vectors can also require potentially expensive changes to trade and other practices.

Balancing these against the expected benefits from fewer invaders is rarely straightforward, especially because the costs often fall on specific industries while the benefits of fewer invaders are more widely dispersed.

Third, it is almost always impossible to accurately determine the ship(s) that delivered a new invasive species. For ballast water releases, invasions are detected some unknown amount of time after the individuals were released. This is in contrast to, for example, most chemical pollution, where blame can often be assigned and compensation sought. For ships, there is no way to assign blame to specific operators, and members of the shipping industry have subsequently never been required to pay for the damage caused by the invasive species that they have spread (Perrings et al. 2005). Indeed, penalties for introducing an invasive species have rarely, if ever, been required or enforced. Thus, as long as the benefits of trade in nonnative species, or the costs of ensuring that invasive species are not transported, remain high, there will be little incentive for industries to change practices.

Managing Purple Loosestrife in North America

The need to better comprehend how science, economics and public perception influence invasive species policy is illustrated in the management history of purple loosestrife (*Lythrum salicaria*). By the mid-1980s, this species was viewed as the invader persona non grata among policy makers, researchers, and managers across much of the United States. Owing to its ability to rapidly spread and to dominate wetlands, purple loosestrife had become the poster child (featured on numerous "wanted" posters) for what can happen when a nonindigenous species is introduced into a new habitat. Wetlands across Eastern North America were transformed into apparent monocultures of purple loosestrife (Fig. 1–2).

In the early 1990s insect biocontrol agents were introduced (Blossey et al. 2001), even though scientific research had yet to fully understand the reasons for the plant's invasion success and how it was impacting ecosystems. The motivation for these introductions was likely increased by the cacophony of news stories using wildly exaggerated language to describe threats associated with purple loosestrife (Lavoie 2010). Millions of individuals from four biocontrol insect species were released, including a flower-feeding weevil (*Nanophytes marmoratus*); a root-mining weevil (*Hylobius transversovittatus*); and two closely related leaf beetles (*Ga-*

FIG. 1–2. Purple loosestrife (*Lythrum salicaria*). GNU free
distribution license.

lerucella calmariensis and *G. pusilla*) (Malecki et al. 1993; Blossey et al.
2001).

The insect releases were quickly criticized. Skeptics argued that the
ecosystem impacts of purple loosestrife had not been fully established,
preventing a full analysis of the benefits of biocontrol against the program
costs plus the potential risk of the biocontrol insects attacking native plants
(Anderson 1995; Hager and McCoy 1998). However, even as the biocontrol insects were still being released and establishing successful populations across purple loosestrife's invaded range, several studies identified
measurable economic and ecological impacts of the purple loosestrife invasion (e.g., Blossey et al. 2001; Brown et al. 2002; Ehrenfeld 2003).

Although the success of biocontrol has varied across different regions
depending on management practices and ecological interactions (McEvoy

et al. 2012), the general consensus is that releasing the biocontrol insects was a good decision. Nonetheless, it is important to note that the consequential and expensive decision to release nonnative insects was based not just on scientific information, but also on news stories, anecdotal evidence from wetland managers, agency reports, and so forth. When the biocontrol was initiated, many unanswered questions remained, which brings us back to our original point. Should policy makers wait for scientific consensus, or are the risks and costs associated with waiting for further information too great? To what extent should public opinion and politically motivated priorities influence policy decisions, if at all? These are difficult questions that include context-dependent factors. Still, the North American experience with purple loosestrife reveals how different interests can come together to initiate meaningful management.

The Ecological Invasion Process

In order to explore how researchers from different areas of inquiry can come together to affect policy, we need to have a common understanding of how a species actually becomes invasive. This occurs through a number of transitions that are collectively referred to as the *invasion process* (Fig. 1–3). Understanding the invasion process is important because it sets out the ecological steps that a species must pass through to become established in its nonnative range. For policy makers, each transition offers different possibilities and options for reducing impacts.

The first step in the invasion process occurs when a species is transported through human agency to an area beyond its native range. The mechanisms that move species are often referred to as *vectors*, and globalization has created many of these. One group of vectors is those that intentionally move species. This includes the pet and ornamental plant trade, the trade in live bait and food, and the movement of species for farming and biological control. For some vectors, such as the trade in plants for outdoor ornamental gardens, the species selected for transportation are those most likely to survive in the new region; this was the case for the introduction of purple loosestrife to North America. Other vectors, such as the trade in aquarium fish or live food, do not consider the environment of the receiving region because the species are kept in artificial environments. As the northern snakehead example illustrates, however, these species can still pose a large risk.

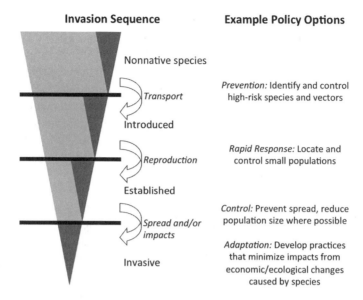

Invasion Sequence   **Example Policy Options**

Nonnative species

Transport

Prevention: Identify and control
high-risk species and vectors

Introduced

Reproduction

Rapid Response: Locate and
control small populations

Established

Spread and/or
impacts

Control: Prevent spread, reduce
population size where possible

Invasive

Adaptation: Develop practices
that minimize impacts from
economic/ecological changes
caused by species

FIG. 1–3. The invasion sequence, illustrating the ecological transitions that a nonnative species can pass through (left), and some of the options available to managers and policy makers at each stage (right). In the left panel, the pool of invasive species is represented by the dark gray wedges. The invasive species pass through each transition, while nonnative species are filtered out.

The other group of vectors is those that move species unintentionally. These are usually associated with human modes of transport that can accidentally entrain species. One such mode, the movement of nonnative species in the ballast water of ships, was described above. Others include the movement of seeds in dirt attached to car wheels, insects and other species that become entrained in cargo containers, and species that arrive as contaminants of intentionally transported live species (e.g., snails attached to aquarium plants).

Species that survive passage in a vector and are released beyond their native range are referred to as *introduced*. If enough individuals are released, and if the environment into which they are released is suitable, such a species may reproduce and form a self-sustaining population. Species doing this are considered to be *established*.

Newly established species have small population sizes and cover small geographic ranges. If the species shows population growth and the ability to spread geographically, it may be considered to be *ecologically invasive.* Ecological definitions of invasive can be as precise as that offered for non-native plants by Richardson et al. (2000), where an invasive species is one that is nonnative, reproducing, and spreading at more than approximately 100m/50 years if by seed, and more than approximately 6m/3 years if by vegetative means. Definitions such as this serve an important purpose in science because their precision can decrease the ambiguity of terms applied. We note, however, that definitions this strict will be difficult to assess for many species because of the requirements to repeatedly measure range size. Additionally, this definition can't be usefully applied to other types of organisms, such as animals.

There has been significant debate within the ecological literature about the appropriate definition of invasive (e.g., Daehler 2001, Davis and Thompson 2000, Blackburn et al. 2011). Some ecologists have even recommended that the "invasion process" be defined without any stage labeled as *invasive* because the term has become so laden with differing ideas (e.g., Collauti and MacIsaac 2004). In general, however, the term *invasive* is reserved by ecologists for nonnative species that have become established and spread widely. That said, whether and how much ecological impact an invasive species should have, how rapidly it should spread, and how dense its population should be, differ among different ecological definitions.

A Definition of *Invasive* for Policy

Discovering that a species meets an ecological definition of invasive does not automatically mean that it should be subject to prevention or control efforts. Many species that rapidly spread through ecosystems are not considered by society to pose sufficient threat that resources should be expended to control them, and many other nonnative species are considered beneficial. For example, nonnative trout and salmon species have been introduced widely across the globe to create new populations for recreational fishing. In many places these populations have become established and spread, and they thus satisfy ecological definitions of *invasive.* Despite this, they were introduced intentionally and in many cases generate large benefits in terms of recreational angling and tourism income. For these reasons, their *ecological invasion* is not cause for action by policy makers.

This leads to an alternative definition of invasive that can be used to inform policy. Probably the most widely accepted such definition is that used by US President Bill Clinton in his 1999 Executive Order 13112, where he stated that an invasive species is "an alien species whose introduction does or is likely to cause economic or environmental harm or harm to human health." This definition is usually interpreted to specify that a species should be considered invasive if the total harm that it causes is greater than any benefits. This makes logical sense as a policy definition; those species that society considers to cause net harm should be the targets of management efforts, while resources should not be expended on nonnative species that are benign or beneficial.

To use this definition a policy maker needs to weigh the environmental, economic, social, and human health costs and benefits of any given species. This is no easy task, as it will often require comparing costs and benefits across different sectors. As an example, many nonnative plant species are introduced for the horticultural trade, and many of these species escape and form populations beyond cultivation. To use the policy definition of invasive, it is necessary to compare the economic benefits of the species in trade, and the benefits to gardeners who enjoy the species, to the costs of controlling the species in the wild, and any attendant impacts on native biodiversity or ecosystem function.

A Definitive Definition of *Invasive*?

It is unfortunate that policy makers and ecologists have adopted the same term—*invasive*—but apply it in different ways. This has created confusion in both ecological and policy circles, and hampers understanding between the two. The situation is made more complex as other disciplines have begun to study nonnative species, and have in some cases introduced modified definitions that meet their own paradigms. A unified definition has not been accepted across these disciplines, meaning that any writer about nonnative species needs to be careful to define his or her use of the term *invasive*, and any reader should be aware of the context in which the term is being used.

Policy for Invasive Species — Managing Harm

The role for invasive species policy and management is to minimize the harm caused by nonnative species. For some types of nonnative species threats, policy has been both successful and well coordinated. Probably the best examples of this are the global institutions created to manage outbreaks of new human diseases (Keller and Perrings 2011). These diseases emerge in one area and spread to others, and are of great concern to the international community. This has led to the creation of the World Health Organization (WHO), which is tasked with identifying outbreaks of new diseases and managing a response that will minimize both the geographical spread and number of people infected by the disease. The effectiveness of this system was tested during the severe acute respiratory syndrome (SARS) outbreak. The international community, led by the WHO, was able to contain the disease within a few months, and fewer than 10,000 people were infected (Mahmoud and Lemon 2004). Although SARS was an enormously costly epidemic, both in terms of human infections and economics, it had the potential to be far more devastating.

Other examples of relatively well-developed policy for invasive species are the efforts to prevent the arrival and spread of nonnative species that harm livestock and crops. The World Organisation for Animal Health (known by its French acronym, OIE) maintains a database of known livestock diseases and requires its 178 member nations to report any outbreaks that occur within their borders. This allows all member nations to conduct trade with other nations in ways that reduce the risk of disease spread. The OIE is widely considered to be an effective institution, and even managed the global eradication of rinderpest, previously a widespread and devastating disease of cattle (Joint FAO/OIE Committee on Global Rinderpest Eradication 2011).

Diseases of crops are also managed aggressively, but most commonly at the national or regional level (as opposed to the international level at which human and livestock diseases are managed). Methods to prevent the introduction and spread of crop diseases and pests are based around quarantine when organisms are imported, and around quarantining infected areas in efforts to prevent species spread.

Two commonalities among human, livestock, and crop diseases help explain why they have been relatively well managed. First, such diseases are widely recognized to pose enormous costs, and these costs are readily quantified—for example, in terms of human mortality and morbidity, or lost trade because agricultural productivity declines and/or other countries refuse to import infected agricultural products. Second, these species are not perceived to have any benefits. The policy consequences of this are two-fold—strong constituencies work to prevent spread of these disease species, and there is no incentive for any party to increase their spread.

The final large category of nonnative species, and the category that has been managed least effectively, is essentially all invaders that are not human, livestock, or crop diseases. This includes species that have enormous economic and environmental impacts, and that have large direct and indirect impacts on human health and welfare. Unlike the better-managed categories of nonnative species, however, these invaders can have benefits as well as costs, and their control may impose costs on industry and/or impede trade. Northern snakehead and purple loosestrife were both introduced through trade and have created economic benefits for importers and retailers. Their impacts, however, are spread widely among society, falling most heavily on the state and federal agencies tasked with controlling the species. This is an example of the "tragedy of the commons," where economic benefits are reaped by a small number of people, but the environmental and economic costs when those species become invasive are the responsibility of society at large.

The lack of regulation for ballast water, also discussed above, is somewhat different from the situation with northern snakehead and purple loosestrife. No benefits accrue from the nonnative species that become established via ballast water, but shipping is a competitive global industry in which companies work hard to keep their costs as low as possible. The installation and operation of ballast water treatment systems by a subset of the industry would put that subset at a financial disadvantage. Hence, action is likely to come only from a strong international mandate that will impose similar financial costs on all shipping companies.

Background to This Volume

This volume is about this final group of species—those that are not already addressed by policy. Successful policy for these species, and the vectors that transport them, could lead to major reductions in the introduction and spread of invasive species. To accomplish this, researchers and policy makers with diverse abilities and interests must come together to produce the multidisciplinary information and understanding that is sorely needed for sound policy, including cost-benefit appraisals, risk assessments, analyses of human motivations, histories of past introductions, and theories about invasion ecology.

With this in mind, the Program on the Global Environment at the University of Chicago brought together researchers and practitioners for a conference in May 2011 to consider how experts from multiple disciplines and perspectives can work together to address the issues presented by nonnative species. The goal was not to create a fully synthetic understanding of nonnative species; that will take thousands of researchers and several decades. Instead, we hoped more modestly to facilitate such synthesizing, by having researchers and practitioners present their work in a way that was accessible to conferees from other disciplines. The conference included historians, legal scholars, economists, ecologists, evolutionary biologists, a children's book author, attorneys specializing in invasive species law, an engineer, and many others.

The current volume has emerged from the conference, but it is not a volume of proceedings. Instead, each chapter was written after the conference, with the multidisciplinary policy focus of the conference in mind. All chapters were reviewed internally by several authors from disciplines different from that of the main author. We collectively believe that better policy for managing invasive species is urgently needed. Because synthetic understanding is necessary for such improvement, each chapter includes a section discussing how the work presented can be applied to policy.

This is not the first multidisciplinary effort to bring together invasive species experts. Economists and ecologists have been working together for several years and have produced a small, but rich, body of work (e.g., Perrings et al. 2001; Keller et al. 2009), and the recent *Encyclopedia of Invasive Species* (Simberloff and Rejmánek 2011) brings together experts from many fields. Additionally, several individual researchers, including a large subset of the authors for this volume, have effectively made themselves multidisciplinary experts in order to pursue their invasive species studies. Despite this, interactions among disciplines remain rare, and we believe that policy suffers for this lack of interaction. Our hope is that this volume will enhance understanding among disciplines and provide a basis and motivation for further integrative invasive species research.

This volume has a focus on policy and authors have generally used the policy definition of *invasive* (see above). Where this is not practical—for example, because it conflicts too severely with the definition used within the authors' discipline—authors have clearly defined their usage of the term. In this way, we trust that readers will be able to interpret the text without being hampered by definitions of *invasive*.

Structure of the Volume

This volume is divided into four sections, each with a short introduction. The first section includes chapters that cover both the science of invasive species and the ways that public perceptions can guide the human responses. Species covered are the Australian invasion of cane toads (Rick Shine, chapter 2), the invasion of gray squirrel in the United Kingdom (Peter Coates, chapter 3), and the ongoing invasions of Asian carp species in the United States (Glenn Sandiford, chapter 4). The final chapter in this section, written by a children's book author who has traveled extensively to educate school children about invasive species, discusses how the impacts of invasive species can best be communicated to this audience (Mark Newman, chapter 5).

Section 2 includes chapters that cover the introduction phase of the invasion sequence. Christina Romagosa (chapter 6) looks at importation records of animals into the United States. Marc Cadotte and Lanna Jin (chapter 7) look at how species evolutionary histories can be used to explain which introduced species become established (and which don't). Michael Springborn, an economist, then discusses how economic modeling can contribute to policy decisions about which species should be allowed for importation (chapter 8).

Section 3 covers management of those nonnative species that become invasive and cause harm. Jon Bossenbroek and colleagues (chapter 9) look at ways to slow the spread of the invasive Emerald Ash Borer in the United States. This is followed by a chapter from Jim Kitchell and colleagues (chapter 10) that discusses the ways that climate change will cause the invasion dynamics of sea lamprey in the Great Lakes to change, most likely causing that invasion to become more damaging. Robert Keller, a civil engineer, has worked to design structures that can exclude invasive carp species from rivers in Australia, and presents this work in chapter 11. Section 2 ends with a meta-analysis of the "enemy release hypothesis," one ecological theory that has been put forward to explain success and failure of species at the introduction stage (Kirsten Prior and Jessica Hellmann; chapter 12).

The final section of the book deals directly with policy for invasive species. First, Stas Burgiel (chapter 13) discusses the international policy environment for invasive species. Next, Clare Shine (chapter 14) addresses issues for policy across the European Union. Marc Miller (chapter 15) reviews the state of policy for invasive species in the United States, and offers suggestions for the way toward more effective legislation. The last chapter of this section is by lawyers Joel Brammeier and Thom Cmar (chapter 16), who look at policy efforts to address aquatic invasions in the Laurentian Great Lakes. Finally, we (the editors) present a concluding chapter to the volume that looks at the way forward for invasive species policy.

Literature Cited

Anderson, M. G. 1995. Interaction between *Lythrum salicaria* and native organisms: a critical review. *Environmental Management* 19:225–231.

Blackburn, T. M., P. Pyšek, S. Bacher, J. T. Carlton, R. P. Duncan, V. Jarošik, J. R. U. Wilson, and D. M. Richardson. 2011. A proposed unified framework for biological invasions. *Trends in Ecology and Evolution* 26:333–339.

Blossey, B., L. C. Skinner, and J. Taylor. 2001. Impact and management of purple loosestrife (*Lythrum salicaria*) in North America. *Biodiversity and Conservation* 10:1787–1807.

Brown, B. J., R. J. Mitchell, and S. A. Graham. 2002. Competition for pollination between an invasive species (purple loosestrife) and a native congener. *Ecology* 83:2328–2336.

Carlton, J. Y., and J. B. Geller. 1993. Ecological roulette: the global transport of nonindigenous marine organisms. *Science* 261:78–82.

Coates, P. 2007. *American Perceptions of Immigrant and Invasive Species: Strangers on the Land*. Berkeley: University of California Press.

Colautti, R. I., and H. J. MacIsaac. 2004. A neutral terminology to define 'invasive' species. *Diversity and Distributions* 10:135–141.

Daehler, C. C. 2001. Two ways to be an invader, but one is more suitable for ecology. *Bulletin of the Ecological Society of America* (January) 2001:101–102.

Davis, M. A., and K. Thompson. 2000. Eight ways to be a colonizer; two ways to be an invader: a proposed nomenclature scheme for invasion ecology. *Bulletin of the Ecological Society of America* (July) 2000:226–230.

Dolin, E. J. 2002. *Snakehead: A Fish out of Water*. Washington, DC: Smithsonian Books.

Ehrenfeld, J. G. 2003. Effects of exotic plant invasions on soil nutrient cycling processes. *Ecosystems* 6:503–523.

Fears, D. 2012. Setting out after snakeheads. *Washington Post*, June 4, B1.

Fields, H. 2005. Invasion of the snakeheads! *Smithsonian* (February):62–70.

Fuller, P. F., A. J. Benson, and M. E. Neilson. 2013. *Channa argus*. USGS Nonindigenous Aquatic Species Database, Gainesville, FL. Accessed December 29, 2013. http://nas.er.usgs.gov/queries/FactSheet.aspx?speciesID=2265.

Gollasch, S., M. David, M. Voigt, E. Dragsund, C. Hewitt, and Y. Fukuyo. 2007. Critical review of the IMO international convention on the management of ships' ballast water and sediments. *Harmful Algae* 6:585–600.

Hager, H. A., and K. D. McCoy. 1998. The implications of accepting untested hypotheses: a review of the effects of purple loosestrife (*Lythrum salicaria*) in North America. *Biodiversity and Conservation* 7:1069–1079.

International Chamber of Shipping. 2012. Accessed July 5, 2012. http://www .marisec.org/shippingfacts/keyfacts/.

Joint FAO/OIE Committee on Global Rinderpest Eradication. 2011. Final Report. Food and Agriculture Organisation of the United Nations (FAO; Rome, Italy); World Organisation for Animal Health (OIE; Paris, France). Accessed October 14, 2013. http://www.oie.int/fileadmin/Home/eng/Media_Center /docs/pdf/Final_Report_May2011.pdf.

Keller, R. P., and J. M. Drake. 2009. Trait-based risk assessment for invasive species. In *Bioeconomics of Invasive Species*, edited by R. P. Keller, D. M. Lodge, M. A. Lewis, and J. F. Shogren, 44–62. New York: Oxford University Press.

Keller, R. P., J. M. Drake, M. Drew, and D. M. Lodge. 2011. Linking environmental conditions and ship movements to estimate invasive species transport across the global shipping network. *Diversity and Distributions* 17:93–102.

Keller, R. P., P. E. S. zu Ermgassen, and D. A. Aldridge. 2009. Vectors and timing of freshwater invasion in Great Britain. *Conservation Biology* 6:1526–1534.

Keller, R. P., and C. Perrings. 2011. International policy options for reducing the environmental impacts of invasive species. *BioScience* 61:1005–1012.

Lavoie, C. 2010. Should we care about purple loosestrife? The history of an invasive plant in North America. *Biological Invasions* 12:1967–1999.

Leung, B., D. M. Lodge, D. Finnoff, J. F. Shogren, M. A. Lewis, and G. Lamberti. 2002. An ounce of prevention or a pound of cure: bioeconomic risk analysis of invasive species. *Proceedings of the Royal Society B* 269:2407–2413.

Leung, B., D. Finnoff, J. F. Shogren, and D. Lodge. 2005. Managing invasive species: rules of thumb for rapid assessment. *Ecological Economics* 55:24–36.

Mahmoud, A. A. F., and A. M. Lemon. 2004. Summary and assessment. In *Learning from SARS: Preparing for the Next Disease Outbreak,* edited by S. Knobler, A. Mahmoud, S. Lemon, A. Mack, L. Sivitz, and K. Oberholtzer, 1–40. Washington, DC: National Academies Press.

Malecki, R. A., B. Blossey, S. D. Hight, D. Schroeder, L. T. Kok, and J. R. Coulson. 1993. Biological control of purple loosestrife. *Bioscience* 43:680–686.

McEvoy, P. B., F. S. Grevstad, and S. S. Schooler. 2012. Insect invasions: lessons from biological control of weeds. In *Insect Outbreaks Revisited*, edited by P. Barbosa, D. Letourneau, and A. Agrawal. Chichester, UK: Wiley.

Miller, M., and R. Fabian, eds. 2004. *Harmful Invasive Species: Legal Responses*. Washington, DC: Environmental Law Institute.

Molnar, J. L., R. L. Gamboa, C. Revenga, and M. D. Spalding. 2008. Assessing the global threat of invasive species to marine biodiversity. *Frontiers in Ecology and the Environment* 6:485–492.

Perrings, C., K. Dehnen-Schmutz, J. Touza, and M. Williamson. 2005. How to manage biological invasion under globalization. *Trends in Ecology and Evolution* 5:212–215.

Perrings, C., M. H. Williamson, and S. Dalmazzone. 2001. *The Economics of Biological Invasions*. Cheltenham, UK: Edward Elgar.

Ricciardi, A. 2006. Patterns of invasion in the Laurentian Great Lakes in relation to changes in vector activity. *Diversity and Distributions* 12:425–433.

Richardson, D. M., P. Pyšek, M. Rejmánek, M. G. Barbour, F. D. Panetta, and C. J. West. 2000. Naturalization and invasion of alien plants: concepts and definitions. *Diversity and Distributions* 6:93–107.

Shine, C., M. Kettunen, P. Genovesi, F. Essl, S. Gollasch, W. Rabitsch, R. Scalera, U. Starfinger, and P. ten Brink. 2010. Assessment to support continued development of the EU Strategy to combat invasive alien species. Final Report for the European Commission. Brussels: Institute for European Environmental Policy (IEEP).

Simberloff, D., and M. Rejmánek, eds. 2011. *Encyclopedia of Biological Invasions*. Berkeley: University of California Press.

Snakehead Scientific Advisory Panel. 2002. First Report to the Maryland Secretary of Natural Resources, July 26, 2002. Accessed August 23, 2012. http://www.dnr .state.md.us/irc/ssap_report.html.

Springborn, M., C. M. Romagosa, and R. P. Keller. 2011. The value of nonindigenous species risk assessment in international trade. *Ecological Economics* 70:2145–2153.

Thomson, C. 2002. It lurks in Crofton's waters. *Baltimore Sun*, June 22.

Thomson, C. 2011. Snakehead infiltration threatens Potomac. *Baltimore Sun*, January 22.

US Fish and Wildlife Service. 2002. Interior Secretary Norton proposes ban on importation of snakehead fish. Press release, July 23. Accessed August 23, 2012. http://www.fws.gov/news/newsreleases/showNews.cfm?newsId=09E74A8D -8E2E-49E9-B0F551156137A349.

Of Toads, Squirrels, Carps, and Kids: How Science *and* Human Perceptions Drive Our Responses to Invasive Species

Ecologists have been studying invasive species for several decades, and this work has built upon a longer history of basic and applied ecological research. This has produced great advances in our understanding of how species are introduced, which species survive and thrive, and how the impacts of invasive species are manifest on ecological and economic systems, and on human health. There are now multiple scientific journals and societies dedicated to the study of invasive species and research efforts continue to grow. In turn, managers have become much more concerned about the impacts of invasive species and spend large amounts of resources controlling and, where possible, eradicating invaders.

Despite the scientific advances, the public remains largely unaware of the problems caused by invasive species. Many people can name one or two high-profile invaders, but few can place those species in the context of the global reshuffling of species that has come about through international trade and travel. While the circumstances and impacts of individual invasions can be enlightening, a full understanding of the underlying processes is required to properly respond to the issues. Without such knowledge, people tend to over- or underreact, and the pressure on policy makers becomes dislocated from the actual issues presented.

This first group of chapters looks at public and policy responses to species invasions. It begins with an ecologist, Rick Shine (chapter 2), discussing his research on the cane toad invasion of Australia, how the perceived impacts have differed from the true harm of the species, and how well-meaning members of the public have taken action. Next, historian Peter Coates (chapter 3) investigates the invasion of the American gray squirrel in the United Kingdom, and discusses how notions of patriotism have colored the response to this species. Glenn Sandiford (chapter 4) examines the history of carp invasions in the United States, and shows that while initial intentions were good when these species were introduced, the appearance of negative impacts has prompted management responses that have little chance of success. Finally, Mark Newman (chapter 5), a journalist and author, describes his work educating school-age children about the risks from invasive species in the North American Great Lakes. Newman, by giving children a broader perspective on invasive species issues, seeks to address the issues raised by the previous chapters.

CHAPTER TWO

The Ecological, Evolutionary, and Social Impact of Invasive Cane Toads in Australia

Richard Shine

Introduction

Invasive species come in all shapes and sizes, belong to a diverse array of evolutionary lineages, and exert their impacts on native fauna and flora by a remarkably varied array of pathways. For example, many of the most damaging invasive organisms are microscopic in size. Pathogens such as the influenza virus decimated native people of many regions during the period of European colonial expansion, and the chytrid fungus has wiped out many species of frogs and toads around the world (Blaustein and Keisecker 2002). At the other extreme of body size, large herbivores such as deer and camels have created ecological havoc in their introduced ranges (Nugent et al. 2001). Clearly, some kinds of organisms are better suited to successful invasion than others, and an extensive literature has explored the features that predict invasion success (Blackburn et al. 2009; Kraus 2009). For example, vertebrate invaders tend to be mobile, and have high reproductive rates (Cassey 2002) and relatively large brains (which may confer the behavioral plasticity needed to meet the challenges of a novel environment: Sol et al. 2008, Amiel et al. 2011).

One group of organisms that at first sight seem poorly suited to become invasive are anuran amphibians (frogs and toads). These small, squat-bodied animals have relatively low mobility, and rely upon frequent access to water to avoid desiccating (Boutilier et al. 1992, Lever 2001). Nonetheless, several amphibian species have proven to be highly successful invaders. The first alleged case of a toad invasion is probably the Old Testament's famed "plague of frogs," the second of ten plagues visited upon the Egyptians by a wrathful God (Exodus 8:1–4). Understandably, empirical data on the anurans involved in that specific invasion, and the details of their ecological impact, are lost in the mists of time. However, we have much more information about some recent anuran invasions, one of which is the focus of the current chapter.

Three species of invasive frogs and toads stand out as major problems, in different parts of the world and for different reasons. The small coqui frog (*Eleutherodactylus coqui*; adults are less than 60 mm in length) does not require water bodies for breeding, and thus has been able to spread through suburban areas in Florida and Hawaii, after reaching these areas as stowaways (Kraus and Campbell 2002). Its high abundance can create significant ecological impacts (Beard et al. 2002). Despite its miniscule size, the coqui frog has had a major social impact because males produce a very loud repetitive call to attract mates—a call so loud and irritating to many people that house prices fall as soon as the coqui invasion front spreads through a new neighborhood (Beard et al. 2009).

The other two troublesome anuran invaders lie at the other end of the body-size spectrum: both are among the world's largest frogs. The North American bullfrog (*Lithobates* [formerly *Rana*] *catesbeiana*) can grow to more than 200 mm in length and has been intentionally spread to many areas outside its native range in eastern North America (notably into the western United States and Canada, and into Asia: Ficetola et al. 2007) where it is a favored food item because of its large body size. Bullfrogs are formidable predators. Many authorities have expressed concern that this invader may affect native frog species (by eating them or competing with them), as well as consuming a wide range of other native taxa (Rosen and Schwalbe 1995; Govindarajulu et al. 2006). Recent research in western Canada, however, suggests that at least some of the concern about ecological impacts of bullfrogs may be misplaced: rather than eliminate native anurans from ponds, they tend to thrive in highly disturbed areas where the native anuran fauna has already been depleted by habitat degradation (Govindarajulu 2004).

The final member of this unloved triumvirate—and the subject of the present chapter—is the cane toad (*Rhinella* [formerly *Bufo*] *marina*: Pramuk 2006), another giant species (to 400 mm long, although most adults are less than half this size; see Fig. 2–1). Native to a large area in

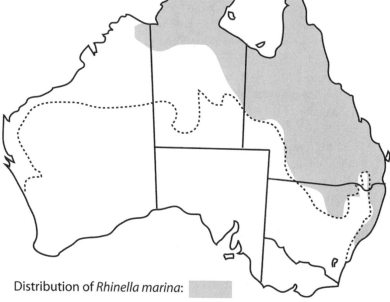

Distribution of *Rhinella marina*:

Potential Habitat:

FIG. 2–1. Upper: the cane toad, *Rhinella marina* (photograph by M. Greenlees). Lower: The current and predicted geographic distribution of invasive cane toads in Australia. Modified from Kearney et al. (2008), with permission.

South and Central America, the cane toad has been introduced intentionally to more than 40 countries worldwide in an attempt to control insect pests of commercial agriculture (Lever 2001; Kraus 2009). The giant toad's alleged success in controlling cane beetles in sugar-cane plantations led to an enthusiastic program to spread the toad to several other cane-growing areas (Lever 2001). The most catastrophic translocation of cane toads occurred in 1935, when agricultural scientists brought 101 toads from Hawaii to northeastern Australia. The toads were bred in captivity, and their offspring (as well as some of the original adults) were then released in sugar-cane plantations along the coast of tropical Queensland.

The toads appear not to have had much effect on the beetles they were meant to control, possibly because the insects spend little time on the ground. However, the toads flourished by eating many other kinds of fauna—mostly small ants and beetles, but occasionally larger items like mice, snakes, and nestling birds (Lever 2001). Within a few years, cane toads were abundant throughout the cane-growing region, and spreading south and west. The southern invasion took them into coastal New South Wales, mostly by direct dispersal but with frequent assistance (mostly inadvertent) in vehicles. These stowaway toads, often hidden in materials for building and gardening, have set up isolated extralimital populations in places well outside the main range of the toad in Australia. For example, in 2010 a thriving population was discovered in Sydney, more than 600 km (370 miles) south of the toad's "southern limit" (Fig. 2–1). Although climate models suggest that the toads are unlikely to be able to spread much farther south (Kearney et al. 2008), so far the toads do not appear to have taken much notice of such predictions.

The toads' westward invasion took them into severely arid areas of western Queensland, a region which many people hoped would limit their expansion. However, the toads continued spreading, rapidly moving through the Northern Territory all the way to the Western Australian border (Urban et al. 2008). Toads now occupy more than a million square kilometers, often in high densities. As a result, the impact of cane toads in Australia has received extensive scientific attention. At the same time, the cane toad has achieved iconic status in Australian society, as the invader that people love to hate. In this chapter I will review our understanding of the impact of invasive cane toads from my own group's work and other research, as well as from more subjective observations.

The Ecological Impact of Cane Toads

Unlike most of the other areas to which *Rhinella marina* has been introduced, Australia lacks any native toads (anurans of the family Bufonidae).
Thus, the Australian fauna was confronted with an invader very different
from any that they had previously encountered. In particular, bufonids
produce distinctive and highly toxic chemicals as a defense against predators. These cardio-active steroids (bufadienolides) can rapidly kill animals
that have not evolved to deal with them (Hayes et al. 2009). Extensive
research suggests that lethal toxic ingestion of toads is the most important
mechanism by which cane toads have affected native species. Toads influence native species in other ways as well, by eating them, competing with
them for food and shelter, and spreading pathogens (Pizzatto and Shine
2009; Pizzatto et al. 2010). Nonetheless, poisoning of predators stands out
as the most important direct effect of toad invasion, and is probably also
the most important reason for indirect effects of toad arrival (i.e., secondary consequences of toad invasion, mediated via effects on some other
species: Shine 2010).

Although popular opinion in Australia holds that the invasion of cane
toads is devastating for most or all native species, the reality has proven
to be less alarming. Cane toad invasion is catastrophic for a small suite of
native predators, but neutral (or indeed, beneficial) for other species. The
taxa that are dramatically impacted, with population crashes of up to 95%
within a few months of toad arrival, include a phylogenetically diverse array of predators with one thing in common: they are large (reviewed by
Shine 2010). Thus, victims of the toad invasion include some of the largest lizards, such as "goannas" (e.g., the yellow-spotted monitor, *Varanus
panoptes*) and the iconic bluetongue skink (*Tiliqua scincoides*), some large
venomous snakes (such as the king brown, *Pseudechis australis*), the cat-
sized marsupial quoll (*Dasyurus hallucatus*), and even some (but not all)
populations of the freshwater crocodile (*Crocodylus johnstoni*). Measuring the exact magnitude of population crashes in such large, mobile, and
often rare predators is not an easy task, but survey data unequivocally
show that such impacts can be substantial. Puzzlingly, though, the impacts
are variable through time and space. For example, some populations of
freshwater crocodiles are unaffected by toad invasion, whereas conspecifics in adjacent river systems die in droves (Letnic et al. 2008). In some
cases, indirect effects of toad invasion may outweigh direct impacts, such

that the toads actually benefit some species. For example, many supposedly toad-vulnerable snake species increase rather than decrease in numbers after toads arrive, probably because toads fatally poison many of the
large varanid lizards that prey upon these snakes (Brown et al. 2011). Similarly, reduced lizard predation on eggs means that toad arrival provides
a strong benefit to survival rates of this life-history phase in freshwater
turtles (Doody et al. 2009). For any species eaten by varanid lizards, the
arrival of cane toads may be good news, not bad news.

Many other predator taxa are potentially at risk of fatal toxic ingestion
when toads arrive, but the population-level effects on these species are
minor. A few individuals die, but the survivors soon learn that toads are
unpalatable (Shine 2010). One of the biggest surprises from our research
program is that native animals like fishes, frogs, and small marsupials are
much smarter than we expected; they are capable of rapid aversion learning (Webb et al. 2008; Greenlees et al. 2010). So long as the first toad
consumed by a predator is not big enough to kill it, that predator rapidly
learns to leave toads alone. On the basis of this observation, we developed
a novel way to conserve vulnerable predators in the face of toad invasion:
we taught them about the dangers of toads. Captive-raised northern quolls
were given a small dead toad that contained a nausea-inducing chemical. The trained quolls thereafter refused to eat toads, and so they were
able to survive after we released them into the wild, even in areas where
toads are common (O'Donnell et al. 2010). Encouragingly, many of our
trained quolls have survived long-term after release, producing offspring
of their own—and hopefully the mothers will be able to pass on this critical knowledge to their offspring.

Another encouraging observation (although less well documented) is
the apparent recovery of predator populations a few decades after toad invasion. Some of the species that crash dramatically when toads first invade
(such as yellow-spotted monitors) eventually rebuild their numbers, and
are common again in areas of Queensland where toads have been present
for many years (Shine 2010). One reason may be that juvenile predators
in such areas have a chance to learn toad-aversion by encountering small
(juvenile) cane toads, unlike the situation at the invasion front where all
of the toads are large adults. Another possibility is that the fauna adapts,
by rapid evolutionary change (see below).

The Evolutionary Impact of Cane Toads

Traditionally, evolution has been regarded as a slow process—too slow to be relevant to ecological and conservation issues. Recent work has challenged this assumption, and shown that many biological systems can respond to intense selection so rapidly that we need to incorporate that adaptive flexibility into management planning (Hendry et al. 2011). The cane toad invasion of Australia provides compelling examples of rapid evolutionary change, both in the toads themselves (in response to the process of invasion, as well as to the novel challenges imposed by Australian conditions) and in the native species impacted by toad arrival (reviewed by Shine 2011).

A biological invasion generates spatially non-equilibrial conditions that can create evolutionary processes not seen in stable populations. For example, mathematical models show that mutations can "surf" an expanding range edge, thus increasing phenotypic diversity (Travis et al. 2010). A long-sustained spread into virgin territory (as has occurred in the east-to-west invasion of cane toads through tropical Australia) introduces another evolutionary process that we have dubbed "spatial sorting" (Shine et al. 2011). The clearest example of this process is the cane toad invasion across tropical Australia, which has been a continuous footrace for more than 75 years. In every generation, the individual toads at the invasion front have moved further from their place of birth than any other toads born in the same place at the same time. These "athletic" individuals may have achieved their rapid dispersal through a high activity level, through speed, by selecting the most efficient path, or by traveling in a consistent direction. When breeding occurs at the invasion front, these athletic toads will be the only ones involved, because slower toads are well behind the front (we've called this interbreeding among superathletic toads "The Olympic Village Effect": Phillips et al. 2010c). By chance, some of the progeny of these fast-dispersing males and females will inherit genes for rapid movement ability from *both* their mother and their father—and thus will travel even further and faster than their parents. The process is a cumulative one, with genes for more rapid dispersal accumulating at the invasion front every generation. The end result will be faster and faster-dispersing toads.

This evolutionary process is very different from Charles Darwin's concept of natural selection. Selection relies upon differential fitness; that is, traits (such as dispersal rate) increase in frequency because they benefit

their bearers in terms of survival or reproductive success. But spatial sorting doesn't require differential fitness—it concentrates genes for faster dispersal within invasion-front individuals even if those genes never increase
an individual toad's chances of surviving and breeding (Shine et al. 2011).
Of course, we could see the same result (an acceleration of invasion front
toads) from natural selection as well, if individuals derived some benefit
from being at the front. Such advantages are perfectly plausible—perhaps
related to lesser competition, because the front has fewer conspecifics with
which individuals have to share their food.

The way to test between these two explanations—spatial sorting versus natural selection—is to see if being at the front actually benefits an
individual toad. Unfortunately that is difficult to measure, because the
toads are so mobile. So far, our comparisons of cane toads at the front
compared with those a few years behind the front suggest a complicated
picture. Toads at the front get more food, and grow faster as a result—but
they rarely reproduce. And they are much more vulnerable to predators,
because the varanid lizards, snakes, etc., that prey upon toads are far more
common than will be the case in a few years' time, when most of these
predators either have been fatally poisoned by eating toads or have learnt
to leave toads alone (Shine 2010).

Although we need more data before we can decide whether the pace
of toad invasion evolves owing to natural selection or spatial sorting (or
both), the rate of invasion clearly *has* accelerated. After toads were first
released in the 1940s and 1950s, the toad front spread at about 10 km
(6 miles) per year. This rate has gradually accelerated ever since, and it is
now around 60 km (37 miles) per year (Urban et al. 2008). Radio-tracking
confirms that invasion-front toads typically travel at least a kilometer
(more than half a mile) every night when conditions are suitable, whereas
toads in long-established populations rarely disperse more than a few me-
ters per night (Alford et al. 2009). Is this due to evolutionary changes in
the toads, or just a case of toads at the current front encountering more
favorable conditions that allow easier dispersal? To distinguish between
these possibilities, Ben Phillips collected toads from across the invasion
history in Australia, and radio-tracked them all in the same area at the
same time. The invasion-front animals still traveled much farther (Phillips
et al. 2008). Does this difference in behavior have a genetic basis, or is it
due to differences in the conditions under which toads develop in different
populations? We have tested this idea also. Toads that were spawned by
parents from different locations were raised in the same place, and radio-

tracked when they reached adulthood. Again, the progeny from invasion-front toads dispersed much faster than did the progeny of toads from older populations (Phillips et al. 2010b). So, the accelerated rate of toad dispersal across tropical Australia is certainly an example of rapid evolution.

Besides evolving rapidly, the cane toads in Australia have also stimulated rapid evolutionary change in their victims among the native fauna. For example, red-bellied blacksnakes (*Pseudechis porphyriacus*) feed mostly on frogs, and are very sensitive to toad toxins. They die in large numbers when the toads arrive. Unlike many native species, these snakes don't seem to learn toad-avoidance. This creates a selective pressure. Genes that code for avoiding toads in the diet leave copies of themselves (i.e., snakes with those genes survive and breed), whereas genes for eating toads disappear from the population (because snakes with those genes are killed by eating toads). As a result, snakes from toad-infested areas are genetically programmed not to eat toads, whereas snakes from toad-free areas will eat a toad if given the opportunity (Phillips and Shine 2006). Other evolutionary changes in the snakes resulting from exposure to toads are increased physiological tolerance of bufotoxins and a smaller head relative to body size (because a small-headed snake can't swallow a toad large enough to be fatal: Phillips and Shine 2006).

These adaptations of the native fauna blunt the ecological impacts of invasive toads over time, and may help explain how impacted species are able to recover despite the continuing presence of toads.

The Social Impact of Cane Toads

Cane toads fascinate the Australian public. The toads are widely reviled, and many people go to considerable effort to kill them (often mistakenly destroying misidentified native frogs as a result: Somaweera et al. 2010). Indeed, several community groups across tropical Australia have sprung up for the avowed purpose of killing cane toads. In some parts of Australia, destroying toads has seemingly been elevated to the status of entertainment or sport. Nonetheless, cane toads play many other roles in Australian society. Stuffed toads holding sporting equipment or musical instruments are icons of the kitsch tourist trade (Fig. 2–2). The Queensland rugby league team is known as the Cane Toads. Children's books are written with cane toads as the villains, or sometimes heroes (Gleitzman 2005). Country music celebrates the toad's resilience in the

FIG. 2–2. An example of the use of cane toads in Australian popular culture (photograph by T. Shine). See also color plate.

face of adversity (Slim Dusty's "Cane toad's plain code"). A documentary movie on cane toads (Lewis 1988) is one of the highest-grossing nonfiction Australian films ever, and likely affected societal attitudes to toads in the same way that children's books have influenced European and American attitudes to other forms of wildlife (see chapter 3, by Peter Coates, in the current volume).

Cane toads probably have had less ecological impact in Australia than other high-profile feral species like rabbits, foxes, cats, camels, and carp. Additionally, the toads' impact has largely been visited on native species unpopular with the general public: venomous snakes, crocodiles, large lizards, and chicken-coop–raiding quolls. Beloved icons such as koalas, kangaroos, and kookaburras have been unaffected. Why, then, do surveys of the Australian public consistently rank cane toads as the worst (or among the worst) invasive species, and why do toads play such a central role in Australian culture? For example, the cane toad was elected as an "icon of Queensland" in a recent popular vote. One reason for the toads' popularity (infamy) involves historical events, such as the influential documentary movie of 1988. Another factor is our ambivalence toward other invaders, with feral mammals (or their domestic cousins) often portrayed as cute and cuddly. In contrast, toads are consistently described as hideously ugly, an evaluation that I simply do not understand—and nor, I suspect, do the millions of people on other continents who see native toads as interesting and attractive components of the local biota, and welcome, useful inhabitants of the local gardens. Cane toads broadly resemble other bufonid species in general morphology and behavior, yet one rarely (never?) sees American, European, Asian, or African toads described as repulsive. Clearly, the Australian public's attitude towards the physical attractiveness of toads is driven largely by a powerful anti-alien sentiment: toads don't belong here, and so we hate them, and so we think they are repulsive. Disturbingly, the language used often has strong analogies with that used by people with xenophobic attitudes towards other groups of humans. The cane toad's special status in Australian folklore also may reflect their abundance in disturbed habitats, and resultant frequent encounters with people. In an increasingly urbanized Australia, with most people living in coastal cities in well-watered areas, it is a cane toad, not a rabbit or a fox, that you are most likely to see (or worse, step on) as you walk around your suburban backyard at night.

Variation in public attitudes towards cane toads can illuminate the reasons for strong opinions about this troublesome pest. The Australian government, during a recent rethink of its policies on dealing with the ecological impact of cane toads, funded a social science study of public attitudes towards toads (Clarke et al. 2009). The survey showed that people living in areas soon to be invaded by the cane toad were deeply apprehensive, and believed that toads posed a major threat to biodiversity. People in areas where toads have been present for many years were less concerned. They generally disliked toads, but did not rate them as a major ecological threat (see Fig. 2–3). In a sense, then, the social impact of cane toads shows the same pattern as the ecological impact on native fauna: most severe at the time of initial invasion, but declining thereafter as the ecosystem (and the local people) adapt to the continuing presence of cane toads in the

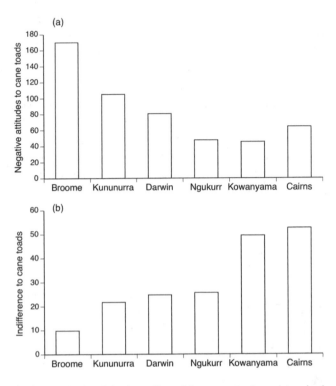

FIG. 2–3. Attitudes of the Australian public to cane toads, as determined by standardized focus-group surveys. The locations of surveys are arranged from west to east across the x-axis of the graph; at the time of testing, cane toads extended from Cairns (on the far eastern edge of their Australian range) to Darwin, but had not reached Kununurra. Responses are shown as normalized comparative response rates, in terms of (a) negative attitudes toward invasive cane toads ($n = 155$) and (b) indifference toward or acceptance of invasive cane toads ($n = 73$). From Clarke et al. (2009), with permission.

neighborhood. To some extent, the greater trepidation of people living in newly invaded (or soon-to-be-invaded) areas is rational. The toads will indeed have substantial ecological impacts on local biodiversity, at least in the short term. However, the response is partly irrational too, reflecting a fear of the unknown and a willingness to believe exaggerated stories about the magnitude of toad impact.

Many of those exaggerated stories, which the media has uncritically promulgated, originated from community groups that were formed to combat the cane toad invasion. The most active and vociferous of these groups are in areas close to the toad invasion front. Some groups claim to have thousands of members, although hard data are scarce. Photographic evidence of their toad-busting activities mostly involves small groups of people. Skepticism about inflated membership numbers is just one manifestation of the intensely politicized debates that have sprung up around toad control (and especially, between rival community groups that compete for funds, membership, and public acclaim). In their attempts to garner publicity and resources, the media-savvy leaders of these groups compete not only with each other, but also with management authorities, politicians at other levels (state, federal, etc.), and scientists. Although interactions have often been positive, at least some community groups seem to view scientists as competitors for public acclaim, rather than collaborators.

From a scientist's perspective, understanding the pressures experienced by community groups and wildlife management authorities can help defuse tensions. For example, it would be political suicide for any community-group leader to publicly agree with scientific assessments that (a) physical removal of toads has little long-term effect, and (b) cane toads have less ecological impact than many other feral species (Shine and Doody 2011). If the allocation of scarce conservation resources is to be based on evidence rather than (or more realistically, as well as) emotion, then scientists must contribute to the public debate about cane toads. When they do so, however, they enter a political minefield with strong competing interests and a diversity of firmly held opinions based on factors other than hard evidence. My only consolation is that interactions with community groups have been even trickier in some other invasive-species scenarios. In Hawaii, some community groups help eradicate invasive coqui frogs, while other groups spread the frogs and prevent eradication (Kraus and Campbell 2002). I have yet to meet an Australian passionately committed to the further spread of cane toads.

Attempts to combat toad invasion illustrate the ways that invasive-species management is intertwined with politics. For example, the imminent arrival of toads in the Darwin area spawned rapid membership growth at a local naturalists' club (Frogwatch NT) which had warned that toad arrival would cause ecological apocalypse. One senior member of that group rose to media stardom as a spokesperson on all issues associated with toads. Some of his contentions about toad impacts, the efficacy of traps as toad-catching devices, and the feasibility of halting the toads' spread ultimately proved inaccurate (e.g., Shine et al. 2009)—but the rapid cycling of news stories meant that such errors had little effect on public opinion. The toad spokesperson was soon elected to high public office, thanks primarily to his media visibility from toad-busting activities. This career change, from leader of a modest naturalist group to elected official, illustrates the potential rewards for high-profile community leaders who can marshal public opinion about invasive species.

In the same way that geographic variation in exposure to toads generates corresponding variation in public attitudes to toads (see Fig. 2–3), beliefs about the feasibility of toad control show pronounced shifts through time as toads arrive. Newsletters of the community groups reveal a consistent shift from confident pronouncements such as "we WILL stop toads" to more realistic "we will reduce toad numbers in the short-term, until scientists can develop a long-term solution." Accepting the impossibility of toad control via direct removal has forced some of these groups to rethink their attitudes to research and scientists. Initial reactions at the time of group formation were largely anti-science because of scientific skepticism about the idea that physical removal could affect toad abundance. The high mobility of these animals (see above) and their enormous fecundity (to >30,000 eggs in a clutch) make it almost impossible to physically remove toads faster than they can replace themselves (McCallum 2006).

As the futility of direct removal in eliminating toads became more difficult to ignore, some community groups supported existing researchers (e.g., the Stop The Toad Foundation donated data, vehicles, and miscellaneous assistance), whereas other groups (e.g., the Kimberley Toadbusters, and Frogwatch NT) set up their own research arms. Although conclusions from this community-based "research" have been promulgated in media releases, much of it appears to lack robust experimental designs and quantitative analysis. None of it has been published in the mainstream scientific literature. Some community groups have criticized scientific research in general, but applauded specific science-based results consistent with their own favored approaches for reducing toad numbers. One group that advocates fencing of water bodies for toad collection enthusiastically acclaimed a research project that supported the effectiveness of that approach (Florance et al. 2011), but ignored research that exposed the futility of trapping for toad control (McCallum 2006; Schwarzkopf and Alford 2006). Most community groups have come to understand that direct removal of toads cannot have any long-term effects, and thus that research offers the only hope for a solution. However, they are frustrated at the slow pace of research outcomes, and in some cases are reluctant to relinquish their own role in combating the toad invasion.

Scientific research on toad control also fits within this social fabric, and has taken several directions. A government research organization (CSIRO) that received $11 million in federal funds initially searched for native-range pathogens of the toad, but that proved unsuccessful and so the agency shifted its focus toward genetic engineering of a self-disseminating virus that would kill toads by disrupting metamorphosis. Neither project produced usable results in terms of toad control, although there were spin-off benefits (Shanmuganathan et al. 2010). The work was discontinued after a review pointed out serious problems in feasibility and environmental risk. For example, a toad-killing virus that escaped from Australia might decimate other bufonid species in their native range, causing far more ecological havoc than the cane toad itself has done in Australia.

Most other control-focused research has had more modest objectives and funding. Some projects have tried to fine-tune existing methods to catch toads—for example, by testing the effects of alternative light sources and sounds associated with traps (Schwarzkopf and Alford 2006), or by measuring the consequences of exclusion-fencing of water bodies (Florance et al. 2011). My own research group has looked for vulnerabilities that could be exploited for selective control of toads, by focusing on ways in which cane toads differ from native Australian frogs or are vulnerable to predators or pathogens (Hagman and Shine 2009; Kelehear et al. 2009; Ward-Fear et al. 2010; Crossland and Shine 2012; Crossland et al. 2012).

In summary, cane toads have had a massive social impact in Australia. Public reactions to toads have been shaped by popular media (books, movies, etc.) as well as by people's own interactions with these animals. The consequent high level of public interest in cane toad control has spawned well-organized, politically adept individuals and groups with passionate beliefs about issues like toad impact and the most effective means of toad control. Perhaps inevitably, the spokespeople for some of these groups have disagreed vigorously with each other, and with the conclusions of scientists.

Implications for Reducing Cane Toad Impact

Broadly, cane toad impacts can be reduced in two ways: either by reducing the numbers of toads, or by somehow assisting the native fauna (and local people) to deal more effectively with this toxic invader.

Most effort has gone into toad control, via direct removal of animals. This effort seems not to have affected rates of toad invasion or long-term toad abundance. Through time, various methods (such as hand-collection and trapping) have been successively promoted as effective, until their respective inadequacies became apparent. Most recently, fencing to exclude toads from water sources has been advocated. This works well in semi-arid regions (Florance et al. 2011), but not in the more well-watered areas that comprise most of the toads' Australian range. Also, collateral effects of water exclusion for native fauna may be a serious problem. The only documented case where fencing a water body enabled cane toad eradication was on Nonsuch Island in Bermuda, where a six-year, $10,000 effort (hopefully) removed all the toads from this tiny (6.5 ha) island (Wingate 2011). We now have several newly developed weapons for toad control, involving native-range parasites (Dubey and Shine 2008; Phillips et al. 2010a), larval pheromones (Hagman and Shine 2009; Crossland and Shine 2012; Crossland et al. 2012), and encouragement of native species (pondside vegetation, native frogs, native invertebrate predators: Hagman and Shine 2006, Ward-Fear et al. 2010, Cabrera-Guzmán et al. 2011). The effectiveness of these ideas has yet to be tested. They may provide local communities with more effective methods, or they may join the long list of toad-control ideas that have been consigned to the dustbin of history.

The other option is to blunt the invader's impact by manipulating responses of the native fauna (and local people). This approach has worked well in the case of one endangered marsupial, the northern quoll (*Dasyurus hallucatus*). We offered captive-raised quolls a small dead toad containing a nausea-inducing chemical, thereby inducing strong and persistent taste aversion (O'Donnell et al. 2010). Many of the "educated" quolls survived after release into toad-infested areas, and they appear to be rebuilding population numbers (unpublished data). By analogy, we might be able to use the same methods to aversion-train vulnerable domestic pets in advance of toad arrival. The same broad approach can work with humans also, in that public education can help people deal with toads—and in particular, ameliorate the anti-toad hysteria in areas close to the invasion front. If people can learn a more realistic view of the scope and duration of toad impact, the social impacts will likely be reduced. Because scientists have direct access to reliable firsthand data on these topics, and training to interpret this evidence, they can play a critical role in public education about the dangers posed by this invasive species, and the likely effectiveness of attempts to combat the toxic invader.

What can the spread of this high-profile invader across Australia tell us about more general issues in invasive-species policy? The cane toad example illustrates how some (but not all) alien species can arouse great concern in the general public, especially if community groups become involved. That public concern can liberate substantial resources to combat the problem—in terms both of money and of volunteer effort. Inevitably, however, the availability of those resources introduces political issues, and dissenting views about the best approaches—and in turn, that dissent can create a media focus whose perspective is very different from that typical of academic debates. To achieve the best outcomes in invasive-species control, scientists need to embed their research programs within the broader cultural context. To identify solutions that are maximally effective in reducing the ecological impact of the invaders, we need to appreciate—and integrate—the rich diversity of public perspectives and opinions aroused by invasive species.

Acknowledgments

I thank the Australian Research Council for funding my work on cane toads, and my postdoctoral fellows, students, and assistants for their efforts. I am also grateful to Reuben Keller and other speakers at the Chicago conference for expanding my own perspectives on alien species.

Literature Cited

Alford, R. A., G. P. Brown, L. Schwarzkopf, B. Phillips, and R. Shine. 2009. Comparisons through time and space suggest rapid evolution of dispersal behaviour in an invasive species. *Wildlife Research* 36:23–28.

Amiel, J. J., R. Tingley, and R. Shine. 2011. Smart moves: effects of relative brain size on establishment success of invasive amphibians and reptiles. *PLoS ONE* 6:e18277.

Beard, K. H., E. A. Price, and W. C. Pitt. 2009. Biology and impacts of Pacific Island invasive species. 5. *Eleutherodactylus coqui*, the Coqui Frog (Anura: Leptodactylidae). *Pacific Science* 63:297–316.

Beard, K. H., K. A. Vogt, and A. Kulmatiski. 2002. Top-down effects of a terrestrial frog on nutrient dynamics. *Oecologia* 133:583–593.

Blackburn, T. M., J. L. Lockwood, and P. Cassey. 2009. *Avian Invasions: The Ecology and Evolution of Exotic Birds*. Oxford: Oxford University Press.

Blaustein, A. R., and J. M. Kiesecker. 2002. Complexity in conservation: lessons from the global decline of amphibian populations. *Ecology Letters* 5:597–608.

Boutilier, R. G., D. F. Stiffler, and D. P. Toews. 1992. Exchange of respiratory gases, ions, and water in amphibious and aquatic amphibians. In *Environmental Physiology of the Amphibians*, edited by M. E. Feder and W. W. Burggren, 81–124. Chicago: University of Chicago Press.

Brown, G. P., B. L. Phillips, and R. Shine. 2011. The ecological impact of invasive cane toads on tropical snakes: field data do not support predictions from laboratory studies. *Ecology* 92:422–431.

Cabrera-Guzmán, E., M. R. Crossland, and R. Shine. 2011. Can we use the tadpoles of Australian frogs to reduce recruitment of invasive cane toads? *Journal of Applied Ecology* 48:462–470.

Cassey, P. 2002. Life history and ecology influences establishment success of introduced land birds. *Biological Journal of the Linnean Society* 76:465–480.

Clarke, R., A. Carr, S. White, B. Raphael, and J. Baker. 2009. Cane toads in communities. Executive Report to the Australian Government. Canberra, ACT: Bureau of Rural Sciences.

Crossland, M. R., T. Haramura, A. A. Salim, R. J. Capon, and R. Shine. 2012. Exploiting intraspecific competitive mechanisms to control invasive cane toads (*Rhinella marina*). *Proceedings of the Royal Society B* 279:3436–3442.

Crossland, M. R., and R. Shine. 2012. Embryonic exposure to conspecific chemicals suppresses cane toad growth and survival. *Biology Letters* 8:226–229.

Doody, J. S., B. Green, D. Rhind, C. M. Castellano, R. Sims, and T. Robinson. 2009. Population-level declines in Australian predators caused by an invasive species. *Animal Conservation* 12:46–53.

Dubey, S., and R. Shine. 2008. Origin of the parasites of an invading species, the Australian cane toad (*Bufo marinus*): are the lungworms Australian or American? *Molecular Ecology* 17:4418–4424.

Ficetola, G. F., W. Thuiller, and C. Miaud. 2007. Prediction and validation of the potential global distribution of a problematic alien invasive species—the American bullfrog. *Diversity and Distributions* 13:476–485.

Florance, D., J. K. Webb, T. Dempster, M. R. Kearney, A. Worthing, and M. Letnic. 2011. Excluding access to invasion hubs can contain the spread of an invasive vertebrate. *Proceedings of the Royal Society B* 278:2900–2908.

Gleitzman, M. 2005. *Toad Rage.* New York: Yearling.

Govindarajulu, P. 2004. Introduced bullfrogs (*Rana catesbeiana*) in British Columbia: impacts on native Pacific treefrogs (*Hyla regilla*) and red-legged frogs (*Rana aurora*). PhD diss., University of Victoria, Victoria, Canada.

Govindarajulu, P., S. Price, and B. R. Anholt. 2006. Introduced bullfrogs (*Rana catesbeiana*) in western Canada: has their ecology diverged? *Journal of Herpetology* 40:249–260.

Greenlees, M., B. L. Phillips, and R. Shine. 2010. Adjusting to a toxic invader: native Australian frog learns not to prey on cane toads. *Behavioral Ecology* 21:966–971.

Hagman, M., and R. Shine. 2006. Spawning-site selection by feral cane toads (*Bufo marinus*) at an invasion front in tropical Australia. *Austral Ecology* 31:551–558.

Hagman, M., and R. Shine. 2009. Larval alarm pheromones as a potential control for invasive cane toads (*Bufo marinus*) in tropical Australia. *Chemoecology* 19:211–217.

Hayes, R. A., M. R. Crossland, M. Hagman, R. J. Capon, and R. Shine. 2009. Ontogenetic variation in the chemical defences of cane toads (*Bufo marinus*): toxin profiles and effects on predators. *Journal of Chemical Ecology* 35:391–399.

Hendry, A. P., M. T. Kinnison, M. Heino, T. Day, T. B. Smith, G. Fitt, C. T. Bergstrom, et al. 2011. Evolutionary principles and their practical application. *Evolutionary Applications* 4:159–183.

Kearney, M. R., B. L. Phillips, C. R. Tracy, K. A. Christian, G. Betts, and W. P. Porter. 2008. Modelling species distributions without using species distributions: the cane toad in Australia under current and future climates. *Ecography* 31:423–434.

Kelehear, C., J. K. Webb, and R. Shine. 2009. *Rhabdias pseudosphaerocephala* infection in *Bufo marinus*: lung nematodes reduce viability of metamorph cane toads. *Parasitology* 136:919–927.

Kraus, F. 2009. *Alien Reptiles and Amphibians: A Scientific Compendium and Analysis*. Dordrecht: Springer Science + Business Media B.V.

Kraus, F., and E. W. Campbell. 2002. Human-mediated escalation of a formerly eradicable problem: the invasion of Caribbean frogs in the Hawaiian Islands. *Biological Invasions* 4:327–332.

Letnic, M., J. K. Webb, and R. Shine. 2008. Invasive cane toads (*Bufo marinus*) cause mass mortality of freshwater crocodiles (*Crocodylus johnstoni*) in tropical Australia. *Biological Conservation* 141:1773–1782.

Lever, C. 2001. *The Cane Toad: The History and Ecology of a Successful Colonist*. Otley, West Yorkshire: Westbury Academic and Scientific Publishing.

Lewis, M. 1988. *Cane Toads: An Unnatural History*. Lindfield, NSW: Film Australia.

McCallum, H. 2006. Modelling potential control strategies for cane toads. In *Science of Cane Toad Invasion and Control*, edited by K. L. Molloy and W. R. Henderson, 123–133. Canberra, ACT: Invasive Animals Co-operative Research Centre.

Nugent, G., W. Fraser, and P. Sweetapple. 2001. Top down or bottom up? Comparing the impacts of introduced arboreal possums and "terrestrial" ruminants on native forests in New Zealand. *Biological Conservation* 99:65–79.

O'Donnell, S., J. K. Webb, and R. Shine. 2010. Conditioned taste aversion enhances the survival of an endangered predator imperiled by a toxic invader. *Journal of Applied Ecology* 47:558–565.

Phillips, B. L., C. Kelehear, L. Pizzatto, G. P. Brown, D. P. Barton, and R. Shine. 2010a. Parasites and pathogens lag behind their host during periods of host range advance. *Ecology* 91:872–881.

Phillips, B. L., G. P. Brown, and R. Shine. 2010b. Evolutionarily accelerated invasions: the rate of dispersal evolves upwards during the range advance of cane toads. *Journal of Evolutionary Biology* 23:2595–2601.

Phillips, B. L., G. P. Brown, and R. Shine. 2010c. Life-history evolution in range-shifting populations. *Ecology* 91:1617–1627.

Phillips, B. L., G. P. Brown, J. M. J. Travis, and R. Shine. 2008. Reid's paradox revisited: the evolution of dispersal kernels during range expansion. *American Naturalist* 172:S34–S48.

Phillips, B. L., and R. Shine. 2006. An invasive species induces rapid adaptive change in a native predator: cane toads and black snakes in Australia. *Proceedings of the Royal Society B* 273:1545–1550.

Pizzatto, L., and R. Shine. 2009. Native Australian frogs avoid the scent of invasive cane toads. *Austral Ecology* 34:77–82.

Pizzatto, L., C. M. Shilton, and R. Shine. 2010. Infection dynamics of the lungworm *Rhabdias pseudosphaerocephala* in its natural host, the cane toad *Bufo marinus*, and in novel hosts (Australian frogs). *Journal of Wildlife Diseases* 46:1152–1164.

Pramuk, J. B. 2006. Phylogeny of South American *Bufo* (Anura: Bufonidae) inferred from combined evidence. *Zoological Journal of the Linnean Society* 146:407–452.

Rosen, P. C., and C. R. Schwalbe. 1995. Bullfrogs: introduced predators in southwestern wetlands. In *Our Living Resources: A Report to the Nation on the Distribution, Abundance, and Health of US Plants, Animals, and Ecosystems*, edited by E. T. La-Roe, G. S. Farris, C. E. Puckett, P. D. Doran, and M. J. Mac, 452–453. Washington, DC: US Department of Interior, National Biological Service.

Schwarzkopf, L., and R. A. Alford. 2006. Increasing the effectiveness of toad traps: olfactory and acoustic attractants. In *Science of Cane Toad Invasion and Control*, edited by K. Molloy and W. Henderson, 165–170. Canberra, ACT: Invasive Animals Co-operative Research Centre.

Shanmuganathan, T., J. Pallister, S. Doody, H. McCallum, T. Robinson, A. Sheppard, C. Hardy, et al. 2010. Biological control of the cane toad in Australia: a review. *Animal Conservation* 13 (Suppl. 1):16–23.

Shine, R. 2010. The ecological impact of invasive cane toads (*Bufo marinus*) in Australia. *Quarterly Review of Biology* 85:253–291.

Shine, R. 2011. Invasive species as drivers of evolutionary change: cane toads in Australia. *Evolutionary Applications* 5:107–116.

Shine, R., G. P. Brown, and B. L. Phillips. 2011. An evolutionary process that assembles phenotypes through space rather than through time. *Proceedings of the National Academy of Sciences USA* 108:5708–5711.

Shine, R., and J. S. Doody. 2011. Invasive-species control: understanding conflicts between researchers and the general community. *Frontiers in Ecology and the Environment* 9:400–406.

Shine, R., M. Greenlees, M. R. Crossland, and D. Nelson. 2009. The myth of the toad-eating frog. *Frontiers in Ecology and the Environment* 7:359–361.

Sol, D., S. Bacher, S. M. Reader, and L. Lefebvre. 2008. Brain size predicts the success of mammal species introduced into novel environments. *American Naturalist* 172:S63–S71.

Somaweera, R., N. Somaweera, and R. Shine. 2010. Frogs under friendly fire: how accurately can the general public recognize invasive species? *Biological Conservation* 143:1477–1484.

Travis, J. M. J., T. Münkemüller, and O. J. Burton. 2010. Mutation surfing and the evolution of dispersal during range expansions. *Journal of Evolutionary Biology* 23:2656–2667.

Urban, M., B. L. Phillips, D. K. Skelly, and R. Shine. 2008. A toad more traveled: the heterogeneous invasion dynamics of cane toads in Australia. *American Naturalist* 171:E134–E148.

Ward-Fear, G., G. P. Brown, and R. Shine. 2010. Using a native predator (the meat ant, *Iridomyrmex reburrus*) to reduce the abundance of an invasive species (the cane toad, *Bufo marinus*) in tropical Australia. *Journal of Applied Ecology* 47:273–280.

Webb, J. K., G. P. Brown, T. Child, M. J. Greenlees, B. L. Phillips, and R. Shine. 2008. A native dasyurid predator (common planigale, *Planigale maculata*) rapidly learns to avoid toxic cane toads. *Austral Ecology* 33:821–829.

Wingate, D. B. 2011. The successful elimination of cane toads, *Bufo marinus*, from an island with breeding habitat off Bermuda. *Biological Invasions* 13:1487–1492.

CHAPTER THREE

A Tale of Two Squirrels: A British Case Study of the Sociocultural Dimensions of Debates over Invasive Species

Peter Coates

From "Over Here" to "Over There"

In his landmark studies of the impact of European flora and fauna either deliberately or unintentionally transplanted to the Americas, Alfred Crosby emphasized the imbalance of biotic interchange between "old" and "new" worlds. Europe had received few "problem" species in return for the bevy of organisms that became notoriously invasive in North America, especially the "English" sparrow and the "European" starling, which had "dispossessed millions of American birds" (Crosby 1972, 1986). Yet, as John MacKenzie later indicated, the flow of troublesome and unwanted trans-Atlantic traffic was not quite so one-way. Crosby's perspective was a "strikingly American one, for anyone who lives in the British Isles is almost daily brought face to face with the fact that ecological colonialism has been a two-way process." Plenty of extra-European species posed difficulties for native wildlife (MacKenzie 2001a, 2001b). Crosby remained insistent. "How often," he rejoined, "have American species swamped and driven to the verge of extinction native species in Great Britain? Do you have over there, for instance, millions of American rats as we do millions of Old World brown and black rats? If there are equivalently successful aliens in the UK, I would like to know about them" (MacKenzie 2001b).

Crosby was right. There are no American fauna whose British exploits compare with those of the Old World's rats. However, various American creatures have flourished and created waves in Britain. This essay shifts attention from European species over here (America) to American species over there (Britain). It challenges the frequent assumption in the US literature on invasive species that Britons are less critical of introduced species and less inclined to regard their activities as invasive (Sagoff 2002). In Britain, many of the most vigorously contested immigrant animals are from North America: gray squirrel, mink, ruddy duck, signal crayfish, and bullfrog. A nonnative species audit by English Nature, the governmental body responsible for wildlife conservation (since renamed Natural England), highlighted the strong American showing among animals with "major negative environmental effects." Of twelve nonnatives wreaking the most havoc, five were North American: signal crayfish, Canada goose, ruddy duck, mink, and gray squirrel (Hill et al. 2005).[1]

Though it was the first American animal introduction—and does not rank top of today's official lists of the unwanted—of all the American animals resident in Britain yet withheld faunal citizenship, it is the gray squirrel (*Sciurus carolinensis*) that still generates the most heated debate. Perhaps because it occupies the highest profile of any wild animal in millions of Britons' daily lives (Harris 2011), it appears to constitute the clearest and most present danger.[2]

This historical perspective on the gray squirrel in Britain has three objectives: first, to examine the relationship between the gray's success and the decline of the native red and how the gray has served as a popular scapegoat; second, to investigate the character and success of efforts to control this "pest" species for economic and ecological reasons; third, to explore the cultural dimensions of British debates over the desirability of the gray squirrel's presence, examining in particular the role of anti-American sentiments and the language through which they are expressed.

Arrival, Expansion, and Reaction

The gray squirrel was introduced during an era of innocent faith in acclimatization. Those responsible, commented the Ministry of Agriculture and Fisheries (MAF) in a publicity leaflet, "no doubt considered it would be an attractive addition to our native wild life. Little did they know what trouble they were laying up for later generations!" (MAF 1953a). The first recorded release took place at Henbury Park, Cheshire (1876). The most consequential release, however, occurred at Woburn Park, Bedfordshire (1890); eight secondary releases were drawn from Woburn's stock. Though importation ended before World War One, internal transplantations continued until 1929 (Middleton 1930).

The "alien rodents" enjoyed early naturalization and rapid spread, assisted by their omnivorous dietary habits (likened to the rat's "very catholic" tastes). The native red (common) squirrel (*Sciurus vulgaris leucourus*) was more specialized in its feeding, largely dependent on pine cones and hazelnuts (MAF 1943b, MAF 1943c, MAF 1943f). The area containing grays in 1930 had virtually doubled by 1935 (Middleton 1935). By the 1950s, its occupation of central England was virtually complete. Prospering across much of the southeast as well, it colonized territory at an annual rate of 2,000 square miles between 1937 and 1945 (Shorten 1953, 1954). For the pioneering invasion biologist, Charles Elton, the American gray squirrel was a prime example of an ecological population "explosion" (Elton 1958).

Opposition emerged in tandem with proliferation. Britons described the larger and more powerful gray as pugnacious, greedy, opportunistic, and fecund. This criticism was directed at both economic and ecological impacts. The belief that the gray's success was inimical to the native squirrel's welfare took hold while it was still being released. Frederick Treves, the author, surgeon, and royal physician, complained (1917) that they evicted reds and "eat everything that can be eaten, and destroy twenty times more than they eat" (Ritchie 1920). Britain lacks American rats. But "tree rat" has been a well-established British soubriquet for the gray squirrel since the early twentieth century (Country Life 1943; Shorten 1954).[3] If not quite "equivalently successful" (MacKenzie 2001b) to the "Old World" rat, the gray squirrel's misdeeds in Britain were often compared to the American exploits of sparrow and starling.

The inter-war era's leading squirrel researcher, A. D. Middleton, based at Elton's Bureau of Animal Population (Oxford University), reported (in his self-described "unprejudiced" reviews) the belief among residents of colonized areas that the red "has been reduced in numbers or driven out by the introduced aliens." He also cited the widespread rumor that the male gray castrated its red rival (Middleton 1930, 1931).[4] Equally fanciful rural folklore held that male grays slaughtered entire red litters and interbred with their mothers to produce hybrid offspring more gray than red (Middleton 1931; Tittensor 1975).

A reviewer of Middleton's book about the gray squirrel in Britain referred to a "little creature" that was "pleasant . . . to look at" but which possessed peerless "obnoxious" habits (Bookman 1931). Contrasting the gray's endearing appearance and manner with its nefarious character and activities was a standard ploy. "Quarrelsome" and "spiteful," the "little ruffian" wrecked bird nests and massacred fledglings. "Long ago," lamented one commentator in 1936, "these woods had been the home of the gentle red squirrels." But since the "American greys" appeared, they had "either fled or died" (Russell 1936). A few months later, 26 county councils distributed more than 5,000 posters beseeching the public to "kill the tree rat" (*Times*, April 7, 1936). The Destructive Imported Animals Act of 1932 (drafted with specific reference to the musk rat and mink, two fur-bearers introduced in the 1920s that had quickly escaped from fur farms) was extended to the gray squirrel in 1937 after it was designated a pest of hardwoods, cereal crops, and fruit trees (Kenward 1987). Direct government action, when it came, was driven by national security considerations and economics rather than ecology. The gray's impact on food production and timber supplies became a serious concern during World War Two, its depredations compared with those of rabbit and rat (MAF 1943c). Government-funded control was first mooted in 1931, when the Ministry of Agriculture and Fisheries (MAF) endorsed the National Anti-Gray Squirrel Campaign, a private initiative under the auspices of *The Field*, a sporting and outdoor magazine (Middleton 1931). The Grey Squirrels Order of 1943, which required occupiers to destroy grays on their property, placed control on a firm footing. Publicizing this order, MAF also emphasized the gray squirrel's undesirable character and impact on indigenous wildlife: they bred "prolifically" and were "as bad or even worse than rats, for they will rob bird's nests of eggs and young, will kill and eat the Red Squirrel, and even devour one another if caught in a trap" (MAF 1943e).[5]

Calls from the shires for a bounty (capitation fee) grew louder, boosted by sportsmen's magazines such as *Gun and Game*. Deploying typically emotive language, Devon's County War Agricultural Executive Committee emphasized the need to curb the "menace" that would soon become "alarming" in the absence of "drastic" measures (MAF 1943d, 1946). A "shilling a tail" bounty was introduced in March 1953 for more effective prosecution of the "war" against "one of the forester's worst enemies" (Forestry Commission [FC] 1953, FC 1955a-c, Sheail 1999, Thompson and Peace 1962).

Pest control officers who sought to galvanize the public with martial and medical metaphors worried that the urbanite and suburbanite did not appreciate the swath of destruction cut through fields and woods by the sprightly denizens of their local parks, which they were inclined to regard as pets rather than pests (FC 1954; Burton 1947). [6] So the campaign against the gray "plague" and the public's "present sentimental attachment" to a squirrel less "standoffish" than the "native Britisher" (Colinvaux 1973) was reinforced by an energetic publicity drive (*Quarterly Journal of Forestry* 1953). A standardized reply to a request for a license to keep grays after 1937 explained that, though "attractive . . . because of the damage they can do to forestry and . . . crops they have to be treated as pests" (Logan 1967). Governmental responses were often freighted with inflammatory language ("nothing can be said in favour of the grey squirrel while it is alive") and liberal use of the designation "alien" (MAF 1943f). Yet, doubtless tempered by the wartime alliance, there was no direct focus on American origins.

"Pest," the most commonly employed term for the gray squirrel, was a designation that paid no heed to national origins (Smout 2011). The avowed intent to "rid the country of this pest" through an "all out effort" was backed by a consciousness-raising exercise that included circulation of "propaganda" materials to schools, police stations, and rural inns. In October 1954, there were 22,000 posters printed to promote the bounty, followed by another run of 3,000 in 1956, when the reward was raised to two shillings (Atkinson 1954; *Nottingham Evening News* 1955; O'Hanlon 1956). These efforts and inducements yielded some ostensibly impressive kill figures: 753,527 squirrels within the bounty's first two years (MAF 1955a and b; FC 1955b; FC 1955c; FC 1956a). Grey Squirrel Clubs (1,202 in total, with a combined membership of 5,774 at the end of May 1953), armed with free cartridges by MAF (1945–55), assisted the cause (Ryle 1952; MAF 1953c; Marjoram 1958).

The bounty, however, barely dented the "virile" and "enterprising" creature's numbers (Burton 1947); in 1958, they were the same as when the scheme started (Waldegrave 1962, MAF 1961b). A few years after termination (March 1958, by which time £100,000 had been paid out), a forester commented on the "alarming progress" of the "grey invasion" (Shorten 1955, Seymour 1961). The Forestry Commission observed that it was simpler to state where grays were not than to describe where they were (FC 1960).

Gray Advance and Red Retreat: A Special Relationship?

UK government officials monitoring the gray's activities rarely focused on its nationality. They usually identified it with reference to geographical origins as the North American Grey Squirrel or the Eastern Grey Squirrel of North America (MAF 1953a, FC 1955a, *Times Weekly Edition* 1952). And when they did indicate the gray's origins, they could get it wrong (South America) (MAF 1943e). At a popular level, however, a nationalistic element has been more or less inseparable from the debate since the start. The "undesirable alien" was not just a gray squirrel, but an American gray squirrel—even when its native terrain was identified as southern Canada as well as the United States (*Meat Trades Journal* 1943, MAF 1943b, MAF 1953b). Nor was it just a tree rat. It was an American tree rat (Watt 1923, *Illustrated London News* 1931, Lancum 1947, Leyland 1955).

Not one of the flurry of books on anti-Americanism published during the presidency of George W. Bush (2000–8) mentioned the phenomenon of ecological nationalism, bio-xenophobia, or a single disruptive animal of American provenance (Sardar and Davies 2003; Hollander 2004; Revel 2004; Ross and Ross 2004; O'Connor and Griffiths 2005; Hollander 2006; Lacorne and Judt 2006; Sweig 2006; Markovits 2007; Berman 2008; Gould 2009). Attitudes toward invasive American animals is the least studied aspect of British anti-Americanism.

"I know of more than one patriotic Englishman who has been embittered against the whole American nation on account of the presence of their squirrels in his garden," remarked Middleton on his book's opening page (Middleton 1931). The editors of the Collins New Naturalist Series that published Monica Shorten's book on squirrels in Britain referred to the problem of controlling the "American cousin" with its "bad reputation" as "almost 'political'" (Shorten 1954). In a representative example of postwar demonization, an eminent geographer denounced it as a "villain of the worst order . . . steadily replacing the lovely little native red squirrel" (Stamp 1955). And in a letter to the editor of the *Times* (21 October 1971), the owner of a woodland garden that the gray allegedly terrorized admitted that "in respect of the alien grey immigrant I plead guilty to racial discrimination."

Vilification did not escape American attention. The *New York Times* reported the Forestry Commission's late 1920s investigation of the gray's impact on timber resources. Alluding to previous American debates over the "English" sparrow, the reporter reckoned prejudice was so strong that the gray was unlikely to receive a fair appraisal. A year later, the *Times* noted the many "outbursts" against the "gray invader" triggered by "patriotic" naturalists, foresters, and agriculturalists (*New York Times* 1929 and 1930).

British opinion on the relationship between gray success and red decline was in fact less monolithic than the *Times* implied. Elton reported in detail to his mother on the situation in his Oxfordshire neck of the woods. Though it had been driven out of various university parklands, and Blenheim Park was "still pure red," waging a "very hard campaign" an "advance guard" of grays had penetrated Bagley Wood (200 hectares of mixed lowland woodland on the edge of Oxford owned by St. John's College). Marking his map with pins to indicate their presence (like a field commander), the young biologist felt "quite warlike" (Elton 1925). Yet in his first book, *Animal Ecology*, he chose his words carefully: the gray was "replacing" the red, but this was not the same as holding the gray responsible. And he could offer no "good explanation" for the unconnected fall in red numbers (Elton 1927 and 1951).

Government views were also split. One official claimed there was no evidence to suggest that the gray had "driven out" the red. Another (who nonetheless objected to the near-ubiquitous "tree rat" label—"quite uncalled for and rather like giving a dog a bad name") disputed this belief, insisting that it had evicted the red over large areas. Refuting this allegation, the previous official (like Elton) cautioned against confusing the gray's "spread" with replacement of the red, whose numbers fluctuated without reference to the gray's presence and activities (MAF 1943a). Moreover, the key variable governing the gray's population levels was not the existence or nonexistence of the bounty but the weather, which governed its food supply and breeding conditions. MAF officials tried to explain this to the Prime Minister, Harold Macmillan, who was mightily exercised by their numbers "swarming again": "What are you doing about grey squirrels?" he demanded, urging reinstatement of the "tail bonus" (Macmillan 1960). As two successive Ministers of Agriculture informed him, gray numbers dropped, despite the termination of the bonus, because of less clement weather conditions (MAF 1960 and 1961a).

The tender feeling of many Britons toward the red squirrel is not self-evident, for the red's reputation was far from unblemished. As various commentators between the 1920s and the 1940s pointed out, the "American stranger" was mischievous and destructive—but no more so than the pesky red (Watt 1923; Smout 2009).[7] The red's taste for the bark and shoots of trees in spring and early summer, and the "very great" "injury" inflicted on young plantations of fir, larch, and Scots pine, were well known (a problem aggravated by gamekeepers' determination to exterminate its natural predators, the marten, sable, and hawk [Rusticus 1850]). The leading late nineteenth-century expert on what was then simply "the squirrel" reported that "unanimously, my correspondents condemn [it] as one of the most destructive animals [of] our forests" and that "nowhere can I find anything said in its favour" (Harvie-Brown 1881). During World War Two, an MAF official acknowledged that the red squirrel itself was once regarded as a pest (though adding that "it is nothing like as bad as the invading grey squirrel, which appears to be . . . hardier and more prolific" [MAF 1943a]).

Integral to the grey's defamation was the rehabilitation of the smaller, gentler, cuter, and more timid red squirrel (Kean 2000). With a little help from one of Beatrix Potter's most beloved animal creations, Squirrel Nutkin (Potter 1903), the red was converted from forester's scourge to national animal emblem.[8] By the 1950s, the red squirrel had become "truly British" (Leyland 1955). Alongside red telephone boxes, warm beer, and cricket bats, the charming, feisty little chap has assumed a hallowed position in the pantheon of Englishness (Kean 2001). Prior to the gray's advent and spread—though its qualities were far from common ("there is nothing in the English landscape so beautiful" [Stillman 1907])—the red was simply the "common" squirrel (Stillman 1899; Finn 1909). But, by 1950, according to one squirrel expert, "our island race" was highly uncommon, confined to "islands" of conifer plantations amidst a "sea of grey squirrel country" (Tittensor 1979).[9] Potter tried to follow Squirrel Nutkin's popularity with a story about a "little fat comfortable" gray squirrel, Timmy Tiptoes (Potter 1911; Shorten 1954; Lear 2007). Timmy's nationality is not mentioned. Nor is the story's setting. But since the cast of characters includes a bear and chipmunk, this is clearly not the Lake District of Cumbria, where there were, as yet, no grays. Though Timmy and his wife, Goody, drink tea like good British squirrels, the story was a flop.

Regardless of whether the associations are positive or unsavory, the assignation of human qualities to nonhuman creatures is not the sole preserve of the authors of animal stories. Humanization (pejorative) of invasive species by natural resource managers, journalists, and members of the public alike is an incorrigible tendency. The UK Environment Agency compiles lists of the "Top Ten Most Wanted Foreign Species," which confirm the high American profile among invasive species in Britain; the signal crayfish and mink rank second and third (Environment Agency [EA] 2006). Signal crayfish and mink, EA explains, have "taken advantage of Britain's welcoming living conditions" and "overstayed their environmental visa" (EA 2006). EA's press office also personifies them as individuals "wanted" for their "crimes." Even the British National Party has pitched in (Monbiot 2009), while an article posted on a related online forum describes the red squirrel as a "victim of immigration, albeit immigration of a different kind to that we nationalists are more accustomed to discussing" (Wakely 2011).

The most familiar rhetorical device, though, has been to humanize invasive species from across the pond through analogy with the "friendly invasion" of American troops during World War Two. GIs stationed in Britain were famously described as "oversexed, overpaid, overfed, and over here" (Reynolds 1995). This celebrated phrase encapsulated British male resentment toward their American counterparts who, bearing gifts of nylon stockings, lipstick, chocolate, cigarettes, and chewing gum, enjoyed a huge competitive advantage in the mating game. From the standpoint of white-clawed crayfish and water vole, signal crayfish and mink are "over sized, over sexed and over here" (EA 2006; Edwards 2001).

The Hegemony of the Gray Squirrel and the Red Crusade

Whereas the population of the "vicious tree pest" (Drummond 1984) in Britain now stands at an estimated 2.5 million (of which 2 million are in England), red numbers have plunged to c. 160,000 (down from c. 12 million on the eve of human colonization) (Skelcher 1997, Carrell 2009, Harris and Yalden 2008).[10] The majority of these survivors (120,000) are in Scotland. The remaining pockets in England are confined to the fringes. The Lake District in the far northwest, home of Potter and Squirrel Nutkin, is one of the besieged English redoubts. Another island in the gray sea south of Scotland is the red squirrel reserve, managed by Red Alert, in the (nonnative Corsican) coastal pinewoods at Formby, Lancashire (Dutton 2004). But the largest surviving population (9,000, representing 80% of mainland England's population) is housed in Kielder Forest, Northumberland, where the Forestry Commission planted moorland with Sitka spruce to establish a strategic timber reserve after World War One. Visiting Kielder recently as part of a review of England's forest resources and values, the Independent Panel on Forestry (IPF) identified the red as a "high profile iconic woodland species" (IPF 2011).

The gray's infamy and the red's dilemma is a matter taken very seriously by the hereditary peers of Parliament's upper chamber (who, perhaps, feel a large dose of empathy with a fellow endangered species). Today's "British" squirrels are mostly descended from continental European stock brought over in the late nineteenth and early twentieth centuries for restocking purposes, as native red populations plummeted. But the "native" squirrel's political champions are immune to this irony. During a squirrel debate in the House of Lords, Lord Inglewood of Cumbria identified the red as the "most lovable and loved of our British native animals" (Hansard 1998). During a more recent squirrel debate, peers of the realm also lost no opportunity to underscore the gray's national origins. Embracing the red as an iconic animal that was firmly British (rather than merely English), Lady Saltoun of Abernethy (Scotland) pronounced its innate superiority. Clothing reds in the admirable qualities of a better class of person, she observed that they are "rather like quiet, well behaved people, who do not make a nuisance or an exhibition of themselves, or commit crimes, and so do not get themselves into the papers in the vulgar way grey squirrels do" (Hansard 2006). The distinction between squirrels in Britain and British squirrels (Squirrel Nutkin's ilk) is subtle but profound. As Lord Rowallan observed in 1998: "we should not encourage them merely because they are beautiful animals ... They ... have another advantage—they are British" (Hansard 1998). By 2006, the red was even more embattled than in 1998. Desperate times demanded drastic measures. Rowallan urged the celebrity chef, Jamie Oliver, who had recently launched a campaign to revolutionize British school dinners, to add the gray squirrel to the canteen menu. For "unless something radical and imaginative is done ... Squirrel Nutkin and his friends and relations are 'going to be toast'" (Hansard 2006; Massie 2007). Lord Redesdale concurred with this elegantly simple policy of "eat a grey to save a red" (Watson 2008; Burnham, 2009) (in other words, "if you can't beat them, eat them").[11] Redesdale, whose Northumberland seat, adjoining Kielder, enjoys a remnant stock of reds, has founded the Red Squirrel Protection Partnership, whose mission is simply to kill grays (Max 2007, Adams 2008). Whereas Redesdale's language is restrained, his right-hand man, Paul Parker, is a plain speaking, self-styled "Verminator" on a crusade: "everything's got its place in nature. This is from America; it shouldn't be here" (Squirrel Wars 2009).[12] Many British countryside lovers and wildlife enthusiasts agree: the "American interloper" is inauthentic (Wright 2011).

Red squirrel defense groups are spreading almost at the rate of the gray itself (Northern Red Squirrels 2009). The Red Squirrel Survival Trust's patron is HRH The Prince of Wales; and Save Our Squirrels, a campaign operated by Red Alert (since renamed Red Squirrels Northern England), received National Heritage Lottery funding (2006) to operate 16 reserves in northern England. Drawings of red squirrels by personalities from the worlds of music, comedy, sport, television and radio, stage and screen, were auctioned on the Internet in September 2000 to raise money for Red Alert (Ingham 2000). The popular case against the gray is so entrenched that it is largely impervious to evidence suggesting that, whatever the position during recent decades and today, its initial spread was not responsible for the red's decline.

Red populations have a long history of fluctuation. Against a background of forest clearance, the native squirrel had virtually vanished from Scotland by the late 1700s (Harvie-Brown 1881; Ritchie 1920). Woodland retreat in England was also accompanied by a fall in numbers. Immediately preceding the gray's importation, however, the reds rebounded; extensive late eighteenth- and early nineteenth-century plantings of nonnative conifers matured between 1860 and 1890, offering congenial conditions (MAF 1959). Nonetheless, owing to disease, reds were under pressure again in many parts of England before the "foreign" squirrel's noticeable spread (probably coccidiosis, but possibly also mange and/or enteritis) (MAF 1959), and subsequent disease outbreaks were unrelated to a gray presence (Reynolds 1981). The early twentieth-century retreat of coniferous woodland, especially under wartime demand, also took its toll. Reds died out in areas without a single gray. Rather than displacing them, grays often advanced into vacated territory (MacKinnon 1978; Usher et al. 1992; Kenward and Holm 1993). And while Crosby conceded that the "relatively large and aggressive" North American counterpart had "replaced" the red over large parts of Britain, he noted that disease had largely cleared the ground (Crosby 1986).[13]

With the exception of islands off the south coast of England such as the Isle of Wight and Brownsea, the remaining sanctuaries of the red, which makes no distinction between native broadleaf trees and nonnative conifers such as Sitka and Norway spruce and Corsican pine, are mainly northern England's and Scotland's coniferous forests (Garson and Lurz 1992; IPF 2011). Ironically, the firm conservationist preference for native broadleaf woodland benefits the grays, because, though Britain's native reds also occupy broadleaf habitat, nineteenth-century (re)introductions from Scandinavia are adapted to coniferous woodlands. Tree diversification at Kielder, for example, must be carefully planned to avoid inadvertent creation of gray invasion corridors, thereby ensuring that the forest remains a "safe haven" for reds (Pratt 2011).

One outlet for dispassionate data on the red-gray relationship (Middleton 1931; Shorten 1954; Gurnell 1987) was a children's book by Woodrow Wyatt, a well-known public figure who became a Conservative Party life peer in 1987. The central character in *The Further Exploits of Mr. Saucy Squirrel* is a dapper squirrel that lives in a beech tree in Wiltshire. Despite his impeccably English manner and pursuits, Mr Saucy Squirrel was in fact of the gray variety. Asked by the shopkeeper, as he buys a bowler hat, if he was "a good English squirrel," he replies: "Dear me. I'm always being asked that question. I'm sorry to have to tell you that my family came from America at the end of the nineteenth century. We are not the old kind of English squirrel people call the red squirrel. But we've been here a long time now." Nor does he suffer from an identity crisis or conflicting allegiance: "We came from America so long ago that we are now properly naturalised English. I don't sing 'The Star-Spangled Banner' . . . I sing 'God Save The Queen.' "

The "English" squirrel, Saucy proceeds to explain, was already in trouble: "We just took over in various places that they'd already left." He also maintains that the two species happily coexist: "We don't quarrel with English squirrels at all. We eat the same food but we don't fight over it" (Wyatt 1977). Though he concedes that grays "probably put the red squirrels off by living in the same areas," the charming picture of Anglo-American harmony is overdone. Grays did not cause the reds' decline, but probably prevented their recovery (Tittensor 1975). Moreover, in common with many other invasive species around the world (not least humans) that spread disease to the detriment of native species, many grays carry a virus, squirrelpox, to which they are immune, that kills reds within two weeks (Rushton et al. 2006, *The Scottish Sporting Gazette* 2006).

The Furry Ugly American

Even-handed researchers frown on the term "tree rat" and suspect that the gray's alien provenance is a major component of objection (Shorten 1954). They also point to the absence of "obvious competition" or note-worthy interaction between the two squirrels (on which the transmission of squirrelpox does not depend) (Colinvaux 1973; Reynolds 1981). Yet the gray still serves as a handy scapegoat. A few recent examples of the construction of an unpalatable Yankee critter illustrate the mentality that feeds what Andrew Tyler, director of Animal Aid, calls the anti-gray "pogrom" (Tyler undated). In his view, the British public believes that "the only good squirrel is a red squirrel" (Tyler 2001). A blog of a self-described Tennessee expatriate and long-term London resident ponders the orthodox view that the gray squirrel is UK Public Enemy Number

One. The immediate prompt was the Lords' debate (2006). At that time, Americans living peacefully in London for years reported their first experience of abuse, whether or not they voted for Bush or supported the invasion of Iraq ("some people just fly off the handle without even talking to me—it's as if they had been waiting to run into an American all day to let their feelings out," a 29-year old woman reported) (BBC 2006). Within this hostile climate of opinion, she wondered whether the assault on the gray was a "further sign of rampant anti-Americanism among the British elite" (Blogspot 2006).

The Tennessean's analysis raises a key issue in discussion of British anti-Americanism: whether it is primarily a trait of the establishment, or essentially a feature of the left. The author of an article in a conservative British magazine that welcomed the tea shop's plucky reappearance in London amid the sprawling American empire of a coffee shop chain confessed that Starbucks is "liable to bring out unworthy feelings of anti-Americanism in me." He portrayed Starbucks as "the Yankee grey squirrel of the high street, which has reduced our native tea and coffee shops to a few redoubts in the North" (Trefgarne 2007).[14]

For some analysts, the anti-Americanism of the ruling class is the only genuine sort (Spiro 1988). They dismiss popular anti-Americanism as a sporadic, nonsystemic phenomenon. But the gray squirrel provides a common enemy against which Britons from all walks of life can unite. A *New York Times* feature about inter-national couples negotiating differences with their respective families supplied another instance of guilt by association. One profiled couple was a London-based American woman and her British husband. Her future mother-in-law made a point, at their first meeting, of bringing up the "awful American gray squirrels" that were "chasing all the lovely English red squirrels out of Britain." This was something she attributed to generic, reflexive anti-Americanism, until she read up about the gray's misdeeds (Emling 2006).

For those who regard anti-Americanism as effectively the preserve of the UK establishment, the "over here" variety exemplifies the largely benign and jocular popular form (Spiro 1988). However, this understates its potency, for the phrase is no innocuous relic. It has proved highly adaptable, surviving the closure of US bases in Britain and the end of the Cold War. This retention of cultural currency has a lot to do with the robust condition of the gray squirrel and other vibrant new species from North America. The transfer from human to nonhuman has been effortless. The richness of existing associations commends the "over here" analogy as the weapon of choice in the environmental manager's rhetorical arsenal.

Today's red squirrel champions tend to forget that their darling, too, once incurred the forester's wrath (MacDonald 1954) and the gamekeeper's ire for taking young pheasants and pheasant eggs (Rusticus 1850; Harvie-Brown 1881). Following nineteenth-century reintroductions based on mainland European stock, Scottish populations rebounded to such an extent that the Highland Squirrel Club, a consortium of large Scottish estates, was able to kill 60,450 reds during the period it administered its own bounty (1903–17) (Ritchie 1920). During World War Two, a forester complained, it was hard to find trees for telegraph poles because reds had extensively damaged suitable specimens (Nature Conservancy 1951). And, as late as the early 1950s, a journalist observed that the red was "only a little less troublesome than the grey, in spite of what has been said to the contrary" and damage to Scots pine plantations was reported (*Illustrated London News* 1953; Shorten 1955).[15]

However, native status provided a buffer and restricted critics' rhetorical options. The threat of an invader that is also a foreigner can be more easily "sexed up" (a phrase connected with those in Tony Blair's government who talked up the military threat posed by Saddam Hussein to justify intervention in Iraq). Even a thoroughly disinterested biologists' account of the two squirrels in Britain was titled "An American invasion of Great Britain" (Usher et al. 1992).

How closely, though, do responses to the gray squirrel and fellow-American fauna mirror attitudes to Americans and American activities? We can chart the peaks and troughs of recent British anti-American feeling with some accuracy. Though we need to distinguish between opposition to specific US policies and general attitudes toward America, the Vietnam War represents a fairly unambiguous high point.[16] And during the Clinton presidency (1992–2000), its profile was comparatively low. On a tour (1997) to promote his book, *A History of the American People*, British historian Paul Johnson, a great admirer of the United States and conservative values, encountered minimal anti-American feeling. He was pleased to report that "overpaid, oversexed, and over here" was merely a "historical curiosity," incomprehensible to Britons under forty (Johnson 1997).

Johnson's verdict on anti-Americanism's imminent death was premature. The George W. Bush presidency and invasion of Iraq precipitated a resurgence. Insofar as how Britons react to the acrobatic raider of their backyard bird feeders is influenced by and sometimes conflated with how they respond to American behavior abroad, British attitudes to the gray tell us as much about British perceptions of Americans as they do about the creature itself. The independent momentum that the dominant British view of the gray squirrel seems to possess fits with the main thrust of recent studies of anti-Americanism in Europe, which contend that the phenomenon is a general orientation with deep cultural and ideological roots (endemic variety), rather than a response to specific events and policies (circumstantial variety) (Kroes 2006). And insofar as a hard core of British anti-Americanism exists—a "sizeable group that can be expected to be anti-American at all times" (Graham 1952)—then it exists with reference to the gray squirrel.

For rhetorical purposes, the "bigger, brasher American cousin" that has "bullied" the red "out of house and home" (Wright 2011) has become a furry embodiment of what some Britons see as domineering US multinationals and chauvinistic foreign policy. (The language of criticism would certainly be far less colorful had Starbucks never crossed the Atlantic.) However, the particularly virulent most recent strain of anti-Americanism in Britain does not explain the high degree of hostility the gray has encountered over the past decade. It is not just about what the gray squirrel stands for. What it does matters too. The gray squirrel would have been as unpopular lately regardless of the global advance of American military and economic might, or presidential popularity, due to its own sustained success (with more than a little help from its unwitting ally, squirrelpox) and the hapless red's continuing decline.

Conclusion: Making Peace with the Gray Squirrel

Research for the English, Welsh, and Scottish governments recently ranked the gray squirrel tenth on the list of costliest invasive species from an economic standpoint (Meikle 2010). It barks trees, nibbles on garden plants and digs up bulbs, raids bird nests, and chews the insulation off electric cables (Marston 1964). It can also cause substantial structural damage to houses by infiltrating lofts. Above all, though, its presence is resented because it is held responsible for the red squirrel's marginalized status. Among government-employed and contracted scientists, the consensus remains that "the main reason for the decline [of the red squirrel] has been the spread of the non-native grey squirrel . . . through direct competition for resources, and also by transmitting squirrelpox virus" (Natural England 2009). As such, Natural England, the Forestry Commission, and the Department of Environment, Food and Rural Affairs (DEFRA)[17] continue to monitor the gray's economic impact and, to maintain biodiversity, support groups like Red Alert.[18] Craig Shuttleworth, national operations director of the Red Squirrel Survival Trust, is moderately optimistic. Citing the example of the northern Welsh island of Anglesey, scene of a successful recent cull of grays and bolstering of the indigenous population, he contends that culls and reintroductions stand a realistic chance of success elsewhere too, particularly on other islands and peninsulas.

However, the UK government has no illusions. Over half a century ago, at the time of the bounty, the nation's leading squirrel researcher doubted that "we shall ever rid ourselves [of the gray]" (Shorten 1954). Today's wildlife managers are even more convinced that the gray is here to stay. Moreover, while many Britons insist that their backyard squirrel "should be the red original, not this grey impostor" (Baker 2005), others have made their peace. A recent Under-Secretary of State at DEFRA pointed out that some Britons enjoy them as "part of our wildlife" (Ruddock 2008). From this standpoint, eradication is not just infeasible; it is undesirable. In another fifty years, whether or not reds are still around, the gray squirrel may have finally shaken off its status as permanent alien, accepted as being of Britain as well as in Britain. If and when this happens, justice will have finally been done, given that many of today's red squirrels are themselves descended from continental European stock.

Policy Implications

British wildlife biologist Stephen Harris has questioned the value of committing further, increasingly scarce resources to what has been a largely fatuous effort to control the gray squirrel population and reinstate reds. Not only is this an essentially unwinnable war (apart from on islands such as Anglesey, Arran, and Brownsea). It is also being waged on behalf of a creature that, beyond Britain, is not at all endangered (Harris 2006). To advance the red cause, a zoological subspecies has effectively been upgraded to a full-fledged, endemic species through casual but potent taxonomic "inflation" (*Economist* 2007). The red squirrel is distributed across Eurasia from Portugal to Japan. The subspecies endemic to Britain is one of seventeen, distinguished from the other sixteen by attributes such as monomorphism (absence of color varieties) and the winter whitening of ear tufts and tail (Tittensor 1970, 1975). But Harris's point that there are still plenty of red squirrels elsewhere in Eurasia (which could provide sources of reintroduced populations at some future point, as they did for earlier restockings [Harris, 1973/4]), is unlikely to console those with a heavy emotional and cultural investment in the British red squirrel's survival. An animal is rarely just an animal. And an animal is particularly precious when wrapped in the flag.

Apart from concentrating our minds on this complicating factor in the formulation of wildlife management strategies, however, it is not clear that lessons can be learned from this particular unintentional ecological experiment and its cultural fallout. Exploring the gray and red squirrels' joint history indicates that native wildlife populations rise and fall without adding invasive species to the mix. This tale of two squirrels also reminds us that the activities of a new species are not necessarily the sole or even the main reason for a native species' plight. Yet this does not mean that efforts to control other invasive species should be relaxed. Like their counterparts elsewhere, British invasion biologists emphasize the need for preventative measures; namely, a rigorous risk assessment exercise on a case-by-case basis for all suggested introductions, to assist in the formulation of "effective control measures *before* they are needed" (Manchester and Bullock 2000).

Though the extirpation of an invasive species is a relatively rare event, two successful campaigns have been executed in Britain. Both targeted creatures from the Americas—one North American (muskrat) and one South American (coypu, or nutria)—and both applied the lessons of the gray squirrel experience. The muskrat was imported to boost fur farming in the early 1920s. Escapees swiftly naturalized and vigorously expanding colonies became an economic pest, preying on crops and undermining the banks of rivers and canals. A short and highly effective eradication campaign by government trappers eliminated the creature from the British Isles (including Ireland) by 1937. Elsewhere in Europe, feral populations continue to thrive. In some regions, they still constitute a serious nuisance. Britain's campaign worked because it was initiated sooner than in continental Europe—at a time when numbers were comparatively low and clustered within a few well-defined areas of infestation. In short, the problem could be nipped in the bud.

Finally, what does the debate over the gray squirrel tell us about how invasive species from other countries are regarded in Britain? Is there a special relationship between American animals and the British? In the absence of fraught relations between the country of origin and the host nation, the debate about an invasive species is likely to operate within much stricter economic and ecological parameters. This is why the gray squirrel is usually associated specifically with the United States rather than North America as a whole. There is far less mileage in criticizing a North American—let alone a Canadian—gray squirrel. Apart from those hailing from North America, most of the invasive species currently causing the greatest perturbation in Britain derive from Asia. So it is not possible to compare responses to American invasive species with attitudes to organisms from European countries with which Britain has often experienced rough relations over recent centuries, such as France and Germany.

At the same time, the gray squirrel's potential for wider damage to British-American relations should not be overstated. Despite the consternation that grays caused Macmillan, they did not spoil his good relations with Eisenhower and Kennedy. The relatively smooth overall course of this political and cultural relationship has remained unruffled by any number of allusions—from citizens, journalists, conservationists, and government employees—to unbridled, supersized, and beastly American beasts.

Acknowledgments

The author thanks Earthscan (now part of the Taylor & Francis group) for permission to reprint some of the material that appeared in "Over here: American animals in Britain," in *Invasive and Introduced Plants and Animals: Human Perceptions, Attitudes and Approaches to Management*, edited by Ian Rotherham and Robert Lambert. London: Earthscan, 2011: 39–54. He also thanks the anonymous reviewers and editors for their comments on earlier drafts, as well as his University of Bristol colleague Stephen Harris for making a series of valuable points on a draft version.

Notes

1. The signal crayfish exemplifies the characteristics that have equipped various American species for competitive success. Brought over in the early 1970s for the restaurant trade, it readily escaped the aquaculture tank. Bigger and more aggressive than its native counterpart, the white-clawed crayfish, it also breeds earlier, lays more eggs that hatch sooner, consumes a wider range of foods, is more tolerant of poor water quality, and spreads a fungal disease to which it is immune. Indige-

nous crayfish have been virtually supplanted south of a line drawn between Severn and Humber. http://www.avonwildlifetrust.org.uk/wildlife/project_crayfish.htm. Invasives from North America also loom large in journalistic attention to alleged British intolerance of nonnative species (Liddle 2002; Johnston 2009).

2. The only continental European country with gray squirrels is Italy (introduced to Piedmont in the northwest of the country in 1948) (Bertolini and Genovesi 2003).

3. A sketch in the MAF records (1952–53), "Get Rid of the Tree Rat," depicts two identical creatures—"grey squirrel without a tail" and "common rat without a tail."

4. A retired gamekeeper from Wiltshire claimed to have witnessed such an emasculating attack during the 1930s (Collyer 1961).

5. In fact, both species typically produced two annual litters of 3–5 kits (FC 1956b).

6. In the early 1970s, Richard Mabey distinguished between the enhancing role of the urban gray squirrel and its less desirable presence in the countryside (Mabey 1973).

7. The red squirrel also preyed on birds (Rusticus 1857).

8. Campaigners Tim and Pat Cook, who coordinate the Patterdale and District Red Squirrel Group in the Lake District, have produced a heavily polemical children's book (profits donated to conservation projects): *Stumpy: Hero of the Lakes*. Stumpy, who lost a front paw and rear foot in combat with Brownfang, leader of the invading grays, is equipped with artificial limbs (complete with sharp metal claws) by a human friend. Leader of the Red Defence Force, Stumpy recruits various local creatures as "warriors" to protect their native Lakeland and repossess lands "stolen" by the voracious, innumerable fast-breeders (pp. 22, 46). The red deer stag (Magnus) refers to grays as "tree-rats" and "greedy" "ugly brutes" that, according to the ravens, "don't belong here" (pp. 6, 20). Another red crusader trumpets the imperative to "push back this hideous grey tide" of "muscular foreign" "grey vermin" (pp. 13, 53, 54).

9. In fact, the red squirrel's habitat in northern England and southern Scotland at that time included broadleaf woodland. And woodland on the Isle of Wight, where the red remains the only variety of squirrel, is also broadleaf. I am grateful to Stephen Harris for pointing this out.

10. There are no available figures for the size of Britain's native squirrel population at the time of the gray squirrel's introduction.

11. American servicemen may have been overfed, but some still craved an authentic taste of home. MAF correspondence confirms the existence of a market among US troops for what, back home, was a prized game species: (F. Winch to F. E. Charlton 1943). Monica Shorten—a keen consumer of squirrel pie and fried squirrel—reports that GIs were prepared to pay up to five shillings a head (Shorten 1954:58). Other Britons boiled them up for wartime dog food (Tallents 1946; *News Chronicle* undated).

12. Parker betrayed no hint of animosity to the United States, erroneously stating that the gray comes from South America.

13. This marked a refinement of Crosby's original stance. Previously, he observed that it had "nearly driven Britain's red squirrels right off the island in the last seventy years" (Crosby 1972).

14. This particular approach may be more firmly rooted in antiglobalization than in anti-Americanism. Those who position themselves as pro-red rather than anti-gray tend to regard their cause as part of a broader effort to preserve national and regional distinctiveness and protect embattled minorities in a world of rampant globalization. A pro-red stance entails the same commitment to the survival of local heritage, community identity, and the ethos of diversity that invests the championing of, say, local cheeses and apples against the tasteless universalism of international agribusiness. Gray squirrels are like Granny Smith apples—examples of the new age of the Homogenocene, toxic to gastronomic, cultural, and ecological pluralism.

15. Given that gray colonies are typically denser than those of reds, it could be that damage levels from grays are also generally higher. I am grateful to Stephen Harris for this observation.

16. Polls for fifteen nations that registered strong general disapproval of US policies in Vietnam also indicated a widely held sympathetic general opinion of the United States (Crespi 1983:720).

17. DEFRA succeeded the Ministry of Agriculture, Fisheries and Food (MAFF) in 2002, which had replaced the Ministry of Agriculture and Fisheries (MAF) in 1955.

18. Page 2009; Dutton 2004:32–36. Some researchers advocate field trials of immunocontraceptive vaccines to control the population of grays (Rushton et al. 2002). At the red squirrel reserve in nearby Formby, University of Liverpool researchers recently recorded the first instance of a wild red squirrel surviving the squirrelpox virus. This glimmer of hope that England's red squirrel population may be developing resistance has received widespread coverage in the British media (Barkham 2013, Johnson 2013, National Trust 2013).

Literature Cited

Adams, T. 2008 (19 October). They shoot squirrels, don't they? *Observer.*

Atkinson, C. 1954. (20 January; 26 October). Records of the Ministry of Agriculture and Fisheries (MAF), MAF 131/53, National Archives, Kew, UK (NA).

Baker, N. 2005. Wild year. *National Trust Magazine* 106:84.

Barkham, P. 2013 (22 November). Red squirrels showing signs of resistance to poxvirus. *Guardian.*

BBC News. 2006. (16 April). Anti-Americanism "feels like racism."

Berman, R. A. 2008. *Anti-Americanism in Europe: A Cultural Problem.* Stanford: Hoover Institution Press.

Bertolini, S., and P. Genovesi. 2003. Spread and attempted eradication of the grey squirrel in Italy, and consequences for the red squirrel. *Biological Conservation* 109:351–358.

Blogspot. 2006. The Vol Abroad. http://thevolabroad.blogspot.com/2006_03_19 _archive.html.

Bookman. 1931. Review of "The Grey Squirrel" by A. D. Middleton. 81:126.

Burnham, N. 2009 (18 March). Eating the enemy. *Guardian.*

Burton, M. 1947 (15 February). The grey squirrel in Britain. *Illustrated London News* 210:211.

Carrell, S. 2009 (10 February). Sure he's cute . . . but not cute enough to save him from the great squirrel cull. *Guardian.*

Crespi, L. P. 1983. West European perceptions of the United States. *Political Psychology* 4/4: 720.

Colinvaux, P. A. 1973. *Introduction to Ecology.* New York: Wiley: 301.

Collyer, A. J. 1961 (3 July). Correspondence: untitled. MAF 131/103.

Cook, T., and P. Cook. 2008. *Stumpy: Hero of the Lakes.* Kendal: Sciurosso.

Country Life. 1943 (30 July). Grey squirrels: 192.

Crosby, A. W. 1972. *The Columbian Exchange: Biological and Cultural Consequences of 1492.* Westport, CT: Greenwood Press: 210–212.

Crosby, A. W. 1986. *Ecological Imperialism: The Biological Expansion of Europe, 900–1900.* New York: Cambridge University Press: 164–165, 193.

Drummond, H. 1984 (1 December). The unequal status of the red squirrel. *Field*: 48.

Dutton, C. 2004. The red squirrel: redressing the wrong. Woodbridge: European Squirrel Initiative: 37–38.

Economist. 2007 (17 May). Species inflation: Hail Linnaeus: Conservationists— and polar bears—should heed the lessons of economics. http://www.economist .com/node/9191545/print.

Edwards, R. 2001(1 July). Beastly aliens taking over the countryside. *Glasgow Sunday Herald.*

Elton, C. S. 1925 (8 November). Charles Sutherland Elton Papers, Ms. Eng. C.3326, Box A/15, Western Manuscripts and Special Collections, University of Oxford.

Elton, C. S. 1927. *Animal Ecology.* London: Sidgwick and Jackson: 28.

Elton, C. S. 1951 (28 March). Quoted in letter from Nature Conservancy to Colonel Boyle (Fauna Preservation Society), FT 1/24 (Records created by and inherited by the Nature Conservancy), NA.

Elton, C. S. 1958. *The Ecology of Invasions by Animals and Plants.* London: Methuen:15, 73, 123, 148.

Emling, S. 2006 (10 January). Meet the family: complexity and stress at holiday time. *New York Times.*

Environment Agency (EA). 2006 (3 August). Top ten most wanted foreign species. http://www.environment-agency.gov.uk/news/1444976.

Finn, F. 1909. *The Wild Beasts of the World*. London: E. C. Jack: 190–191.

Forestry Commission (FC). 1953 (10 March). War on the grey squirrel. MAF 130/53.

FC. 1954 (5 May). Committee on Grey Squirrels. Minutes of meeting. MAF 131/53.

FC. 1955a. Hints on controlling grey squirrels. MAF 131/53.

FC. 1955b. Committee on Grey Squirrels. Minutes of meeting on 4 May. MAF 131/53.

FC. 1955c. Committee on Grey Squirrels. Minutes of meeting on 26 October 1955. MAF 131/53.

FC. 1956a (12 January). The campaign against the grey squirrel. MAF 131/53.

FC. 1956b. Traps for grey squirrels. MAF 131/88.

FC. 1960. The grey squirrel: a woodland pest. Leaflet no. 31. MAF 149/116.

Garson, P. J., and P. W. W. Lurz. 1992. The distribution of red and grey squirrels in northeast England in relation to available woodland habitats. *Bulletin of the British Ecological Society* 23:133–139.

Gould, C. 2009. Don't tread on me: anti-Americanism abroad. London: Social Affairs Unit.

Graham, M. D. 1952. Anti-Americanism, British garden variety. *Antioch Review* 12:217–228.

Gurnell, J. 1987. The natural history of squirrels. London: Christopher Helm: 160–164.

Hansard (Lords). 1998 (25 March). Session 1997–98. 587, Column 1318, http://www.parliament.the-stationery-office.co.uk/pa/ld199798/ldhansrd/vo980325/text/80325-10.htm#80325-10_head0.

Hansard (Lords). 2006 (23 March). Session 2005–2006. 680, Column 362,

http://www.publications.parliament.uk/pa/ld200506/ldhansrd/vo060323/text/60323-04.htm#60323-04_head1.

Harris, S. 1973/4. The history and distribution of squirrels in Essex. *Essex Naturalist* 33/2:65.

Harris, S. 2006 (September). The red has lost—so accept the grey. *BBC Wildlife* 24/9:38–39.

Harris, S. 2011. Our findings about mammals (Wildlife to Work survey). *BBC Wildlife* 29/11:72.

Harris, S., and D. W. Yalden, eds. 2008. *Mammals of the British Isles: Handbook*. Southampton: Mammal Society: 10, 13, 63, 70.

Harvie-Brown, J. A. 1881. *The History of the Squirrel in Great Britain*. Edinburgh: McFarlane and Erskine: 97, 103–105.

Hill, M., R. Baker, G. Broad, P. J. Chandler, G. H. Copp, J. Ellis, D. Jones, et al. 2005. *Audit of Nonnative Species in England*. Peterborough, UK: English Nature: 30.

Hollander, P., ed. 2004. *Understanding Anti-Americanism: Its Origins and Impact at Home and Abroad*. Chicago: Ivan R. Dee.

Hollander, P. 2006. *Anti-Americanism: Irrational and Rational.* New York: Transaction.

Illustrated London News. 1931 (14 November). Indicted as egg-thief and tree-damager: the grey squirrel: 773.

Illustrated London News. 1953 (18 April). A pest of the British countryside. 222: 624–625.

Independent Panel on Forestry (IPF). 2011. Visit of the independent panel on forestry to Northumberland (including Kielder) on 26 July 2011: 3, 5, 9, 13.

Ingham, J. 2000 (15 September). Stars go nuts to help save the red squirrel. *Daily Express.*

Johnson, M. 2013 (2 November). Supersquirrels! Survival hope for our furry friends. *Liverpool Echo.*

Johnson, P. 1997 (1 November). The decline and fall of anti-Americanism in Britain. *Spectator* 279: 30.

Johnston, I. 2009 (26 January). UK accused of "racism" towards invaders from across the pond. *Independent.*

Kean, H. 2000. Save "our" red squirrel: kill the American grey tree rat. In *Seeing History: Public History in Britain Now*, edited by H. Kean, P. Martin, and S. Morgan, 51–64. London: Francis Boutle .

Kean, H. 2001. Imagining rabbits and squirrels in the English countryside. *Society and Animals* 9:164.

Kenward, R. E. 1987. Bark stripping by grey squirrels in Britain and North America: why does the damage differ? In *Mammals as Pests*, edited by R. J. Putman, 144–154. London: Chapman and Hall.

Kenward, R. E., and J. L. Holm. 1993. On the replacement of the red squirrel in Britain: a phytotoxic explanation. *Proceedings of the Royal Society B* 251:187–194.

Kroes, R. 2006. European anti-Americanism: what's new? *Journal of American History* 93/2:417–431.

Lacourne, D., and T. Judt. 2006. *With Us or Against Us: Studies in Global Anti-Americanism.* Basingstoke: Plagrave Macmillan.

Lancum, F. H. 1947. *Wild Animals and the Land.* London: Crosby Lockwood: 69.

Lear, L. 2007. *Beatrix Potter: A Life in Nature.* London: Allen Lane: 245.

Leyland, E. 1955. *Wild Animals.* London: Edmund Ward: 63.

Liddle, R. 2002 (4 September). Our hatred of certain furry foreigners. *Guardian.*

Logan, V. M. 1967 (4 May). Correspondence: application for a squirrel licence. MAF 131/250.

Mabey, R. 1973. *The Unofficial Countryside.* London: Collins: 76–77.

MacDonald, P. J. 1954 (8 March). Letter to Mrs Vizoso (Monica Shorten). MAF 131/103.

MacKenzie, J. M. 2001a. Editorial. *Environment and History* 7: 253.

MacKenzie, J. M. 2001b. Editorial. *Environment and History* 7: 380.

MacKinnon, K. 1978. Competition between red and grey squirrels. *Mammal Review* 8:185–190.

Macmillan, H. 1960 (3 January, 29 February and 15 August). PREM (Prime Minister's Office Files) 11/3196, NA.

Ministry of Agriculture and Fisheries (MAF).1943a (20 January; 4 and 9 February). Minute sheets. MAF 44/45.

MAF. 1943b. Grey squirrels. Advisory leaflet 58. London. MAF 44/45.

MAF. 1943c (10 September). Minute sheet. MAF 44/45.

MAF. 1943d (15 September). Memorandum to executive officers of County War Agricultural Executive Committees in England and Wales, destruction of grey squirrels: Grey Squirrels Order. MAF 44/45.

MAF. 1943e (18 October). F. E. Charlton to Minister of Agriculture. MAF 44/45.

MAF. 1943f (4 January). Press notice. MAF 44/45.

MAF. 1946 (17 October). MAF 44/45.

MAF. (1952–53). "Get Rid of the Tree Rat." Infestation Control Division. MAF 131/87.

MAF. 1953a (February). The grey squirrel. MAF 131/83.

MAF. 1953b (April). You versus pests: the grey squirrel menace. MAF 131/53.

MAF. 1953c (31 May). Grey squirrel campaign: progress report to 31st May, 1953. MAF 131/83.

MAF. 1955a (21 June). The campaign against the grey squirrel: more than 750,000 "kills" in two years. MAF 131/53.

MAF. 1955b (October). Grey squirrel campaign: progress report for six months ended 30th September, 1955. MAF 131/53.

MAF. 1959 (January). Squirrel film script. MAF 131/53.

MAF. 1960 (5 January; 17 August and 5 October). PREM 11/3196.

MAF. 1961a (29 March and 15 May). PREM 11/3196.

MAF. 1961b (9 January). Grey squirrels. MAF 131/90.

Manchester, S. J., and J. M. Bullock. 2000. The impacts of nonnative species in UK biodiversity and the effectiveness of control. *Journal of Applied Ecology* 37/5:859.

Marjoram, T. J. 1958 (28 March). Grey squirrels. MAF 149/116.

Markovits, A. S. 2007. *Uncouth Nation: Why Europe Dislikes America*. Princeton: Princeton University Press.

Marston, C. (MAF, Infestation Control Divisions). 1964 (7 May). Letter to W.A. Williams (Industrial Pest Control Association). MAF 131/250.

Massie, A. 2007 (9 October). The ugly American abroad: animal version. http://blogs.spectator.co.uk/alex-massie/2007/10/the-ugly-american-abroad-animal-version/.

Max, D. T. 2007 (7 October). The squirrel wars. *New York Times*.

Meat Trades Journal. 1943 (January). Grey squirrel pelts (clipping). MAF 44/45.

Meikle, J. 2010 (15 December). Rabbits named Britain's most costly invasive species. *Guardian*.

Middleton, A. D. 1930. The ecology of the American grey squirrel (*Sciurus carolinensis* Gmelin) in the British Isles. *Proceedings of the General Meetings for Scientific Business of the Zoological Society of London* 54/2:812, 839, 837.

Middleton, A. D. 1931. *The Grey Squirrel: The Introduction and Spread of the Grey Squirrel in the British Isles*. London: Sidgwick and Jackson: v, 1–2, 7, 9, 71–75.

Middleton, A. D. 1935. The distribution of the grey squirrel in Great Britain in 1935. *Journal of Animal Ecology* 4: 274–276.

Monbiot, G. 2009 (1 October). How British nationalists got their claws into my crayfish. *Guardian*.

National Trust. 2013 (25 November). Formby's red squirrel population recovering. http://www.nationaltrust.org.uk/article-1355813511001/

Natural England. 2009 (16 September). Review of red squirrel conservation activity in northern England: 1.

Nature Conservancy. 1951 (13 March). Letter to Colonel Boyle (Fauna Preservation Society), FT 1/24.

News Chronicle. Undated. This is no time to be little grey squirrel (clipping). MAF 131/87.

New York Times. 1929 (11 November). American squirrel on trial for his life in England: he is charged with killing his English cousins and with destroying bird life.

New York Times. 1930 (3 August). Gray squirrel in England gains faster than in American home.

Northern red squirrels. 2009. *Northern News* 3:1.

Nottingham Evening News. 1955 (15 December). Attack on squirrels stepped up. MAF 131/88.

O'Connor, B., and M. Griffiths, eds. 2005. *The Rise of Anti-Americanism*. Abingdon: Routledge.

O'Hanlon, S. R. 1956 (17 January). Correspondence: grey squirrel bounty scheme. MAF 131/53.

Page, R. 2009 (15 October). Red squirrels fight for survival. *Daily Telegraph*.

Potter, B. 1903. *The Tale of Squirrel Nutkin*. London: Frederick Warne.

Potter, B. 1911. *The Tale of Timmy Tiptoes*. London: Frederick Warne.

Pratt, M. 2011. Kielder Forest, multipurpose forestry in action: consequences for biodiversity. Visit of the Independent Panel on Forestry to Northumberland (including Kielder) on 26th July 2011: 9.

Quarterly Journal of Forestry. 1953 (January). Grey squirrel clubs. 47:55.

Revel, J.-F. 2004. *Anti-Americanism*. London: Encounter.

Reynolds, D. 1995. *Rich Relations: The American Occupation of Britain, 1942–1945*. London: HarperCollins: xxiii.

Reynolds, J. C. 1981. The interaction of red and grey squirrels. PhD diss., School of Biological Sciences, University of East Anglia: 1, 187, 200.

Ritchie, J. 1920. *The Influence of Man on Animal Life in Scotland: A Study in Faunal Evolution*. Cambridge: Cambridge University Press: 181, 290.

Ross, A., and K. Ross, eds. 2004. *Anti-Americanism*. New York: New York University Press.

Ruddock, J. 2008 (27 February). Letter from Ruddock (Parliamentary Under-Secretary of State, Climate Change, Biodiversity and Waste, DEFRA) to Michael Jack (Chairman, Environment, Food and Rural Affairs Committee). http://www.parliament.uk/documents/upload/080227-ruddock-on-grey-squirrels.pdf.

Rushton, S. P., J. Gurnell, P. W. W. Lurz, and R. M. Fuller. 2002. Modeling impacts and costs of gray squirrel control regimes on the viability of red squirrel populations. *Journal of Wildlife Management* 66:683–697.

Rushton, S. P., P. W. W. Lurz, J. Gurnell, P. Nettleton, C. Bruemmer, M. D. F. Shirley, and A. W. Sainsbury. 2006. Disease threats posed by alien species: the role of a poxvirus in the decline of the native red squirrel in Britain. *Epidemiology and Infection* 134/3:522, 531.

Russell, D. 1936 (22 February). A rogue in grey. *Saturday Review of Politics, Literature, Science and Art* 161:243.

Rusticus. 1850 (20 April). The ways of the squirrel. *Chambers' Edinburgh Journal* 329:247.

Rusticus. 1857 (9 April). The squirrel: second paper. *Leisure Hour* 276:229.

Ryle, G. B. (FC, England). 1952. Letter to J. Marjoram (MAF, Pest Control Division), MAF 131/83.

Sagoff, M. 2002. What's wrong with exotic species? In *Philosophical Dimensions of Public Policy*, edited by V. Gehring and W. Galston, 327–340. New Brunswick, NJ: Transaction.

Sardar, Z., and M. W. Davies. 2003. *Why Do People Hate America?* Thriplow: Icon Books.

Scottish Sporting Gazette. 2006. Red under threat. 25:130.

Seymour, W. 1961. Grey squirrels. *Quarterly Journal of Forestry* 55:293.

Sheail, J. 1999. The grey squirrel (*Sciurus carolinensis*)—a UK historical perspective on a vertebrate pest species. *Journal of Environmental Management* 55:145–156.

Shorten, M. 1953. Notes on the distribution of the grey squirrel (*Sciurus carolinensis*) and the red squirrel (*S. vulgaris leucourus*) in England and Wales from 1945 to 1952. *Journal of Animal Ecology* 2:134–140.

Shorten, M. 1954. *Squirrels*. London: Collins: ix, x, 1, 5, 38, 58, 93–96, 172–176, 185–186.

Shorten, M. 1955. Damage caused by squirrels in Forestry Commission areas in 1954 and 1955: 9–11. MAF 130/82.

Skelcher, G. 1997. The ecological replacement of red by grey squirrels. In *The Conservation of the Red Squirrel: Sciurus vulgaris L.*, edited by J. Gurnell and P. Lurz. London: People's Trust for Endangered Species: 76.

Smout, T. C. 2009. The alien species in twentieth-century Britain: inventing a new vermin. In *Exploring Environmental History*, edited by T. C. Smout. Edinburgh: Edinburgh University Press: 172.

Smout, T. C. 2011. How the concept of alien species emerged and developed in 20th-century Britain. In *Invasive and Introduced Plants and Animals: Human Perceptions, Attitudes and Approaches to Management*, edited by I. D. Rotherham and R. A. Lambert. London: Earthscan: 59.

Spiro, H. J. 1988. Anti-Americanism in western Europe. *Annals of the American Academy of Political and Social Science* 497:124.

Squirrel Wars: Red versus Grey. 2009 (9 July). Channel 4. http://www.channel4.com/programmes/squirrel-wars-red-vs-grey.

Stamp, L. D. 1955. *Man and the Land*. London: Collins: 210.

Stillman, W. J. 1899 (23 September). The squirrel (letter to the editor). *Outlook* 4:235.

Stillman, W. J. 1907 (31 August). A chat about squirrels. *Estates Gazette*.

Sweig, J. E. 2006. *Friendly Fire: Losing Friends and Making Enemies in the Anti-American Century*. New York: PublicAffairs.

Tallents, S. 1946. Letter to the editor. *Times*, December 2.

Thompson, H. V., and T. R. Peace. 1962. The grey squirrel problem. *Quarterly Journal of Forestry* 56:38–39.

Times, April 7, 1936.

Times Weekly Edition. 1952 (14 February). Alien squirrels' tricks (clipping). FT, 1/24.

Tittensor, A. M. 1970. The red squirrel (*Sciurus vulgaris* L.) in relation to its food resources. PhD diss., Department of Forestry and Natural Resources, University of Edinburgh.

Tittensor, A. M. 1975. *Red Squirrel*. London: Forestry Commission: 16.

Tittensor, A. 1979 (25 October). What future for the reds? *Country Life* 166:1394–1395.

Trefgarne, G. 2007 (20 January). Is it time for tea? *Spectator* 303:65.

Tyler, D. 2001 (29 January). BBC Radio 4.

Tyler, D. (Undated). Scapegoating the aliens. Animal Aid special report. http://www.animalaid.org.uk/images/pdf/factfiles/aliens.pdf.

Usher, M. B., T. J. Crawford, and J. L. Banwell. 1992. An American invasion of Great Britain: the case of the native and alien squirrel (*Sciurus*) species. *Conservation Biology* 6:108.

Wakely, C. 2011 (13 August). British fauna threatened by habitat loss and invasive species. http://nationalistunityforum.co.uk/index.php/british-fauna-threatened-by-habitat-loss-and-invasive-species/.

Waldegrave (Earl). 1962 (31 January). Oral reply. Grey Squirrels. House of Lords. MAF 131/54.

Watson, L. 2008 (17 October). Game dealer rustles up recipe for red squirrel conservation. *Cumberland News*.

Watt, H. B. 1923. On the American grey squirrel (*Sciurus carolinensis*) in the British Isles. *Essex Naturalist* 20:189–205.

Winch, F., to F. E. Charlton. 1943 (21 October). MAF 44/45, NA.

Wright, J. 2011 (9 February). Hope for Formby's grey squirrels. *Formby Times*.

Wyatt, W. 1977. *The Further Adventures of Mr. Saucy Squirrel*. London: Allen and Unwin: 78–80.

CHAPTER FOUR

Fish Tales: Optimism and Other Bias in Rhetoric about Exotic Carps in America

Glenn Sandiford

Introduction

Policy about invasive species and other introduced biota is a "trans-scientific" issue (Weinberg 1972). Such matters are "questions of fact that can be stated in the language of science but are, in principle or in practice, unanswerable by science" (Majone 1989:3). Scientific knowledge can serve only as a guide. Ultimately, subjective judgment determines our individual and collective choices about whether the threats and impacts of exotic species are (un)acceptable and merit policy action (Larson 2007).

Trans-science issues thus transcend science and extend into politics. Stakeholders must persuade others of their particular position on invasive species. A persuasive argument is not necessarily a logical demonstration but involves rhetoric that inspires commitment (Majone 1989; Weber and Word 2001). Argumentation is central to the policy process. Whoever determines the ruling narrative about nature determines how society manages nature (Soulé and Lease 1995), as reflected in the ebb and flow of ideologies that have shaped US policy about biotic introductions since colonial times (Pauly 1996; Coates 2006).

For the past two decades, rhetoric about introduced species has been under scrutiny, particularly discourse about invasive biota. Critiques of science journals, government reports, and media stories uncovered narratives that were variously deemed alarmist, militaristic, demonizing, overgeneralized, and outdated (Eskridge 2010; Chew 2009; Larson 2005; Coates 2006). The criticism sparked strong rejoinders from conservation biologists and environmentalists (Simberloff 2003; Soulé and Lease 1995; Perry and Schueler 2003). However, largely absent from the ongoing debate are analyses of rhetorical tendencies that facilitated some of history's most costly and unwanted invasions by species *initially assessed as desirable*. Such introductions, invariably made with government involvement or sanction, show a historical pattern of rhetoric that begins with an overly optimistic assessment of the species *prior to importation*. After the introduced plant or animal becomes problematic, its attendant narrative undergoes a drastic evolution that cascades through denial, blaming, revisionism, and metaphorical reinvention of the now unpopular species. Particularly striking is how traits initially earmarked as virtues—fecundity, hardiness, etc.—become negative qualities (Fine and Christoforides 1991; Luken and Thieret 1996).

Many flora and fauna have undergone such transformations, including fish. During the past 130 years, four foreign carps have invaded the Mississippi Valley, the world's third largest river basin and encompassing all or part of 31 states. A fifth carp is becoming invasive there. The first invasion, which began in the 1880s, involved common carp. The US Commission of Fish and Fisheries (USFC) imported common carp from Europe for stocking nationwide as a food fish. Within two decades, it had become one of America's most widespread freshwater fishes—and, in sharp contrast to the pre-introduction hype of its "expert" boosters, widely condemned as inedible and harmful to native nature. Twentieth-century efforts to eradicate, control, and harvest common carp collectively cost hundreds of millions of dollars, remain ongoing in many states, and cemented *Cyprinus carpio* as one of the most oft-cited examples of a "bad" introduction. Yet despite this well-known history, the 1960s, 1970s, and 1980s saw a succession of four more foreign carps—collectively known as Asian carp—arrive in America. Their facilitator was a group of federal, state, and commercial aquaculturists centered in Arkansas who extolled the utility of Asian carps as biocontrols for aquatic weeds, algae, and parasites. The fish had long been reared as food too. In America, however, Asian carps not only fell short of their proponents' boosterism, today they are headline news for their massive invasion of Midwestern rivers, where they constitute 90% of fish biomass in some waters and injure boaters in collisions. In a replay of the saga with common carp, federal and state agencies are now spending tens of millions of dollars to limit further spread of Asian carp, incentivize its harvesting, and persuade Americans to eat it.

The nineteenth-century introduction of common carp is often characterized as emblematic of that era's rudimentary understanding of nature. The implication is that such disastrous intentional introductions could never occur in the "modern" age of information and environmental awareness. Yet such introductions did occur contemporarily, with Asian carps, whose infamy as invasive species has surpassed even that of common carp.

The striking similarities in rhetoric about foreign carps, spread across three centuries and huge advances in science, offer important lessons for policy makers. Above all, invasive species policy must better account for *human nature*, especially the tendency for excessive optimism and flawed foresight in advocates' narratives about foreign biota.

Rhetoric and Policy

"Optimism bias" is the universal human tendency to expect things will turn out better than actually happens. People are more optimistic than realistic (Kahneman and Tversky 2000). In public policy, optimism bias manifests itself as a systematic tendency for exaggerated hopes about megaprojects (urban rail, highways) and large-scale technologies (ethanol, hydropower), including acclimatization of plants and animals (Tenner 1996). Whether overestimating the likelihood of positive events or underestimating the risk of negative events, optimism bias can result in unwanted outcomes like cost overruns and benefit shortfalls (Flyvbjerg et al. 2005; Lovallo and Kahneman 2003).

"Framing" is the tailoring of narratives by policy participants, whether scientist, bureaucrat, fish farmer, environmentalist, or journalist. Potential influences on framing include values, emotions, education, job, social and institutional context, political agenda—and optimism bias. At worst, framing can distort data and exclude alternate viewpoints from consideration. Narrow framing minimizes the scope of policy "problems" in hopes of avoiding controversy, whereas broad-scale framing treats controversy as an unavoidable risk in the pursuit of a resolution that is more inclusive and long-term (Lakoff 2010; Ethridge 2008).

Optimism bias and framing occur unconsciously, but they can also be deliberate efforts toward strategic misrepresentation.

Idealizing Common Carp

In 1831, the owner of a transatlantic ship transported "two or three dozen" common carp from France to his estate beside the Hudson River, sixty miles north of New York City. Capt. Henry Robinson's carp were soon spawning in his ponds, so each year he released some of his excess fish into the Hudson River. Floods washed additional specimens into the river. Within two decades, common carp had become the first commercially important fish from abroad to establish reproducing populations in US waters (Anonymous 1851).

Common carp are large members of Cyprinidae, the world's biggest family of freshwater fish. In nineteenth-century America, cyprinids ranked low as table fish, being viewed as "coarse or insipid, and in this country are eaten only when other fish cannot be had" (Norris 1868:217). Yet in China and Europe, a handful of cyprinids including common carp had been reared as food for centuries. A "carp belt" had materialized in medieval Western Europe where generations of pisciculturists bred strains of *Cyprinus carpio* for pond culture, giving rise to folklore and holiday customs centered around the fish. Feral offspring of escaped carp further reshaped Europe's wild aquatic ecosystems (Balon 1995; Hoffman 1995).

In New York, Hudson River fishermen were netting common carp by the early 1840s, although anyone keeping their carp risked a fine. Capt. Robinson had persuaded state lawmakers to protect the piscine newcomer with a five-year ban on carp fishing (Herbert 1850).

Some New Yorkers applauded the captain's importation of carp. He had established "the practicability of introducing foreign fishes into our waters," wrote state zoologist James DeKay in inviting "other patriotic individuals" to attempt such endeavors (DeKay 1842:189). But popular outdoors writer Henry William Herbert belittled *Cyprinus carpio* as tasting like "the sole of a reasonably tender India-rubber shoe." Herbert complained that New York legislators had long ignored calls to protect declining native fauna, but "no sooner is a coarse, watery, foreign fish accidentally thrown into American waters, than it is vigorously and effectively protected"—and for no good reason in Herbert's mind. Besides his insults about its taste, the English-born writer deemed carp "very miserable sport" (Herbert 1850:199–200).

In 1872, with the Hudson River now yielding wagonloads of carp, *Cyprinus carpio* arrived on the West Coast. California dairy farmer Julius

Poppe returned from his native Germany with a handful of carp finger-lings for his ranch. Within a year, Poppe's young carp had bred (Poppe 1880). As he began selling the offspring, first locally and then abroad, his success was heralded in periodicals and government reports. Some of his yearlings weighed eight pounds, a growth rate far surpassing those of American fishes. In 1874, a storm flood washed some of Poppe's carp into a creek. Soon San Francisco fishmongers were selling carp, area piscicul-turists were breeding it—and local streams were "well stocked with this European stranger" (Anonymous 1875; Anonymous 1879:180).

Whether on East Coast or West Coast, enclosed pond or open river, common carp had bred easily and without need of human intervention, rapidly become abundant, and spread fast. And by the late 1870s, this precedent on both sides of the continent was well known.

Such a fish seemed manna for US advocates of pisciculture. Since the 1850s, in mainstream periodicals, books, and government reports, they had preached sermonic promise for fish stocking. Their utopian vision entailed native and foreign fishes being artificially propagated in huge numbers for planting as piscine seed to replenish existing fisheries and create new ones. Much like agriculture was converting America's woods and prairies into field, public pisciculture would transform US waters into a vast fish farm (Taylor 1999; Kinsey 2006). Boosters envisioned the Great Lakes and Mississippi Valley "alive with shining shoals" of salmon, shad, and other introduced fish (Scott 1869:379).

Their prophecies were persuasive to a fast-growing nation politically disinclined against regulation. America's fisheries were declining in the face of heavy commercial fishing, dams, and pollution. In the 1860s, sev-eral states began fish stocking programs, mostly with locally available spe-cies (Pisani 1984). The scale and scope of public pisciculture increased exponentially after Congress created the federal fish commission in 1871. Inaugural USFC chief Spencer Baird opened the floodgates to "an era of unprecedented transfers, introductions, and enthusiasms for fish of whatever species" (Thompson 1970:2). Collaborating with fish culturists and state fish commissions, Baird oversaw a national movement for fish stocking that became one of the century's most popular governmental programs. During the 1870s, the USFC functioned as a "biological clear-inghouse" that shipped tens of millions of ova and fry by train around the country for large-scale indiscriminate stocking (Allard 1978; Taylor 1999:77). During this honeymoon period public discourse about pisci-culture, shaped by Americans' naive faith in scientific experts like Baird, seldom extended beyond subjective arguments about which fishes were best.

Baird imported several species from Europe, but consignments rarely survived the transoceanic voyages. No foreign fishes were established in US waters during the 1870s, even though the USFC imported and stocked, without success, Atlantic salmon (*Salmo salar*) and whitefish (probably *Coregonus lavaretus*) from Germany, and sole (*Solea solea*) and turbot (*Psetta maxima*) from England. Results with American fishes were only marginally better. At decade's end, federal and state fish commissions had recorded only two self-sustaining successes of commercial significance—a Pacific Coast fishery for Atlantic shad (*Alosa sapidissima*), and white catfish (*Ameiurus catus*) in California (Towle 2000).

Enter common carp. In 1879, many Americans had high expectations about *Cyprinus carpio*, thanks to six years of boosterism in the mainstream press. The hype had started after correspondence in 1872 between Baird and a carp culturist in Germany named Rudolph Hessel, who lauded *Cyprinus carpio* as "a fish of great value for table-use and for feeding other fish." In central Europe, carp was common in large rivers and lakes, but "especially adapted for ponds." Hessel had seen 30-lb. specimens in Germany, where most villages and estates had carp ponds—"their propagation being left to nature" (Sandiford 2009; Hessel 1874:567).

In 1873, Hessel emigrated to America. Baird lacked sufficient funds to employ the German, but he did include Hessel's carp praise in his second USFC report to Congress (Hessel 1874), along with his own five-paragraph assessment of the cyprinid. According to Baird, "no other species . . . promises so great a return in limited waters." He portrayed *Cyprinus carpio* as a primarily herbivorous fish that grew much larger than trouts and basses, without the expense of supplemental feeding. "It is on this account that its culture has been continued for centuries." As for edibility, carp meat was "firm, flaky, and in some varieties almost equal to the European trout" (Baird 1874:lxxvi–lxxvii).

The production ethos that framed Baird's inaugural endorsement of *Cyprinus carpio* dovetailed with conventional wisdom about harnessing aquatic nature (Steinberg 1991). However, other elements of the Bairdian carp narrative are unexplainable by cultural norm or scientific ignorance. In particular, despite not having tasted carp meat himself, Baird repeatedly dismissed criticism of its edibility as uninformed. "Our great object is to increase the amount of animal food in the country, and it is purely a matter of cookery to make [carp] palatable" (Baird 1877:69).

Another hallmark of Baird's narrow-framed carp rhetoric was his rationalization that no American fish had the combination of attributes offered by *Cyprinus carpio*. In his third USFC report, published in 1876, Baird summarized the "good qualities of the carp":

1. Fecundity and adaptability to the processes of artificial propagation.
2. Living largely on a vegetable diet.
3. Hardiness in all stages of growth.
4. Adaptability to conditions unfavorable to any equally palatable American fish and to very varied climates.
5. Rapid growth.
6. Harmlessness in its relations to other fishes.
7. Ability to populate waters to their greatest extent.
8. Good table qualities (Baird 1876:xxxii–xxxviii).

If viewed through a different lens, five of carp's attributes (1, 3, 4, 5, 7) could be deemed undesirable. Some Americans had recently come to such realization about English sparrow (*Passer domesticus*). In 1874, two decades of enthusiasm about the imported bird had quickly evaporated after ornithologists described it ousting native avifauna by dint of aggressiveness and abundance—traits that had previously been cast as sparrow virtues (Coates 2006; Fine and Christoforides 1991). Baird, one of America's leading ornithologists, was unavoidably privy to the "sparrow war," but it no more persuaded him to reevaluate his framing of carp than did critics' persistent skepticism about the fish's edibility. Instead, Baird hired Hessel to procure carp from Europe for a federal breeding program.

On May 19, 1877, after failing with two earlier trips, Hessel sailed into New York City with 345 live carp from Germany. He and his precious cargo were transferred to a state hatchery in Baltimore, where Hessel began preparations to breed the fish. He also wrote an essay that framed *Cyprinus carpio* as a piscatorial all-rounder that rapidly turned plant matter into meat in almost any water and climate. Hessel's 35-page endorsement contained not one negative word about carp, except for a mention of occasional moldy taste in pond specimens, which seemed a trivial anomaly given Hessel's assurance that only trout and salmon fetched higher prices than carp in Europe. His narrative offered something for everyone—farmer, fisherman, pisciculturist, fish commissioner, urban resident. One fish fitted all, which helps explain why Hessel's essay was widely reprinted after its publication in a USFC report during 1878 (Sandiford 2009; Hessel 1878).

That same year, with big-city newspapers now embracing the Bairdian carp narrative, Hessel and his fish relocated from Baltimore to the
nation's capital, where the USFC had built a 12-acre carp hatchery beside
the Washington Monument (Baird 1880).

Common Carp Fulfills the Hype

In October 1879, the USFC finally began distribution of *Cyprinus carpio*.
Its inaugural effort saw 12,265 fingerlings shared among 300 applicants
in twenty-five states as far west as California. Most of the carp were put in
large cans and shipped by rail (Smiley 1884).

The young fish thrived in their new surroundings, mostly farm ponds.
A carp craze ensued. The USFC, deluged with requests, shipped a half-
million carp during the next three annual distributions (Cole 1905). Yet
demand still exceeded supply, such was the nation's enthusiasm. *Cyprinus
carpio* grew faster in America than in Europe. In warmer states, where
carp often did not enter winter dormancy, some specimens spawned at the
age of 16 months—a remarkably fast maturation. "They will, if well cared
for, grow almost as fast as pumpkins," declared one recipient in South
Carolina (Smiley 1886:815).

By 1883, millions of carp were breeding in rural ponds and small
lakes—and small numbers were appearing in the nets of commercial
fishermen (Smiley 1886), having escaped into open waters like their piscine predecessors in New York and California. Public reaction to this
rapid colonization of the continent was strongly positive. In news story
and agency report, *Cyprinus carpio* was embraced as a "naturalized" fish
(Hart 1885:29).

Transformation of Common Carp

Amidst the celebratory chorus, anti-carp narratives persisted. Critics' lead
topic of contention was edibility. In 1883, the USFC conducted a survey
of carp recipients. Remarkably, three-fifths of the 600-plus respondents
had not eaten carp—because their fish had either perished, remained too
small to eat, or were too precious to kill. However, of the 242 people who
had tasted carp meat, 85% rated it as favorable (Smiley 1883a). Baird and
his staff interpreted the survey results as proof that the edibility issue was

a matter of ignorance. They launched a campaign to teach Americans to properly rear, prepare, and cook *Cyprinus carpio*. Information and recipes appeared in USFC circulars, state fish commission reports, and the press (Smiley 1883b; Anonymous 1883).

Yet the anti-carp narratives gained momentum. "The carp is a humbug," declared one Texas newspaper in January 1885, calling for protection of native fish that were "far superior" to it (Burr 1950:27–28). Months later, the *Chicago Tribune*, which had long promoted carp, revealed its sudden unease about the "foreign interloper." "It is to be hoped the carp will nowhere develop into a nuisance, as has another importation, the English sparrow" (*Chicago Tribune* 1885:4).

In 1886, a new USFC survey showed "a much larger percentage of dissatisfaction" with carp edibility than the first poll (Smiley 1886). Yet Baird and other carp defenders continued to frame the problem as one of education, with a new twist—cost-effectiveness. *Cyprinus carpio*, declared an Illinois fish commissioner, was "the cheapest food for the greatest number of people, for the least amount of money" (American Fisheries Society 1886:91).

Inexpensive, and increasingly common in rivers and lakes, carp was sold at many fish markets by the time Baird died in August 1887. New USFC chief Marshall McDonald expanded federal fish stocking, but he focused on shad, long revered for its savory meat. Although the carp program continued, its rhetorical prominence faded as policy prioritized other species. Nonetheless, the country's first commercial carp fishery arose in Lake Erie, with most of the harvest going to Eastern cities ("Homerus" 1891).

In 1890, however, carp's abundance in California was deemed problematic by that state's fish commission, which became the first agency to frame carp as undesirable (Sandiford 2009). The new narrative not only spread quickly in the national press, its tone harshened. The *Chicago Tribune* and *New York Times*, reporting that California's carp had invaded salmon spawning waters and de-vegetated a hundred-mile stretch of marsh prized by duck hunters, characterized *Cyprinus carpio* in terms like "scavenger," "greedy," and "obnoxious" (Anonymous 1891a; Anonymous 1891b).

As carp populations mushroomed in other states during the early 1890s, anglers complained that the fish muddied waters and ate ova and fry of game species (*Forest and Stream* 1895). More generally, critics rued carp traits that had once enamored boosters—fecundity, adaptability, fast growth, hardiness. Those attributes now stymied carp control efforts that were launched in accordance with the dominant new policy narrative, namely invasion by a "piscatorial pariah" (Anonymous 1895).

Yet when the USFC finally ceased carp distribution in 1896, *Cyprinus carpio* had become common fare for thousands of rural and urban poor. These people, often immigrants from Europe, needed only a fishing pole or a couple of cents to get fresh carp. They inspired another metaphor for carp loyalists—"poor man's fish." This narrative had particular resonance in Illinois, where a major buffalofish fishery had crashed in the 1870s because of overfishing. By the early 1900s, *Cyprinus carpio* had transformed the Illinois River into America's most productive freshwater river fishery, prompting fishermen there to crown carp as their "bread-winner" (Sandiford 2009).

Nationally, however, prices for carp would remain so low that the fish never became commercially significant in America. By 1920, the upper third of the Illinois River was so polluted with Chicago's sewage that not even *Cyprinus carpio* could survive there (Thompson 1928), while wartime "eat carp" campaigns failed to change the nation's overall poor opinion about carp edibility (McCrimmon 1968). Anglers disdained it as a trash fish. Research gradually unveiled it as an environmental engineer whose activities impacted entire ecosystems (Cahn 1929; Threinen and Helm 1954). The findings provided a scientific rationale for carp control and eradication programs in many states.

America's experience with *Cyprinus carpio* would sour US fishery managers on introductions of foreign fish for a half-century. Stocking continued unabated, but programs mostly involved transplanting of North American species, primarily salmonids (Stroud 1975).

After World War II, state and federal agencies regained enthusiasm for foreign fish, beginning with a group of warm-water species from Africa called tilapia (*Oreochromis spp.*). Hype about tilapia's potential for fish farming, control of pond weeds, and sport fishing echoed the production-oriented rhetoric about common carp three-quarters of a century earlier. By the time "tilapiamania" waned in the 1960s, several tilapia species were becoming invasive in warmer states—and interest had shifted to another foreign fish, this time a cyprinid from Asia.

Idealizing Grass Carp

During the 1950s, the four-state region of Alabama, Arkansas, Louisiana, and Mississippi had become home to an aquaculture industry. Besides fish farms, its infrastructure included university departments and state agencies (Parker 1989).

North America lacked large plant-eating fishes. So in 1957, when Auburn University fish biologist Homer Swingle briefly reviewed eleven herbivorous fishes that might control pond weeds for fish farmers, his candidates were all foreign species. Among them was grass carp, a native of large rivers in China. Grass carp reportedly exceeded 10 pounds. Its tolerance of cold water would enable it to overwinter in northerly states, unlike most of Swingle's other fishes, which were tropical species. Grass carp's one flaw—in the eyes of aquaculturists—was that it was not known to spawn outside its native range. Citing scientific reports from abroad, Swingle said grass carp "appears to be one of the most promising for biological control of rooted aquatics if methods to induce reproduction under controlled conditions can be devised" (Swingle 1957:13).

Two years later, a Chinese aquaculture researcher stationed in Washington, DC, with the United Nation's Food and Agriculture Organization (FAO), also recommended grass carp for US waters, although with a qualifier. Shao-Wen Ling cautioned the US Fish and Wildlife Service (USFWS) that

> the unforeseen danger of careless introduction of exotic species could be tremendous. It is, therefore, obvious that any fish (no matter how good it is in their native country) to be introduced should be carefully investigated and thoroughly studied and experimented under strictly controlled conditions.
>
> Undoubtedly the grass carp is the most efficient aquatic plant eating fish so far known to fish culturists, and since its native habitat is quite similar to some of the major river systems in the U.S.A., it should be able to adapt to American waters well. But the possibility of having it become another major problem fish like the common carp is so great that unless the fish can become acceptable to the Americans its introduction should not be done hastily. (Ling 1959:205)

In August 1963, Ling and a handful of fish experts from the USFWS, Auburn University, and Arkansas Game and Fish Commission attended a two-day meeting at the USFWS's new aquaculture research center in Stuttgart, Arkansas. The group made an unpublicized decision to import some grass carp for research (Stroud 1972; Mitchell and Kelly 2006).

Four months later, on November 16, 1963, America's first grass carp arrived from Malaysia at the Stuttgart laboratory. Auburn University soon received a separate consignment of 12–13 grass carp, this time from Taiwan (Mitchell and Kelly 2006).

Press coverage of the importations was sparse, and framed around proponents' optimism (Anonymous 1964). Early scientific reviews from

Stuttgart and Auburn University were similarly constructed. At both institutions, the young grass carp consumed prodigious amounts of vegetation. Researchers evaluated these results only within the context of aquaculture production. Implications for nature beyond the pond were not discussed. Acknowledgments of ignorance and risk were formulaic, as evident in the absence of concern about grass carp's tendency to jump as a possible means of escape, and reminders to readers that other countries already had grass carp programs (Stevenson 1965; Avault 1965).

As with early Bairdian assessments of common carp, initial expert reviews of grass carp were narrowly framed, self-justifying in terms of institutional agenda and funding, and predisposed toward approval for stocking.

Grass Carp Becomes Controversial

Unlike foreign fishes, birds and mammals from abroad had remained a staple of twentieth-century wildlife management in America. While popular with sportsmen and conservation agencies, imported fauna consistently drew criticism from a minority of scientists and writers. Exotic introductions were "unscientific, economically wasteful, politically short-sighted, and biologically wrong," wrote one zoologist weeks before the inaugural importation of grass carp (Hall 1963:518). In worst cases, exotic species became invasive. Researchers had begun to recognize a worldwide acceleration in the rate of human-related biological invasions, such as recent irruptions of sea lamprey and alewife in the Great Lakes (Elton 1958).

The post-WWII enthusiasm for foreign fishes rekindled concern about their invasiveness (American Fisheries Society 1955). Concern became controversy after 1966, when USFWS researchers at the Stuttgart laboratory in Arkansas learned to spawn grass carp. Within two years, more than 100,000 young grass carp had been shared among institutions and agencies in the region, notably the Arkansas Game and Fish Commission (Mitchell and Kelly 2006).

The indiscriminate distributions caused widespread alarm outside Arkansas. Critics viewed the region's aquaculture community—government biologists, agency bureaucrats, fish farmers—as intent on stocking grass carp, without regard for the thirty states hydrologically connected to Arkansas within the Mississippi River basin (Reiger 1976). In 1969, almost five dozen scientists, fish farmers, policy makers, and aquarists gathered in Washington, DC, for a conference about introductions of exotic fishes.

Organizers hoped to bridge "communications gaps" that had created confusion and suspicion, and develop policy guidelines. The conferees' recommendations included a proposed moratorium on open-water releases of grass carp, pending further study (Stroud 1969:1; Lachner et al. 1970).

The recommendation went unheeded by the Arkansas Game and Fish Commission and other grass carp enthusiasts. Instead, they began a double-barreled program of stocking and persuasion. In June 1969, three months after the conference, the commission made the first known public planting of grass carp, stocking a reservoir in Arkansas (Mitchell and Kelly 2006).

A year later, in Arkansas, grass carp were found for the first time in open waters within the Mississippi basin. Commercial fishermen netted several adult specimens in the White River, just a few miles from the Stuttgart aquaculture center. Personnel there were informed of the discovery, but it was not publicized. Months later, in February 1971, the *St. Louis Post-Dispatch* reported the capture of three more adult grass carp, this time in Illinois waters of the Mississippi River (Mitchell and Kelly 2006). Amidst speculation about the origin of these free-swimming fish, the Arkansas Game and Fish Commission hatched one million grass carp and began stocking them in open waters. During the next eighteen months, grass carp became "a routine management tool" for aquatic weed control in Arkansas (Henderson 1979:27). The commission put almost 300,000 in several dozen ponds and lakes, and also shipped consignments to at least sixteen other states for stocking and commercial use (Guillory and Gasaway 1978; Mitchell and Kelly 2006). By the end of 1972, grass carp had been stocked in forty states (Pflieger 1978).

The Arkansas commission and its allies had also launched a promotional campaign that one critic described as "justifying the fish's presence and promoting grass carp as the greatest thing since blackstrap molasses" (Reiger 1976:109). To avoid Americans' disdain for *Cyprinus carpio*, grass carp was renamed by its boosters as white amur, after the Amur River in its native range in Asia. Proponents also tapped into societal fears about herbicides and other chemicals, a hot-button issue rarely mentioned in their early rhetoric. They reframed white amur as a benign "green" tool for cleaning not just fish farm ponds, but thousands of weed-infested lakes, reservoirs, and ditches (Sneed 1971; Bailey and Boyd 1972; Sills 1970). In this new narrative, a fish originally imported to improve aquaculture's profitability was transformed into a social good.

The boosterism percolated into mainstream press. White amur was "the perfect fish," according to a *Time* reporter who interviewed an Arkansas Game and Fish Commissioner (Anonymous 1972). Other commentators mocked the utopian rhetoric. A *Washington Post* journalist lampooned grass carp as the "great white shmoo," after a cartoon character in the popular comic strip *Li'l Abner* (Willard 1971). The shmoo was the world's most amiable creature, eager to serve humanity. But like a fertility myth gone awry, shmoos reproduced so prodigiously in their efforts to please people that they eventually caused disaster.

Framing of Grass Carp Broadens

Besides satire, a different narrative had gained traction in the controversy. Whereas boosters of grass carp focused on its advantages, other scientists viewed the fish through a broader lens that encompassed potential *dis*advantages too. "The risks are too great," warned ichthyologist David Greenfield in his 1973 recommendation to the American Fisheries Society that grass carp be studied much more thoroughly before any decision about stocking. Greenfield worried that young grass carp would compete with similarly-aged native fish for food, while adult grass carp would consume vegetation utilized by native fish and waterfowl for shelter, food, and spawning. Greenfield also noted multiple assumptions, omissions, and flaws in the early reviews, including insufficient acknowledgment that grass carp would likely become difficult to control if natural reproduction occurred. "One need only to look at the common carp, *Cyprinus carpio*, for a graphic example" (Greenfield 1973:51).

By now, grass carp was spreading through the Mississippi basin and Arkansas fish farmers were openly marketing it to residents in states where it was banned (Vance 1975; Guillory and Gasaway 1978). Nonetheless, the policy debate had shifted, thanks to persuasive argument by critics of grass carp. Invoking new metaphors like biological pollution, they had nudged some of the burden of proof onto the cyprinid's advocates, who now faced pressure to disprove its potential disadvantages. In Florida, proponents of grass carp had to argue their case in court (Anonymous 1976).

At a 1978 conference on grass carp, an Arkansas Game and Fish Commission biologist defended the fish—and his agency—in a brief paper that mixed sarcasm, defiance, and frustration. Much like nineteenth-century expert advocates of *Cyprinus carpio*, Scott Henderson was dismissive of anyone who did not share his favorable view of grass carp. His two-pronged framing of the problem was classically narrow—his commission's mandate was "optimum sustained harvest" of Arkansas fisheries, and control of aquatic weeds was essential because they impacted all water uses. Complaining that 100% certainty about grass carp's harmlessness was an impossible expectation, Henderson said controlled field tests were the only way to determine the cyprinid's effects and whether it would naturally reproduce (Henderson 1979).

Two years later, a research team announced it had been catching grass carp larvae in the southern Mississippi River since 1975 (Conner et al. 1980). However, the news that grass carp were spawning naturally was not only unsurprising, it was eclipsed by a new narrative of optimism, this time about techniques for producing nonfertile grass carp. During the 1980s and 1990s, many of the 35 states that had banned grass carp relaxed regulations so genetically altered "triploid" specimens could be stocked in thousands of weed-choked reservoirs, private lakes, and golf course ponds. The controversy reached a narrative stalemate, with supporters lauding reductions in nuisance plants and critics complaining of reductions in native flora and fish (Spear 1994; Bauers 1995). Scientific evidence supported both narratives. Meanwhile, *fertile* grass carp continued their gradual invasion of the Mississippi Valley and elsewhere.

Black Carp

In the late 1990s, sudden controversy about another large Asian cyprinid affirmed a further shift in policy discourse about exotic species. Black carp closely resembles grass carp, but it grows bigger and eats snails and mussels. Native to rivers in eastern Asia, black carp had long been reared by Chinese aquaculturists for food and medicine. In the early 1980s, Southern fish farmers and researchers imported the species to control snails that carried a fish parasite. Commercial use was limited to triploid black carp. But in 1999, after a new parasite devastated Louisiana's catfish industry and began to spread, an alleged shortage of the triploids led Mississippi to grant its catfish farmers a one-year window to purchase fertile "diploid" black carp from aquaculturists in Arkansas, the only state where black carp was bred commercially (Nico et al. 2005).

The move dismayed other states within the Mississippi Valley. They feared a third invasion by a foreign cyprinid, this one a carnivorous species known to surpass five feet and 150 pounds in its native waters. In February 2000, twenty-eight wildlife agencies took a collective step to preempt a black carp invasion *before* any specimens were found in open waters. They petitioned the USFWS to list black carp as "injurious," citing its potential impacts on endangered mussels (Ferber 2001). Under federal law, "injurious" species cannot be transported across state lines or imported/exported.

Although an industry official said most fish farmers did not use black carp (Kilborn 2002), a minority insisted the fish was essential to their operations. They premised their narrative on a lack of proven alternatives for snail control. Forecasting economic hardship for the industry and higher fish prices, the group lobbied against the proposed listing, led by an Arkansas Game and Fish Commissioner who sold small numbers of black carp at his 1,200-acre fish farm (Bennett 2002; Meersman 2004). "There has not been a single black carp that escaped into the wild," he told the *New York Times* in August 2002 (Kilborn 2002).

Seven months later in Illinois, a 12¾-pound black carp was netted in a lake that connects to the Mississippi River in times of flood (Chick et al. 2003). In subsequent years, more free-swimming specimens would be caught within the river basin, including dozens of fertile diploids in Louisiana (Nico and Neilson 2012). But the black carp "war" had become a side note to a far bigger controversy involving two other Asian cyprinids that had arrived unannounced three decades earlier—in Arkansas.

Bighead Carp and Silver Carp

In 1973, at the height of the grass carp controversy, an Arkansas fish farmer received a consignment of live fish from Asia. He had ordered grass carp, but found among them black carp and two filter-feeding cyprinids—bighead carp and silver carp (Kelly et al. 2011). Voracious eaters of plankton, bighead carp and silver carp are effective cleaners of pond water. In China, where they have been reared as food fish for centuries, bighead carp sometimes exceed 100 pounds, while silver carp reach 60 pounds (Kolar et al. 2007).

The Arkansas fish farmer kept his grass carp, but offered his other Asian cyprinids to the Arkansas Game and Fish Commission, on condition some would be returned to him if the agency learned to spawn the new species. The commission accepted his offer, and began breeding trials with the five dozen fish. The black carp quickly died, so FAO aquaculturist Ling was invited back to Arkansas. During the summer of 1974, Ling hatched a million silver carp and 20,000 bighead carp for the commission (Egan 2006; Kelly et al. 2011).

A year later, with research barely underway into potential uses for the two cyprinids, fishermen discovered adult silver carp in Arkansas tributaries of the Mississippi River (Kolar et al. 2007). By 1981, silver carp had spread into the river's main stem, and a bighead carp had been caught in Kentucky waters of the Ohio River, 60 miles from its confluence with the Mississippi River. The source of these free-swimming specimens was presumed to be Arkansas fish farms, which were already propagating both species (Freeze and Henderson 1982). During the next two decades, silver carp and bighead carp spread northward through the Mississippi Valley, advancing at an annual average of 35 miles. By the late 1990s, both species were abundant and breeding in major tributaries, including the Illinois River (Laird and Page 1996), whose once-polluted waters now supported a premier sport fishery for bass anglers. Occasional warnings by biologists worried about the cyprinids garnered little media attention. Ironically, southern fish farmers had abandoned use of silver carp because of its unexpected habit of jumping when disturbed, while US consumption of bighead carp remained negligible despite periodic marketing.

During this same era, national consciousness about invasive exotics had mushroomed, partly because of disastrous large-scale invasions by species like zebra mussel. Unlike earlier decades of research about the benefits of foreign species, scientists now focused on *undesirable* ecological and economic impacts. Preservation of native biodiversity became a policy imperative, while the increasing coverage of invasive exotics in the popular press featured dark narratives about an America "under assault by foreign plants, fish and other animals" (Swanson 2001:1).

Asian Carp Invade American Minds

In 2001, amidst this climate of concern, bighead carp and silver carp surfaced from their quarter-century of relative obscurity. Biologists reported in *Science* that not only had Illinois populations of the two cyprinids exploded, their northward invasion of the Illinois River was nearing Chicago canals that accessed the Great Lakes. A $1.6-million electric barrier being built in the Chicago Sanitary and Ship Canal (CSSC) might not have sufficient voltage to deter fish as big as Asian carp (Chick and Pegg 2001).

By the following summer, the experimental barrier was operational, but the prospect of an Asian carp invasion of the Great Lakes had become front-page discourse. Yet again, traits that were once viewed as an asset—in this instance an appetite for plankton—had become problematic. US and Canadian fishery managers, scientists, environmentalists, tourism leaders, and anglers collectively predicted devastation of a fishing industry worth $4–7 billion. They urged immediate action. Commercial fishing could perhaps reduce carp numbers in waters already invaded, but the preferred option to prevent invasion of Lake Michigan was a second CSSC barrier, this time designed to be permanent. Otherwise Asian carp "could turn the Great Lakes into a carp pond," one policy maker warned in the *Chicago Tribune* (Deardorff 2002).

Within a couple of years, bighead carp and silver carp constituted two-thirds of fish life in stretches of the Illinois River. Even more newsworthy, however, Asian carp were injuring people. Specifically, silver carp jump as high as six feet when approached by boats and other watercraft. Collisions with the soaring fish are akin to being hit with a bowling ball. On the Illinois River silver carp left researchers and recreationists with neck injuries, broken noses, and concussions—and in the new Internet era, provided remarkable visual drama. Photographs showed boats surrounded by scores of airborne silver carp, while YouTube videos of the fish smashing into people went viral. One video won $10,000 on *America's Funniest Home Videos*. Such imagery helped the four Asian carps collectively become the 21st-century poster child for invasive species.

That status, along with their threat to the iconic Great Lakes, gave the cyprinids major political currency. As they neared Chicago and the controversy intensified, narratives about Asian carp materialized at every government level up to the White House, in multiple states and provinces, as well as from industry, science, environmentalists, tourism, commercial and sport fishermen, and media. The fish even became attached to agendas beyond Asian carp. In 2007, a Midwest power company narrowly framed its proposed increase of wastewater discharge into the CSSC as a way to combat invasive fish, when really it was attempting to avoid compliance with water pollution standards (Ethridge 2008). At the other end of the framing spectrum was proposed closure of the CSSC to sever connection between the Mississippi Valley and the Great Lakes. Pursued in court by Great Lakes states outside Illinois, the concept of hydrologic separation enmeshed Asian carp in a multibillion-dollar vision for replumbing Chicago's outdated sewer/storm water system (Metropolitan Planning Council 2011).

Asian carp have generated thousands of news stories, from local to international. In the logjam of narratives competing for policy dominance, rival stakeholders recruit science to enhance their persuasiveness. They invoke research to affirm or undermine virtually every core assertion about Asian carp, from basic matters like their alleged presence in the CSSC to complex questions about the seriousness of their threat to the Great Lakes (Hood 2010a; Egan 2010). The controversy has become the epitome of trans-science, as evident in December 2010 when a judge rejected a multi-state lawsuit seeking closure of two Chicago-area locks in hopes of sealing Lake Michigan from the advancing carp. The judge acknowledged that earlier that year, despite a new $9-million electric barrier in the CSSC, environmental DNA (eDNA) of Asian carp had been discovered in Lake Michigan and a live bighead carp had been caught in a lake upstream of the CSSC with unblocked access to Lake Michigan. Nonetheless, said the judge, the plaintiff states and their expert witnesses hadn't shown that Asian carp were "anywhere near, much less on the verge of, establishing a population in Lake Michigan" (*Michigan v. US Army Corps of Engineers* 2010:51).

Status Quo

In the three years since the judge's ruling, eDNA of bighead and silver carp has been detected in Lake Erie (Jerde et al. 2013). However, actual specimens of either species have yet to be caught in any of the Great Lakes. This fact is variously framed as a triumph of speedy preventative measures, evidence that the threat was overblown, or a lull before the storm.

Meanwhile, the populations of Asian carp established elsewhere in US waters continue to spread. In late 2013, researchers announced that grass carp were breeding naturally in Lake Erie (Chapman et al. 2013), confirming an earlier study indicating that some of the lake's tributary rivers were potential spawning sites for grass carp as well as other Asian carps (Kocovsky et al. 2012). Bighead carp and silver carp, along with black carp, have been federally listed as "injurious" species. Anglers, commercial fishermen, and biologists implicate the two filter-feeders in declines in native fish and other undesirable changes—ecological, economic, and cultural—that are affecting thousands of people, especially along the Illinois River (Lampe 2003; Hood 2010b; Irons et al. 2007).

Illinois is leading a multistate push to create domestic and foreign markets for Asian carp, in hopes that commercial fishing will suppress their abundance. The effort far exceeds any previous promotion of a foreign cyprinid. State agencies offer Asian carp to prisons and food banks (Gallagher 2011), while celebrity chefs feature them in haute cuisine and rebrand them with names like silverfin and Asian bass (Bergin 2011). Recipes and deboning tips abound on the Web.

However, despite this latest upsurge of optimism-saturated rhetoric, America's appetite for cyprinids remains negligible. Of the millions of pounds of Asian carp being harvested, most are exported abroad—to Asia.

Policy Implications

As America approaches two hundred years of experience with foreign carps, much has changed in science, media, and policy. Yet too much remains the same. Argumentation's crucial role in policy transcends time— and narratives about the cyprinids remain afflicted with optimism bias, narrow framing, and revisionism.

Grass carp is still marketed as an "aquatic cow," even though "aquatic goat" is a more accurate metaphor for its catholic voraciousness. Knowledge about its ecological impacts remains incomplete. Of 2,000 published studies, most have focused on grass carp's biology or plant consumption rather than broader questions of ecology and ecosystem (Dibble and Kavalenko 2009).

Defenders of Southern fish farmers blame the Asian carp invasion on government agencies (Mitchell and Kelly 2006; Kelly et al. 2011). But their revisionist accounts ignore the close ties between public agency and private industry and between scientist and fish farmer, and exclude the fish farmers' own initial enthusiasm for each carp and subsequent resistance to stricter controls. Similarly selective are economic valuations of black carp that exclude potential adverse effects of escaped black carp to the mussel industry (Wui and Engle 2007).

Such flaws are problems of human nature. They occur regardless of historical context. Invasive species policy must accordingly guard against such tendencies. Above all, policy must reflect better foresight, crafted around a precautionary principle. Aquatic species reared in captivity invariably find their way into open water, accidentally or otherwise. If they become established in rivers and lakes, eradication is virtually impossible. Thus proposed importations of foreign fish should be treated as proposals for *introduction*, even if the fish are intended only for controlled experiments. For the same reasons, assessments must evaluate exotic fish not just for potential benefits but also for potential disadvantages (Naylor et al. 2001).

Creating markets for invasive biota as a means to control them is risky policy. Commercial utility must not subsume biological heritage in management of such species. That would undermine the nation's commitment to restoration of native nature, as well as hinder cultivation of "a sense of biological history" that has been lacking in humanity's centuries-long enthusiasm for exotic species (Todd 2001:253).

Such mindfulness still seems lacking in contemporary aquaculture. In 2009, at the height of the Asian carp controversy, two aquaculturists idealized silver carp as "a prize rather than a problem" for America (Buck and Schroeder 2009:13). Their narrative, in its optimism, minimal concern for native nature, and myopic emphasis on harvest, echoes Rudolph Hessel's "one fish fits all" endorsement of *Cyprinus carpio*. But whereas Hessel penned his nineteenth-century essay *before* common carp became problematic, the contemporary article was written in the aftermath of five carp invasions spread across 180 years—a history that is conspicuously absent from the article.

"The only perfect science known to man is hindsight," declared Scott Henderson of the Arkansas Game and Fish Commission in 1978, while criticizing opponents of grass carp stocking (Henderson 1979:29). America's history with invasive foreign carps suggests otherwise. Hindsight can be made so imperfect by optimism bias and other human tendencies that it perpetuates flawed policy for centuries.

Literature Cited

Allard, D. C. 1978. *Spencer Fullerton Baird and the US Fish Commission: A Study in the History of American Science*. New York: Arno Press.

American Fisheries Society. 1886. Untitled. *Transactions of the American Fisheries Society* 15:91–92.

American Fisheries Society. 1955. Untitled. *Transactions of the American Fisheries Society* 83:376–380.

Anonymous. 1851. Untitled. *Transactions of the American Institute of the City of New-York, for the Year 1850*, 397–398. Albany, NY.

Anonymous. 1875. Carp. *Forest and Stream* 3 (January 7):340.

Anonymous. 1879. Carp breeding in California. *Chicago Field* 12 (November 1): 180.

Anonymous. 1883. The flavor of carp. *Forest and Stream* 21 (December 6):368.

Anonymous. 1891a. Readable topics. *Chicago Tribune*, November 14, 4.

Anonymous. 1891b. Fish in irrigating ditches: the carp become a nuisance to the California farmers. *New York Times*, October 18, 8.

Anonymous. 1895. The German carp in disfavor. *New York Times*, July 2, 2.

Anonymous. 1964. Can grass carp spawn here? *Washington Post*, October 11, C12.

Anonymous. 1972. Man's best friend? *Time*, January 31, 50.

Anonymous. 1976. 2 Florida agencies in power fight over Asian fish. *New York Times*, December 26, 22.

Avault, J. 1965. Preliminary studies with grass carp for aquatic weed control. *Progressive Fish-Culturist* 27:207–209.

Bailey, W., and R. Boyd. 1972. Some observations on the white amur in Arkansas. *Hyacinth Control Journal* 10:20–22.

Baird, S. 1874. Report of the Commissioner. In: *US Commission of Fish and Fisheries, Report of the Commissioner for 1872 and 1873*, i–xcii. Washington, DC: GPO.

Baird, S. 1876. Report of the Commissioner. In: *US Commission of Fish and Fisheries, Report of the Commissioner for 1873–75*, vii–xlvi. Washington, DC: GPO.

Baird, S. 1877. Untitled. *Transactions of the American Fisheries Society* 6:67–69.

Baird, S. 1880. Report of the Commissioner. In *US Commission of Fish and Fisheries, Report of the Commissioner for 1878*, xv–lxiv. Washington, DC: GPO.

Balon, E. 1995. The common carp, *Cyprinus carpio*: its wild origin, domestication in aquaculture, and selection as colored nishikigoi. *Guelph Ichthyology Reviews* 3:1–55.

Bauers, S. 1995. Grass-eating fish: farmers' blessing, biologists' curse. *Houston Chronicle*, June 11, A48.

Bennett, D. 2002. Black carp: threat to aquaculture. *Delta Farm Press*, December 13.

Bergin, M. 2011. Carpe diem? *E Magazine* 22:20–21.

Buck, H., and G. Schroeder. 2009. The silver carp: a fish for the future. *World Aquaculture* 40:12–13.

Burr, J. G. 1950. The beginning of Texas fish conservation work: chapter 7 (1879–1885). *Texas Game and Fish* (January):10.

Cahn, A. 1929. The effect of carp on a small lake: the carp as a dominant. *Ecology* 10:271–274.

Chew, M. 2009. The monstering of tamarisk: how scientists made a plant into a problem. *Journal of the History of Biology* 42:231–266.

Chapman, D., J. Davis, J. Jenkins, P. Kocovsky, J. Miner, J. Farver, and R. Jackson. 2013. First evidence of grass carp recruitment in the Great Lakes Basin. *Journal of Great Lakes Research* 39:547–554.

Chicago Tribune. 1885. Untitled [editorial]. September 21, 4.

Chick, J., R. Maher, B. Burr, and M. Thomas. 2003. First black carp captured in US [letter to editor]. *Science* 300:1876–1877.

Chick, J., and M. Pegg. 2001. Invasive carp in the Mississippi River basin [letter]. *Science* 292:2250–2251.

Coates, P. 2006. *American Perceptions of Immigrant and Invasive Species: Strangers on the Land.* Berkeley: University of California Press.

Cole, L. 1905. The German carp in the United States. In *Report of the Bureau of Fisheries, 1904,* 523–641. Washington, DC: GPO.

Conner, J. V., R. P. Gallagher, and M. F. Chatry. 1980. Larval evidence for natural reproduction of the grass carp (*Ctenopharyngodon idella*) in the lower Mississippi River. In *Proceedings of the Fourth Larval Fish Conference,* edited by L. A. Fuiman, 1–19. Ann Arbor, MI: US Fish and Wildlife Service.

Deardorff, J. 2002. Giant carp ready to eat way through Great Lakes. *Chicago Tribune,* July 18, A1.

DeKay, J. 1842. *Zoology of New York, or the New York Fauna—IV: Fishes.* Albany, NY: White and Visscher.

Dibble, E., and K. Kavalenko. 2009. Ecological impact of grass carp: a review of the available data. *Journal of Aquatic Plant Management* 47:1–15.

Egan, D. 2006. Chaos uncorked. *Milwaukee Journal Sentinel,* October 15, A1.

Egan, D. 2010. Explanation for carp may be a fish story. *Milwaukee Journal Sentinel,* August 15, B1.

Elton, C. 1958. *The Ecology of Invasions by Animals and Plants.* London: Methuen.

Eskridge, A. 2010. Alien invaders, plant thugs, and the Southern curse: framing kudzu as environmental other through discourses of fear. *Southeastern Geographer* 50:110–129.

Ethridge, T. 2008. Framing environmental policy: the case of invasive species. Annual meeting, Midwest Political Science Association, Chicago. April 3. http://www.allacademic.com/meta/p268500_index.html.

Ferber, D. 2001. Will black carp be the next zebra mussel? *Science* 292:203.

Fine, G. A., and L. Christoforides.1991. Dirty birds, filthy immigrants, and the English sparrow war: metaphorical linkage in constructing social problems. *Symbolic Interaction* 14:375–393.

Flyvbjerg, B., M. K. S. Holm, and S. L. Buhl. 2005. How (in)accurate are demand forecasts in public works projects? The case of transportation. *Journal of the American Planning Association* 7:131–146.

Forest and Stream. 1895. Untitled [editorial]. (August 17):133.

Freeze, M., and S. Henderson. 1982. Distribution and status of the bighead carp and silver carp in Arkansas. *North American Journal of Fisheries Management* 2:197–200.

Gallagher, J. 2011. Carp-rich Illinois has a plan for Asian invader: feeding the poor with a pest. *St. Louis Post-Dispatch,* July 14, A1.

Greenfield, D. 1973. An evaluation of the advisability of the release of the grass carp, *Ctenopharyngodon idella,* into the natural waters of the United States. *Transactions of the Illinois Academy of Sciences* 66:47–53.

Guillory, V., and R. Gasaway. 1978. Zoogeography of the grass carp in the United States. *Transactions of the American Fisheries Society* 107:105–112.

Hall, R. 1963. Introduction of exotic species of mammals. *Transactions of the Kansas Academy of Science* 66:516–518.

Hart, Lane S. 1885. In *Pennsylvania State Commissioners of Fisheries, Report of the State Commissioners of Fisheries, for the Years 1883 and 1884*, 29. Harrisburg, PA.

Henderson, S. 1979. Grass carp: the scientific and policy issues. In *Proceedings of the Grass Carp Conference, Gainesville, Florida, January 1978*, edited by J. Shireman, 25–29. Gainesville: Institute of Food and Agricultural Sciences, University of Florida.

Herbert, H. W. 1850. *Frank Forester's Fish and Fishing of the United States, and British Provinces of North America*. New York: Stringer and Townsend.

Hessel, R. 1874. Method of treating adhesive eggs of certain fishes, especially of the Cyprinidae, in artificial propagation. In *US Commission of Fish and Fisheries, Report of the Commissioner for 1872 and 1873*; 567–570. Washington, DC: GPO.

Hessel, R. 1878. The carp and its culture in rivers and lakes: and its introduction in America. In *US Commission of Fish and Fisheries, Report of the Commissioner for 1875–76*; 865–900. Washington, DC: GPO.

Hoffman, R. 1995. Environmental change and the culture of common carp in medieval Europe. *Guelph Ichthyology Reviews* 3:57–85.

"Homerus." 1891. Carp in Lake Erie. *Forest and Stream* 36 (January 29):29.

Hood, J. 2010a. Impending catastrophe questioned by carp experts. *Chicago Tribune*, September 12, A18.

Hood, J. 2010b. River town on front line of Asian carp invasion. *Chicago Tribune*, February 14, 24.

Irons, K. S., G. G. Sass, M. A. McClelland, and J. D. Stafford. 2007. Reduced condition factor of two native fish species coincident with invasion of nonnative Asian carps in the Illinois River, U.S.A. Is this evidence for competition and reduced fitness? *Journal of Fish Biology* 71(sd):258–273.

Jerde C., L. Chadderton, A. Mahon, M. Renshaw, J. Corush, M. Budny, S. Mysorekar, and D. Lodge. 2013. Detection of Asian carp DNA as part of a Great Lakes basin-wide surveillance program. *Canadian Journal of Fisheries and Aquatic Sciences* 70:522–526.

Kahneman, D., and A. Tversky, eds. 2000. *Choices, Values and Frames*. New York: Cambridge University Press.

Kelly, A., C. Engle, M. Armstrong, M. Freeze, and A. Mitchell. 2011. History of introductions and governmental involvement in promoting the use of grass, silver, and bighead carps. In *Invasive Asian Carps in North America*, edited by D. Chapman and M. Hoff, 163–174. Bethesda: American Fisheries Society.

Kilborn, P. 2002. A huge, boat-hurdling carp is no Mississippi fish story. *New York Times*, August 26, A1.

Kinsey, D. 2006. "Seeding the water as the earth": the epicenter and peripheries of a western *aqua*cultural revolution. *Environmental History* 11:527–566.

Kocovsky, P., D. Chapman, and J. McKenna. 2012. Thermal and hydrologic suitability of Lake Erie and its major tributaries for spawning of Asian carps. *Journal of Great Lakes Research* 38:159–166.

Kolar, C., D. Chapman, W. Courtenay, C. Housel, J. Williams, and D. Jennings. 2007. *Asian Carps of the Genus* Hypophthalmichthys *(Pisces, Cyprinidae)—A Biological Synopsis and Risk Assessment.* Washington, DC: US Fish and Wildlife Service.

Lachner, E., R. Robins, and W. Courtenay. 1970. Exotic fishes and other aquatic organisms introduced into North America. *Smithsonian Contributions to Zoology* 59:1–29.

Laird, C., and L. Page. 1996. Nonnative fishes inhabiting the streams and lakes of Illinois. *Illinois Natural History Survey Bulletin* 35:1–51.

Lakoff, G. 2010. Why it matters how we frame the environment. *Environmental Communication* 4:70–81.

Lampe, J. 2003. Illinois River becoming carp central. *Peoria Journal-Star*, November 2, D12.

Larson, B. 2005. The war of the roses: demilitarizing invasion biology. *Frontiers in Ecology and the Environment* 3:495–500.

Larson, B. 2007. An alien approach to invasive species: objectivity and society in invasion biology. *Biological Invasions* 9:947–956.

Ling, S. W. 1959. Report to the US Fish and Wildlife Service on fish-farming development in the U.S.A. Cited in Stevenson, James. 1965. Observations on grass carp in Arkansas. *Progressive Fish-Culturist* 27(4) (October):203–206.

Lovallo, D., and D. Kahneman. 2003. Delusions of success: how optimism undermines executives' decisions. *Harvard Business Review* 81:56–63.

Luken, J., and J. Thieret. 1996. Amur honeysuckle, its fall from grace. *BioScience* 46:18–24.

Majone, G. 1989. *Evidence, Argument and Persuasion in the Policy Process.* New Haven, CT: Yale University Press.

McCrimmon, H. 1968. *Carp in Canada.* Ottawa: Fisheries Research Board of Canada.

Meersman, T. 2004. These carp have connections. *Minneapolis Star Tribune*, June 14.

Metropolitan Planning Council. 2011. A Fork in the River. Accessed May 14 2012. http://www.metroplanning.org/multimedia/video/488.

Michigan v. US Army Corps of Engineers, 2010 WL 5018559 (N.D. Ill. Dec. 2 2010). http://www.asiancarp.us/documents/Judge-Dow-Order.pdf.

Mitchell, A., and A. Kelly. 2006. The public sector role in the establishment of grass carp in the United States. *Fisheries* 31:113–121.

Naylor, R., S. Williams, and D. Strong. 2001. Aquaculture—a gateway for exotic species. *Science* 294:1655–1656.

Nico, L. G., and M. E. Neilson. 2012 *Mylopharyngodon piceus*. Gainesville, FL: USGS Nonindigenous Aquatic Species Database. Revised March 15, 2012. http://nas.er.usgs.gov/queries/factsheet.aspx?SpeciesID=573.

Nico, L. G., J. D. Williams, and H. L. Jelks. 2005. *Black Carp: Biological Synopsis and Risk Assessment of an Introduced Fish*. Bethesda, MD: American Fisheries Society.

Norris, T. 1868. *American Fish-Culture*. Philadelphia: Porter and Coates.

Parker, N. 1989. History, status, and future of aquaculture in the United States. *CRC Critical Reviews in Aquatic Sciences* l:97–109.

Pauly, P. 1996. The beauty and menace of the Japanese cherry trees: conflicting visions of American independence. *Isis* 87:51–73.

Perry, G., and F. Schueler. 2003. Metaphors, misuse, and misconceptions. *Science* 301:1480–1481.

Pflieger, W. 1978. Distribution and status of the grass carp (*Ctenopharyngodon idella*) in Missouri streams. *Transactions of the American Fisheries Society* 107:113–118.

Pisani, D. 1984. Fish culture and the dawn of concern over water pollution in the United States. *Environmental Review* 8:117–131.

Poppe, R. 1880. The introduction and culture of the carp in California. In: *US Commission of Fish and Fisheries, Report of the Commissioner for 1878*, 661–666. Washington, DC: GPO.

Reiger, G. 1976. The white amur caper: making careers with carp. *Audubon* 78:108–111.

Sandiford, G. 2009. Transforming an exotic species: nineteenth-century narratives about introduction of carp in America. PhD diss., University of Illinois.

Scott, G. C. 1869. *Fishing in American Waters*. New York: Harper and Brothers.

Sills, J. 1970. A review of herbivorous fish for weed control. *Progressive Fish-Culturist* 32:158–161.

Simberloff, D. 2003. Confronting introduced species: a form of xenophobia? *Biological Invasions* 5:179–192.

Smiley, C. 1883a. Notes on the edible qualities of German carp and hints about cooking them. *Bulletin of the United States Fish Commission, III*, 305–332. Washington, DC: GPO.

Smiley, C. 1883b. Answers to 118 questions relative to German carp. *Bulletin of the United States Fish Commission, III*, 241–248. Washington, DC: GPO.

Smiley, C. 1884. Report on the distribution of carp to July 1, 1881, from young reared in 1879 and 1880. In: *US Commission of Fish and Fisheries, Report of the Commissioner for 1882*, 943–1008. Washington, DC: GPO.

Smiley, C. 1886. Some results of carp culture in the United States. In *US Commission of Fish and Fisheries, Report of the Commissioner for 1884*, 657–890. Washington, DC: GPO.

Sneed, K. 1971. The white amur: a controversial biological control. *American Fish Farmer* 2:6–9.

Soulé, M., and G. Lease, eds. 1995. *Reinventing Nature? Responses to Postmodern Deconstruction*. Washington, DC: Island Press.

Spear, K. 1994. Lakes lose in carp war on hydrilla. *Orlando Sentinel*, August 4, 11.

Steinberg, T. 1991. *Nature Incorporated: Industrialization and the Waters of New England*. New York: Cambridge University Press.

Stevenson, J. 1965. Observations on grass carp in Arkansas. *Progressive Fish-Culturist* 27:203–206.

Stroud, R. 1969. Conference on exotic fishes and related problems. *Sport Fishing Institute Bulletin* 203:1–4.

Stroud, R. 1972. White amur in Arkansas. *Sport Fishing Institute Bulletin* 239:4–5.

Stroud, R. 1975. Exotic fishes in United States waters. *Sport Fishing Institute Bulletin* 264:1–4.

Swanson, S. 2001. Foreign species seize share of US habitat. *Chicago Tribune*, April 15, 1.

Swingle, H. S. 1957. Control of pond weeds by the use of herbivorous fishes. *Proceedings of the Southern Weed Conference* 10:11–17.

Taylor, J. 1999. *Making Salmon*. Seattle: University of Washington Press.

Tenner, E. 1996. *Why Things Bite Back: Technology and the Revenge of Unintended Consequences*. New York: Alfred Knopf.

Thompson, D. 1928. The "knothead" carp of the Illinois River. *Illinois Natural History Survey Bulletin* 17:285–320.

Thompson, P. 1970. The first fifty years. In *A Century of Fisheries in North America*, edited by N. Benson, 1–11. Washington, DC: American Fisheries Society.

Threinen, C. W., and W. T. Helm. 1954. Experiments and observations designed to show carp destruction of aquatic vegetation. *Journal of Wildlife Management* 18:247–251.

Todd, K. 2001. *Tinkering with Eden: A Natural History of Exotics in America*. New York: W. W. Norton.

Towle, J. 2000. Authored ecosystems: Livingston Stone and the transformation of California fisheries. *Environmental History* 5:54–74.

Vance, J. 1975. Amur is a four-letter word. *Field and Stream* 79:12–20.

Weber, J., and C. Word. 2001. The communication process as evaluative context: what do nonscientists hear when scientists speak? *BioScience* 51:487–495.

Weinberg, A. 1972. Science and trans-science. *Minerva* 10:209–222.

Willard, H. 1971. Can the world stand a real shmoo? *Washington Post*, December 23, D1.

Wui, Y. S., and C. Engle. 2007. The economic impact of restricting use of black carp for snail control on hybrid striped bass farms. *North American Journal of Aquaculture* 69:127–138.

CHAPTER FIVE

"Sooper" Impact: Drawing the Attention of Kids to the Dangers of Invasive Species

Mark Newman

Fantasy is a natural part of childhood as kids learn to deal with a variety of psychological fears, their young minds struggling to process emotions and feelings. The popularity of action movies, anime, comic books, and video games may be reflections of kids learning to deal with "life." Or maybe not. That's a debate best left for psychotherapists and educators.

Regardless, pop culture shows that monsters and horror movies remain popular among youth. And whether one is a fan of Superman, Batman, or Spiderman, kids love superheroes. They learn at an early age that the world can be a dangerous place—"don't talk to strangers," "look both ways before you cross the street"—so seeking the solace of superheroes seems only natural.

Now you may be asking yourself, what does this have to do with invasive species?

Science has identified more than 180 nonnative species in the Great Lakes. While foreign exotics like zebra mussels and round gobies may look harmless, their introduction to the Great Lakes and surrounding waterways has had an enormous impact on the freshwater ecosystem—a multimillion-dollar nightmare that calls for action. Indeed, the scope of the problem has reached such a level that many believe it will take superheroic efforts to restore and preserve our waters.

Sooper Yooper: Environmental Defender (Fig. 5–1) is a children's picture book that tells the story of Billy Cooper, an ex-Navy SEAL who lives

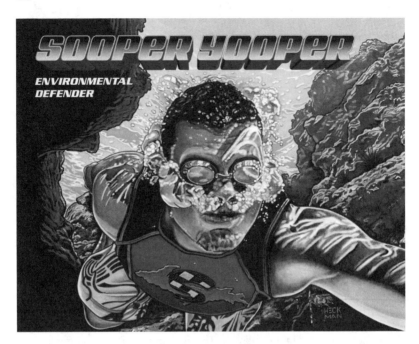

in Michigan's Upper Peninsula and guards the Great Lakes against invasive species. I wrote the story with the late artist Mark Heckman, my closest friend. To understand why we created this book, it might help to provide a little background.

Mark and I first met in the late 1980s. At the time, I was the managing editor of a regional sports magazine and the publisher invited Mark to create a cover for one of the issues. I was just making the switch from newspaper reporter, having covered music and entertainment after getting my start in sports, and Mark was just beginning to establish himself as an artist of some repute.

Many artists are content to work in relative obscurity, brushing off dreams of fame and fortune for the simple pleasure of pursuing their passion without any pretense. But Mark was hardly an ordinary artist. He had a larger than life personality and it was no coincidence that his preferred medium was the 14-foot-by-48-foot billboard canvas.

At the time, Mark was just beginning to build his reputation through self-promotional billboards that playfully proclaimed his prowess at painting (e.g., "World's Best Artist"; see http://www.markheckmanart.com for examples of Heckman's artwork) or his tongue-in-cheek desire to be the "first artist in space." Mark and I learned we had a lot in common. We liked the same sports, same music, same movies, and, most importantly for what would come later, the same desire to make a difference. Initially I helped Mark promote his work with press releases, but we soon discovered we were kindred spirits and began collaborating on projects.

I think Mark realized he had found a creative compatriot when I was the first person to suggest that he wasn't crazy to consider erecting an AIDS billboard in our ultraconservative hometown of Grand Rapids, Michigan. Acquired immune deficiency syndrome had only recently entered the public conscience and Mark was moved to act after reading *And the Band Played On*, Randy Shilts' nonfiction book about the human immunodeficiency virus (HIV).

Mark wanted to paint the word AIDS in giant letters across the billboard and then cover them with 2,001 condoms dipped in paint. I thought it was an idea of unadulterated genius. The end result was a statement that was heard throughout the world and covered by CNN, MTV, and most major news magazines and newspapers around the world. We were off and running.

Having successfully drawn the attention of people to the need for a greater awareness of AIDS and HIV, we set our sights on other targets.

Less than a year before the Rodney King riots in Los Angeles, we addressed the topic of racism with a billboard for the imaginary all-black Afro Country Club, "where only the ball is white." We erected boards on all types of topics, from homelessness ("The Bum Rap") to gun control ("Deadman's Curve"), but the vast majority of our work was related to the environment, more specifically water issues.

Living in the midst of the largest group of freshwater lakes in the world (by total surface area), Mark and I were proud to be counted among the 40 million people who depend upon the Great Lakes for recreation, food, and drink. Bordered on its sides by four of the five lakes, Michigan is known as the "Great Lakes State" and residents are said to be never more than six miles from water. And so we confronted a host of water issues during the course of our work, doing our small part to raise public awareness and encourage good stewardship of our natural resources.

Over time Mark painted more than 200 billboards, a large body of work which put him into contact with many movers and shakers, including a number of noted environmentalists. One of these was Peter Wege, a local philanthropist whose father had founded Steelcase, one of the nation's leading office furniture manufacturers. Mark developed a friendship with Mr. Wege, who lent generous support to many of his artistic endeavors, including a number of green-themed billboards.

In our quest for new worlds to conquer, Mark and I talked about the need to reach younger audiences with our ideas. We both agreed that a children's book could provide an ideal platform for a pro-environment story and, with Mr. Wege's blessing, Mark and I began discussing various concepts.

From the start, Mark wanted the main character to be a superhero. He was a big fan of superheroes, not only for what they represented in their roles as crime fighters and upholders of strong moral codes, but also for their highly stylized appearance, with their distinctive costumes and muscular physiques. His office was filled with superhero paraphernalia, from Spiderman movie posters to Incredible Hulk figurines and Superman collectibles.

We gravitated toward the idea of putting our main character in the Upper Peninsula, where people are not only down-to-earth but also proudly protective of the outdoors. Since natives there are known as Yoopers, our hero would naturally be known as Sooper Yooper; in fact, he is Billy Cooper the Sooper Yooper, and that's the only rhyme in the book because I'm no Dr. Seuss. While there is a rhythm to some of the story, I made a conscious decision to avoid any singsong text.

Superheroes typically have extraordinary powers. We contemplated all manner of unique superpowers—the ability to communicate with fish, the ability to swim superhuman distances, even underwater sonar-like ability that could be utilized in some untold fashion. We talked about dozens upon dozens of potential strengths that might come in handy while protecting the Great Lakes.

In the end, we reached the conclusion that the vitality of our natural resources rests upon common men and women, not a single superhero, and so we decided to make him unique. He is a superhero without a specific superpower. We bowed somewhat to the conventions of the superhero genre by making him an ex-Navy SEAL, but he is more a common Joe than a G.I. Joe for the simple reason that we wanted to underscore the idea that anyone can be a superhero when it comes to protecting the earth.

So Billy Cooper had no special powers, but we weren't going to deny him a sidekick. A dog is a man's best friend, of course, and Mark had a special affinity for English bulldogs. Tank had recently become his second

FIG. 5–2: Billy Cooper, the hero from *Sooper Yooper: Environmental Defender*, with a sea lamprey. ©Mark Heckman. Used with permission. See also color plate.

bulldog after the death of Zane, the pet that he owned when we first met. In the book we dubbed him Mighty Mac, the nickname for the Mackinac Bridge, the longest suspension bridge in the Western Hemisphere, which connects Michigan's peninsulas.

With a hero and a sidekick in place, we needed a villain. A good drama, after all, needs an antagonist who represents a threat or obstacle to the main character. Well, the Great Lakes are filled with villains—more than 180 species, fish and plants that may not be inherently villainous but nonetheless are not native to the Great Lakes and thus pose a danger to the health of the waters. Since a large number of invasive species can be traced to the ballast water of ocean freighters, we personified the threat in the form of Eye-Gore, a Russian shipping magnate with an oversized ocular condition.

Mark had certain images in mind that he wanted to paint before the story was even developed. He knew he would animate sea lamprey (Fig. 5–2) and zebra mussels, but he also had plans for the Mackinac Bridge, Pictured Rocks National Lakeshore, and the Detroit skyline. When I asked him how Detroit fit into our story, he didn't know. "You'll figure it out," he replied, reflecting his confidence in the nature of our collaboration, which had been working for the better part of 20 years.

He began painting well before the story was finished, which was a blessing in disguise. Mark was meticulous in his approach to his artwork, almost "paint-staking" in the way that he applied acrylics from his brush to the canvas or, in this case, on board. It took him a month, sometimes two months, to finish each painting and with more than two dozen illustrations, it was a process that took him over two years.

That process was made even longer by the fact that Mark was not feeling well during the time that we worked on the book. Although Mark was a fastidious eater, exercised and worked out regularly, never smoked, and rarely drank as much as a beer, it was clear that his health was suffering. He had been plagued with various digestive tract problems over the years, but the pain in his midsection was more serious this time. There were times when I visited him in his office and he would be literally on the floor, writhing in pain. Something was clearly wrong, although his doctors were having trouble diagnosing his precise ailment.

Midway through the process of creating the book, Mark was diagnosed with non-Hodgkin lymphoma. At Stage III, his prognosis was not promising, but he fought his condition valiantly, crisscrossing the country in an attempt to find a cure. He started at the University of Michigan Hospital in Ann Arbor, but his illness would take him to the National Cancer Institute in Bethesda, Maryland; the MD Anderson Cancer Center in Houston, Texas; and the Mayo Clinic in Rochester, Minnesota.

He did his best to keep painting through endless chemotherapy sessions and countless trips to the hospital. His illness weakened him considerably, but he continued to fight until the end came on May 6, 2010. Sadly, he never got to hold the actual book in his hands. We had completed the primary work on *Sooper Yooper* shortly before his death, but the publisher was still shopping the book to printers when he passed away. He was only 49.

It's a sad story, but in a way it has a happy ending.

Mark and I had talked about promoting the book by visiting schools in various cities across the state of Michigan. We made a list of about a dozen places and made plans to barnstorm from one city to the next. When it became clear that he would probably be too ill to attempt such an undertaking, he called me. "You have to promise me that you'll still go if I can't," he begged. In truth, he didn't have to ask. I had been excited about the prospect of promoting the book and yet I had to admit that I was a little apprehensive. Without Mark at my side, what was I going to talk about? Kids loved Mark because he was the class clown, the practical joker in the crazy clothes. Compared with Mark, I was Mr. Conservative. The more I thought about it, the more I realized I had my work cut out for myself.

Ultimately, I decided that the best course of action was to create a program that would tell the story behind the story. With the help of the Wege

Foundation, I would put together a show-and-tell presentation that would include real specimens, including a sea lamprey, zebra mussels, Eurasian ruffe, and spiny water fleas. I wanted the program to be engaging and entertaining as well as educational. I resolved to make it appealing to both students and teachers. My rule of thumb was: would I find it interesting if I were sitting in the audience?

When Mark and I wrote *Sooper Yooper*, we didn't really think too much about defining our potential audience. Mark's colorful illustrations would capture the imagination of kids and my wordplay, hopefully, would entice the adults. We figured it was a book that was more likely to be read to kids by a parent or a teacher, so we didn't worry about writing down to children. In fact, we decided to include a "Sooper Glossary" in the back of the book to explain some of the phrases and terminology. I'm no science expert just as I'm no Dr. Seuss. To make up for my educational deficiency, I quickly devoured everything I could find about exotic species in the Great Lakes. It was like late-night cramming for college—research on the run—as I had only a couple of short months before the school year began in earnest. *Pandora's Locks*, journalist Jeff Alexander's excellent book about the opening of the Great Lakes–St. Lawrence Seaway and the resulting invasion of exotic species, became my primer, while the Internet provided infinite sources for my research. With more than 180 nonnative species, it became evident that I would have to narrow my search.

Opting to reduce the list to a Top 10, I decided to focus my presentation on sea lamprey, zebra mussels, Eurasian ruffe, round goby, rusty crayfish, spiny water flea, New Zealand mud snail, purple loosestrife, Eurasian watermilfoil and bloody red shrimp. Digging through facts, I tried to retrieve information that might interest a general audience. In the meantime, Terri McCarthy, vice president of programs for the Wege Foundation, was working to put together a traveling trunk of invasive species to help illustrate my presentation. Her search was rewarded when she discovered that the Minnesota Sea Grant people had just what we were seeking. It included museum-quality preserved specimens of zebra mussels, Eurasian ruffe, and spiny waterfleas. Disappointed by the "realistic" rubber sea lamprey included in the kit, Terri was later able to secure an actual "plasticized" sea lamprey from the University of Michigan. Its one-foot size may have been a little substandard but it had a great set of teeth.

Part of my program needed to explain how ballast water is a major source for introducing invasive species into the Great Lakes ecosystem and it occurred to me that I better make sure that my audience first understood the interconnectedness of the Great Lakes. I figured it would be a good idea to present a few facts about each lake, focusing primarily on Lake Superior since it forms the northern coast of the Upper Peninsula.

I learned that most (but not all) schools use the acronym of HOMES as a mnemonic device for helping students learn the names of the Great Lakes. I discovered there were similar tools for remembering other facts. If you wanted to recall the largest to the smallest, you could use the phrase Sam's Horse Must Eat Oats and the first letter of each word would provide the clue. Similarly, if you wanted to remember the order of the Great Lakes from west to east, you could refer to the phrase of Sam's Mother Hates Eating Onions.

From the beginning, it was my intention to make the program as interactive as possible. I had no interest in standing in front of a classroom and spouting endless facts and details about the Great Lakes and invasive species. It was important to me to involve the students as much as possible. This would help ensure that my presentation was more of a vibrant, breathing program than a simple resuscitation of micromanaged minutiae. My module on Lake Superior provides a good example. Superior is not only the largest of the Great Lakes, it is also the coldest. Why is it the coldest? Younger students will offer answers like it's the furthest north, closest to Canada or the North Pole, or the furthest from the equator. All good answers, I say, but there's an even better one. Usually there will be a student who will explain the inability of the sun to warm such a large body of water before someone will answer that it is the deepest. Bingo! How deep is it? Often, the guesses will start in numbers equated with ocean depths. Sometimes I'll try to get the students to use their mathematic skills to arrive at the correct number. Divide your number by three, I will say, or multiply it by four. It's fun to see their wheels spinning. The answer of 1,332 feet is only a number; I want them to visualize how deep that really is. I'll ask, who has been to Chicago? Many students usually raise their hands. I then show a picture of the Sears Tower (now the Willis Tower), one of the tallest buildings in the world, and explain how it would be almost completely underwater at the deepest point of Lake Superior.

Now I have them thinking. Lake Superior is so deep, I explain, that it holds 3 quadrillion gallons of water. (For younger kids, it becomes 3 quadrillion milk jugs, since I learned they couldn't picture a gallon). With lower elementary students, I'll ask how many zeroes in quadrillion. On the board, I draw nine zeroes. Is that enough? No! I add three more. Is that enough? No! I add three more. Is that enough? I usually hear a yes or two among the chorus. I pause, egging them to change their minds. Yes! But 3 quadrillion is only a number. I ask them to picture all of North America and all of South America, waving my hand high and low. If you took Superior and spread it over both continents, everything would be under one foot of water. Wow! In fact, Superior is so deep that you could take all of the other Great Lakes, plus three more Lake Eries, and all of them would fit into the volume of Superior.

So Superior is very deep. It is also very cold. How cold is it? The average temperature is only about 40 degrees Fahrenheit and, of course, water freezes at 32 degrees, so it's almost like ice. But it's rare that the entire lake completely freezes over, not only because it is so deep but also because it is relatively stormy. Superior has seen more than 550 shipwrecks. There's an even a song about a famous one, *The Wreck of the Edmund Fitzgerald.*

And so it goes. I assembled various facts about the lakes. Michigan used to be called (in French) "the Lake of the Stinking Water"; Huron is the second largest by volume but has the longest coastline and is home to 30,000 islands; Erie is the shallowest, warmest, and as a result, most biologically active, with more fish caught for human consumption than the other Great Lakes combined; Ontario gets its name from the Iroquois tribe.

Having established the Great Lakes as an aquatic ecosystem on which 40 million people depend, I explain how our waters were compromised by the arrival of ocean freighters. I talk about how roughly a thousand ships annually make the 2,340-mile journey through the St. Lawrence Seaway that extends all the way to Duluth, Minnesota, the furthest western point and the busiest inland freshwater port in the world. To help them understand the concept of ballast water, I explain that ships would bob like a cork in the ocean without the appropriate weight. Ballast water from a ship's previous ports provides the necessary balance, but that water must be discharged before the ship takes on new weight in the form of cargo. To give historical perspective, I pose the question. Why couldn't big ships enter the Great Lakes more than 100 years ago? It usually takes a little coaxing,—there was a natural barrier, I'll remind them—but eventually a student will provide the answer of Niagara Falls, giving me the opportunity to explain the complex system of channels, locks, and canals that makes shipping possible and which opened the floodgates so to speak for the arrival of many invasive species.

The first presentation of my program came in October 2010. Having tested out my information on family and friends beforehand, I had high hopes that it would be well received, but the response exceeded my wildest dreams. Request for the Sooper Yooper tour seemed to grow exponentially. It certainly helped that the program was offered at no cost, thanks to the generous support of the Wege Foundation. Originally, Diane Heckman, Mark's wife, and I talked about keeping our visits to Fridays, but that was quickly amended to include Wednesdays and by the end of the school year, I was fitting schools into my schedule whenever I could. By the close of the 2010–11 school year, I had spoken to nearly 18,000 students in 45 cities in Michigan, Illinois, and Indiana. A feature on National Public Radio as well as local media coverage (from the *Grand Rapids Press* to the *Chicago Tribune*) helped publicize the tour, which was booked on a first come, first served basis.

Initially, the thought was that the program would appeal mostly to third and fourth graders, but it wasn't long before I was fielding requests from middle schools and eventually high schools that wanted me to visit. More and more schools were asking if they could include first graders or even kindergartners. I adjusted the program, which filled the better part of an hour, according to the audience, stressing some elements more than others. A variety of visuals and video clips helped hold the interest of the young and old alike.

Over the course of the school year, the program evolved as students raised questions or whenever I uncovered new pieces of information. For example, kids want to know if they can be attacked by sea lamprey. "We're warm-blooded and they're cold-blooded, so they don't attack people because we're not on their menu," I explain. But I proceed to tell them how I met a man and a woman who worked in sea lamprey control and how they wanted to know what it felt like to have a lamprey attached to their arm (a thought that naturally horrifies most listeners). Of course, I say, I had to ask, "Did it hurt?" At this point, you could probably hear a pin drop in the classroom. It was odd, I continue, because they both gave almost the same word-for-word reply. "It didn't really hurt. It kind of tickled. But it was more of a creepy feeling knowing that the thing was attached to their arm"—to which students breathe a sigh of relief.

In regard to zebra mussels, I talk about how they will stick to just about anything—boats, docks, rocks, and even water pipes—and how they managed years ago to shut down the water system in the city of Monroe, Michigan, by clogging the intake valves. A principal at one school helped confirm the "stickiness" issue, relating how his niece did a research paper on zebra mussels at Michigan State University and how she wanted to see if there was anything they wouldn't stick to. She tried all manner of materials, even no-stick Teflon, and they stuck to everything. I hear lots of stories. When I showed the sharp edges of the zebra mussel at one school, a boy said he had to get 15 stitches after he fell while running on a beach with zebra mussels.

Once when I talked about how fishermen dislike round goby, a boy replied, "My dad and I hate them so much, we cut off their heads!" A girl next to him said, "That's kind of harsh." Undeterred, the boy retorted. "They're invasive species! They don't belong here!" But some students, particularly girls, think the goby is kind of cute, which surprised me. I started asking, "Who thinks he's cute?" Usually there are at least a few hands. "Who thinks he's ugly?" More hands. "Well, if you think he's ugly, let me show you ugly!" On the projector screen, I click to a ghastly microscopic view of the spiny water flea. "Eeeeewww!" The slide never fails. Moments later, I show another slide displaying a fishing line with hundreds of spiny water fleas. I had wondered whether the slide was staged until a teacher at one Upper Peninsula school told me his line looked exactly as depicted after he went fishing in Lake Superior for the first time. "I didn't know what they were, but they stuck to my line like Velcro," he said. "They were like bristly gobs of jelly. Gross!"

I learned to be quick on my feet early during my school tour. Talking about purple loosestrife, I explained how scientists wanted to see if they could find an insect that would eat only the invasive plant. I explained how more than a decade of research in Europe revealed there were five varieties of beetles that would eat only purple loosestrife and that we had started to import beetles in an effort to control the plant. "That's awful," a teacher at one of the first schools replied. "What are we going to do about the beetles?" I paused at the question. "We don't have to worry about the Beatles," I answered. "The Beatles broke up 40 years ago!" Yes, it's a bad joke but it's become a regular part of my routine.

Not content to confine my program to the Top 10, I conclude my presentation with two potential invasive species: *Hydrilla* and Asian carp, the latter eliciting the most awe of anything in my program. The sheer size of the bighead carp and the wild, leaping antics of the silver carp seem to fire the imagination and bring out stories. A teacher at a Catholic school in Chicago related the tale of her niece who got hit in the mouth by an Asian carp while tubing on the Mississippi River. She had her front teeth knocked out and required surgery to repair the damage to her lip. Imagine my surprise when I later related that story at other schools and kids laughed. That was not the desired response, so I added a bit. "Imagine getting hit in the mouth by just a 30-pound carp. That's like getting hit in the mouth by two bowling balls. Would it hurt? Of course!"

After a couple of months of the tour, I met with the Wege Foundation to let them know that I was very thankful for their support and that they had more than fulfilled their promise to Mark, his wife, and me. On behalf of the schools, I could not thank them enough. They assured me that they wanted to continue supporting the program, as education is the first branch of the foundation's mission. With the backing of the Wege Foundation, the program was presented to more than 33,000 students at 109 schools in five states during its second year (Michigan, Illinois, Indiana, Ohio, and Wisconsin). There is now a growing waiting list as the word of mouth spreads. As an example, a visit to a Cleveland Heights, Ohio, school in 2012 yielded a request to bring the program to seven other schools in 2013. Many schools have asked to be an annual stop on the tour. Michigan City High School in Indiana, for example, has asked me to speak to their sophomore class each year. I guess I'm becoming an invasive species of sorts in the schools, but I try to do more good than harm. Your school system, unlike our ecosystem, is safe. I promise.

November 2012 saw the arrival of the sequel, *Sooper Yooper: The Quest of the Blue Crew*. Inspired by the work of real scientists, the new book follows the exploits of the environmental CSI team known as the Blue Crew, an elite group of scientific superheroes assembled by Billy Cooper in his defense of the Great Lakes. Like the first book, the new work relies on clever wordplay and interesting illustrations to carry the story. I debated about finding an artist who could paint in Mark's style, but in the end I felt it better if I chose a different route and went it alone. The new illustrations, which adopt a style more along the lines of oil painting, are all based on photographs I took of scientists from several Midwestern schools: Grand Valley State University, Michigan State University, the University of Notre Dame, and the University of Wisconsin. There are even a few rhymes in the book, courtesy of the dapper rapper MC Sirius, the "unflappable scientist in a dog-eared hat." I'm not sure, but I may have broken new ground with the first hip-hop scientist. I also developed a second program in conjunction with the new book, still based on the Great Lakes and invasive species, but centered on the themes of teamwork and collaboration.

Policy Implications

So why do kids need to know about the Great Lakes and invasive species? At the conclusion of my Sooper Yooper program, I underscore the moral or the message of the "story." I tell students that Mark and I never heard about things like spiny waterfleas or Asian carp when we were growing up. Nobody talked about them. When the students grow up, there may be a new invasive species—a new plant, fish, or other animal—that they had never heard about when they were growing up but now threaten the waters. "You are the future," I say. "You have to guard our Great Lakes. I don't want to be an old grandparent with grandkids someday and say, 'C'mon kids, we're going to see the Not-So-Good Lakes. They used to be called the Great Lakes, but now they're called the Not-So-Good Lakes.' You've got to keep them Great! And that's why Mark and I wrote the book *Sooper Yooper: Environmental Defender*."

Current research indicates there is a dire need to address environmental issues in children's literature. A recent study (Williams et al. 2012) found that fewer children's books are set in nature and, surprisingly, animals are also disappearing as featured subjects. Researchers at several universities reviewed about 8,100 images in 296 children books that had been honored as Caldecott Medal winners from 1938 to 2008. Their work found that use of natural environments declined from 40 to 25 percent, and images of both wild and domestic animals declined dramatically (Williams et al. 2012).

Chris Podeschi of Bloomsburg University of Pennsylvania, who is one of the authors of that study, suggested that their findings could spell dire consequences. "This is just one sample of children's books, but it suggests there may be a move away from the natural world as the population is increasingly isolated from these settings. This could translate into less concern about the environment" (quoted in Hemlich 2012). Richard Louv, the author of *Last Child in the Woods*, an influential work about the increasing divide between children and the outdoors, pointed to the study as further evidence that today's wired generation is becoming disconnected with the "real world." "Nature experience isn't a panacea, but it does help children and the rest of us on many levels of health and cognition," Louv told *USA Today*. "I believe that as parents learn more about the disconnect, they'll want to seek more of that experience for their children, including the joy and wonder that nature has traditionally contributed to children's literature" (quotes from Hemlich 2012).

Sooper Yooper: Environmental Defender and its sequel uphold that tradition by pitting superhero protagonists against the invasive species threatening the Great Lakes. The books and supporting programs have been designed to educate and entertain kids about the importance of protecting our natural resources. My hope is that my work is a starting point for discussion about invasive species and that students will be open to other experiences and viewpoints. If nothing else, I want to engage kids about science and get them to care about the environment and the living world around them. My message is simple: when it comes to the environment, everyone can be a superhero!

Literature Cited

Alexander, J. 2009. *Pandora's Locks: The Opening of the Great Lakes–St. Lawrence Seaway*. East Lansing: Michigan State University Press.

Heckman, M., and M. Newman. 2010. *Sooper Yooper: Environmental Defender*. Holt, MI: Thunder Bay Press.

Hemlich, N. 2012. Study: new children's books lack reference to nature, animals. *USA Today*, February 28, 4D. http://usatoday30.usatoday.com/news/health/wellness/story/2012-02-27/Study-New-childrens-books-lack-reference-to-nature-animals/53275082/1.

Mark Heckman website. Accessed October 30, 2013. http://www.markheckmanart.com.

Newman, M. 2012. *Sooper Yooper: The Quest of the Blue Crew*. Grand Rapids, MI: Green Junction Press.

Sooper Yooper website. Accessed December 3, 2012. http://www.sooperyooper.com.

Williams, J. A., Jr., C. Podeschi, N. Palmer, P, Schwadel, and D. Meyler. 2012. The human–environment dialog in award-winning children's picture books. *Sociological Enquiry* 82:145–459.

Here They Come: Understanding and Managing the Introduction of Invasive Species

For a species to become invasive, it must first be transported and released beyond its native range. Globalization has created many "vectors" that do this, from the intentional movement of species as pets to the unintentional transport of species as hitchhikers on planes, trucks, and boats. While many nonnative species are valuable in trade, and the costs of keeping species out of vectors may be high, the available evidence suggests that the introduction stage is the best point at which to manage invaders. That is, implementation of policies that prevent invaders from arriving is much cheaper than dealing with invaders after they are established.

The first step to managing vectors of introduction is to properly understand their scale and how they operate. Particularly problematic vectors move a large number of individual organisms, move them rapidly and in relatively good conditions so that they are released in good health, and move organisms between similar environments having conditions to which they are adapted. Once the risk presented by a vector is assessed, the actual species being transported must be investigated to identify those whose characteristics make them likely to become established and invasive.

In chapter 6, ecologist Christina Romagosa analyzes data about the number of vertebrate species being introduced to the United States through trade in live organisms. Many of these species are imported for sale as pets, and care is taken to ensure that they arrive rapidly and in good health. From work such as Romagosa's we discover which species are arriving, and these can then be assessed for the likelihood that they will become invasive. In chapter 7, ecologists Marc Cadotte and Lanna Jin investigate one factor—taxonomic relatedness to known invasive species—that can be used to assess the risk posed by introduced species. Finally, economist Michael Springborn (chapter 8) combines themes from the previous two chapters to show how knowledge of the actual species being introduced can be combined with predictions of invasiveness to determine rational policy. Springborn shows that although predictions of invasiveness are not perfect, they are good enough that using them to decide which species should be imported could lead to large overall economic benefits.

CHAPTER SIX

Patterns of Live Vertebrate Importation into the United States: Analysis of an Invasion Pathway

Christina Romagosa

Introduction

G lobal trade has increased substantially since the 1950s (Hulme 2009), and higher demand for wildlife is intrinsically linked with this trend (Nijman 2010). The wildlife trade creates global movement of millions of individuals annually. This anthropogenic transport of wildlife is a major threat to biodiversity by depleting wild populations, homogenizing distinct flora and fauna, and introducing invasive species, disease, and parasites (Wilcove et al. 1998, Hanselmann et al. 2004). The intensity at which wild-life is currently traded will further accentuate these cascading ecological consequences. For these consequences to be fully recognized, the scope of global trade in wildlife needs to be better understood.

With respect to biological invasions, the trade in live specimens is the most important pathway for vertebrate introductions worldwide (Kraus 2009). Much current research activity and management attention focuses on risk assessment and the aftereffects of these introductions rather than how trade specifically contributes to the biological invasion process. The biological invasion process has been divided into several stages and transitions (Kolar and Lodge 2001, Williamson and Fitter 1996) that allow for a more precise analysis of species invasions and their primary pathways (Fig. 6–1). In order to disrupt a pathway by management actions, scientists and policy makers must first understand its functioning: how, when, and

Biological Invasion Process

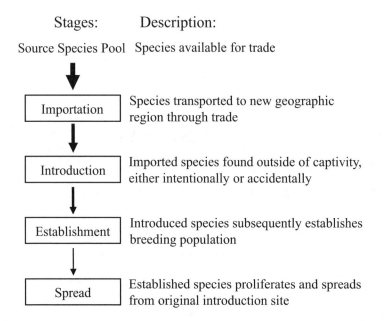

Stages: Description:

Source Species Pool   Species available for trade

Importation   Species transported to new geographic
                region through trade

Introduction   Imported species found outside of captivity,
                either intentionally or accidentally

Establishment   Introduced species subsequently establishes
                breeding population

Spread   Established species proliferates and spreads
                from original introduction site

FIG. 6–1. Stages of the biological invasion process, modified to relate to trade.

from where organisms are delivered, and how propagule supply changes over time (Ruiz and Carlton 2003). This understanding necessarily focuses on the first stages of the invasion process, when species are first entrained by a particular pathway, but also how these same species progress through the remaining stages. The most cost-effective efforts for invasive species management are at these early stages (Lodge et al. 2006).

The primary purpose for this book is to link our increasingly globalized world to biological invasions, and develop policy recommendations aimed at reducing this global threat. This chapter frames the live vertebrate trade pathway within this overarching purpose. Assessing global trends in vertebrate trade is challenging, because detailed information on the movement of vertebrates is difficult to obtain. The United States is one of the largest global markets for live vertebrates, with over 185 million individuals imported from and 30 million individuals exported to almost every country in the world each year (Defenders of Wildlife 2007). Because import records are maintained by US Fish and Wildlife Service (USFWS), a record of legal trade can be generated. Several studies have examined US trade in live vertebrates by using these records (e.g., Nilsson 1990, Hoover 1998, Franke and Telecky 2001, Schlaepfer et al. 2005, Defenders of Wildlife 2007, Rhyne et al. 2012), but few have assessed long-term changes in vertebrate trade patterns and how they relate to biological invasions. Gaps exist in current international regulation of the live animal trade (Convention on Biological Diversity 2010), and the United States, as a top importer, should make an effort to reduce introductions through this pathway.

This chapter reviews the US vertebrate trade and its relationship to biological invasions by synthesizing import data collected by the USFWS over 30 years for six vertebrate groups (tetrapods: amphibians, turtles, lizards, snakes, birds, and mammals). I break this chapter into three sections following an introduction to the USFWS import data. First, I assess specific vertebrate trade dynamics by summarizing the cumulative number of species imported over time, comparing the magnitude of individuals imported for available time periods, and summarizing geographic patterns to identify past and emerging trading partners and donor regions. Second, each of these analyses is considered within the biological invasion process framework described in Fig. 6–1. Finally, I consider policy implications and recommendations based on these analyses of the vertebrate trade data.

USFWS Import Data

Since 1966, the United States has required that a declaration form (Form 3-177) accompany any wildlife transported into or out of the United States (Title 50 CFR Part 14.61). The intention for this mandatory documentation was to improve and strengthen reporting requirements for the importation of all species.

The USFWS maintained these forms only as hard copy until 1983 when select information from these forms was also entered into a computerized database (Law Enforcement Management Information System, LEMIS). An updated version of LEMIS became fully operational in 2000, by which time, a centralized data entry group had been tasked with entering information from Form 3-177 (USFWS 2001). According to the USFWS (2001), the information entered into the LEMIS database is used to develop statistics on many aspects of the international wildlife trade. The LEMIS database assists the USFWS in criminal investigations, monitoring trade in species listed by the Convention on International Trade in

TABLE 6.1 **Availability of importation data from USFWS Form 3-177 by taxonomic group and year**

Taxonomic group	Years available	Reference
Amphibian	1970–1971	Busack 1974
	1998–2010	This study (LEMIS)
Reptile (turtles, lizards, snakes)	1970–1971	Busack 1974
	1989–1997	Franke & Telecky 2001
	1998–2010	This study (LEMIS)
Bird	1968–1972	Banks 1970, Banks & Clapp 1972, Clapp & Banks 1972, Clapp & Banks 1973, Clapp 1975
	1977–1980	Nilsson 1977
	1983–1988	Nilsson 1990
	1998–2010	This study (LEMIS)
Mammal	1968–1972	Paradiso & Fisher 1972, Clapp 1973
	1998–2010	This study (LEMIS)

Endangered Species (CITES), as well as assessing the major ports of entry and manpower needed to run them (Rhyne et al. 2012).

Until recently, the agency destroyed hard copies and computerized storage of data from these forms every 5–7 years. Therefore, to date, the only sources of these data prior to 1999 are from previous compilations. Unfortunately, not all information from Form 3-177 is available in these compilations. Recent data (after 1999) entered into LEMIS can be obtained through a Freedom of Information Act (FOIA) request filed through the USFWS.

As these data consider only legal trade, they must be considered a minimum estimate of trade by the United States. Moreover, data derived from the LEMIS database are incomplete and not fully accurate. Declaration information (e.g., invoices that list species in the shipment) that may accompany each shipment is not always entered into LEMIS, so the database may not contain species-level identification for all individuals reported to USFWS (see also Rhyne et al. 2012). Additionally, because typographical errors are present, it is necessary to examine each record to identify outliers as well as incorrect taxonomy. A thorough treatise of flaws in LEMIS data is provided in Schlaepfer et al. (2005) and Reaser and Waugh (2007). These USFWS data are the only information available to assess US trade patterns in both CITES-listed and CITES-unlisted species.

Beginning in 2002, I obtained LEMIS import data for live individuals from 6 taxonomic groups (amphibians, turtles, lizards, snakes, birds, and mammals). The 200,000 records cover the period 1998–2010 and document approximately 70,000 shipments and at least 85 million individuals. Import data prior to 1998 were obtained from previous compilations (Table 6–1), and updated to current taxonomy where necessary (after Romagosa et al. 2009). Imported species were categorized as either indigenous or nonindigenous to the continental United States.

Characterizing Trade: Importation Patterns

The goal for this section is to characterize live vertebrate importation to the United States and the dynamics of this pathway. Each subsection covers several analyses of importation patterns and concludes with a related discussion. Data used for each analysis below may vary by year owing to availability and quality; these differences are noted within the text.

Quantitative and Taxonomic Patterns

I used data derived from USFWS Form 3-177 to estimate the magnitude of individuals imported to the United States for available years from 1968 to 2010. For this assessment, both indigenous and nonindigenous species were considered because not all individuals are identified to the species-level in these data. For example, individuals in a shipment of amphibians may be identified in the LEMIS database only as "nonCITES amphibian," making it impossible to discern which individuals are native. For each taxonomic group, the average number of individuals imported per year was calculated for five time periods: 1968–1972 (all taxa), 1983–1987 (birds), 1989–1993 (reptiles), 1998–2002 (all taxa), and 2006–2010 (all taxa). The top 5 species imported by quantity, indigenous and nonindigenous, were summarized for periods 1968–1972 and 2006–2010.

In order to explore further specific trends among and within each taxonomic group, I determined the total number of nonindigenous species imported per group for available years (1970–2010). To determine the percentage of global biodiversity traded, the total number of nonindigenous species per taxonomic group imported from 1970–2010 was compared with the global diversity per taxonomic group (excluding species indigenous to the continental United States). Finally, I used a paired t-test to compare

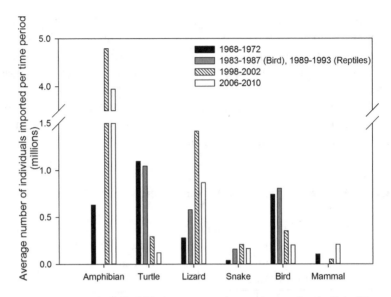

FIG. 6–2. Average number of individuals per taxonomic group imported to the United States: 1968–1972, 1983–1987 (birds), 1989–1993 (reptiles), 1998–2002, 2006–2010.

the average number of nonindigenous species imported in 1970–1971 and 2008–2009.

At least 125.7 million animals were imported to the United States during 1968–2010. Approximately 3.5 million amphibians, reptiles, birds, and mammals were imported annually during 1968–1972; this number doubled to 7 million individuals during 1998–2002, and declined to 5.5 million during 2006–2010 (Fig. 6–2). The difference between 1968–1972 and 1998–2002 was driven mostly by a substantial increase in imported amphibians. For turtles and mammals, the annual averages decreased between the same two time periods. As a result of these trends, the most heavily imported taxonomic group shifted from turtles to amphibians. Data available for the 1980s and 1990s for reptiles and birds suggest that the greatest change in the annual averages for lizards, turtles, and birds occurred after the early 1990s (Fig. 6–2). An increase in snakes occurred mostly between the 1970s and early 1990s. Mammals were the only group that increased in the most recent time period, 2006–2010 (Fig. 6–2), thanks to increased importation of small rodents, primarily desert hamsters (*Phodopus* sp., Table 6–2). Declines in the quantity of live vertebrate imports between 1998–2002 and 2006–2010 suggest that the trade is susceptible to the general economic climate. The decline in vertebrate imports is attributed to the economic

TABLE 6.2 **Top 5 species imported in 1968–1972 and 2006–2010 listed by average number of individuals for each time period; introduction, establishment, and spread status for each species within the continental United States; and geographic region of origin as reported by USFWS import data. An asterisk indicates species indigenous to the United States**

Taxon	Scientific Name	Average number individuals imported		Invasion stage			Geographic region of origin as reported by Form 3-177
		1968–1972	2006–2010	Introd	Estab	Spread	
Amphibian	Lithobates[Rana] pipiens*	812,600	9,208	X	X		Central and South America
	Xenopus laevis	102,871	159,513	X	X	X	Eastern Asia
	Cynops pyrrhogaster	30,938	16,972	X			Eastern Asia
	Rhinella marina [Bufo marinus]	20,210	673	X	X	X	Caribbean
	Triturus cristatus	10,115	23				Europe
	Lithobates catesbeianus* [Rana catesbeiana]	1,992	1,859,940	X	X	X	Eastern Asia
	Hymenochirus curtipes		831,623				Eastern Asia
	Bombina orientalis		323,217	X			Eastern Asia
	Lithobates [Rana] forreri		203,920				Central and South America
Turtle	Trachemys scripta*	913,720	15,396	X	X	X	North America, Central and South America
	Chelonoidis [Geochelone] denticulata	57,888	567	X			Central and South America
	Malayemys subrijuga	18,783	61	X			Southeast Asia
	Pelodiscus sinensis	6,092	34,985	X			Southeast Asia

(continued)

TABLE 6.2 (continued)

| Taxon | Scientific Name | Average number individuals imported | | Invasion stage | | | Geographic region of origin as reported by Form 3-177 |
		1968–1972	2006–2010	Introd	Estab	Spread	
	Chelonoidis [Geochelone] carbonaria	5,825	3,224	X			Central and South America
	Testudo horsfieldii	57	22,596	X			Central Asia
	Pelomedusa subrufa	324	4,349				Africa
	Testudo graeca	715	3,193	X			Africa
Lizard	Iguana iguana	176,486	166,920	X	X	X	Central and South America
	Cnemidophorus lemnis-catus	23,763	866	X	X		Eastern Asia
	Chamaeleo jacksonii	3,878	1,070	X	X		Africa
	Gekko gecko	3,134	22,070	X	X	X	Southeast Asia
	Agama agama	3,083	8,514	X	X	X	Africa
	Takydromus sexlineatus		94,657	X			Southeast Asia
	Physignathus cocincinus	61	88,540	X			Southeast Asia
	Varanus exanthematicus	138	22,975	X			Africa
	Pogona vitticeps		22,826	X			Southeast Asia

Snake	Boa constrictor	19,681	17,676	X		X	Central and South America
	Epicrates cenchria	3,108	414	X			Central and South America
	Broghammerus [Python] reticulatus	1,684	2,291	X			Africa
	Calloselasma rhodostoma	1,548	13				Southeast Asia
	Eunectes murinus	1,477	622	X			Central and South America
	Python regius	817	107,282	X			Africa
	Xenochrophis vittatus		3,701				Southeast Asia
	Python molurus	523	2,741	X	X	X	Southeast Asia
	Elaphe taeniura	58	2,393				Eastern Asia
Bird	Serinus canaria	99,480	44,497	X			Eastern Asia
	Estrilda troglodytes	66,563	1,496				Africa
	Amandava amandava	59,159	19	X			Southeast Asia
	Brotogeris versicolurus	52,165		X		X	Central and South America
	Lonchura malacca	46,645	4	X			Southeast Asia
	Phasianus colchicus	1,019	40,578	X	X	X	North America (Canada)
	Serinus mozambicus	28,265	12,776	X			Africa
	Uraeginthus bengalus	42,686	9,978	X			Africa
	Carduelis carduelis	888	8,767	X			Oceania

(continued)

TABLE 6.2 (continued)

Taxon	Scientific Name	Average number individuals imported		Invasion stage			Geographic region of origin as reported by Form 3-177
		1968–1972	2006–2010	Introd	Estab	Spread	
Mammal	Saimiri sciureus	34,610	97				Central and South America
	Macaca mulatta	25,444	1,326	X	X		Eastern Asia
	Aotus trivirgatus	4,008	41				Europe
	Chlorocebus aethiops	3,890	321				Caribbean
	Lepus americanus*	3,655		X	X		North America (Canada)
	Phodopus roborovskii		87,164				Europe
	Phodopus sungorus		31,522				Europe
	Macaca fascicularis	1,824	23,185				Eastern Asia
	Mesocricetus auratus	3	7,636	X			Europe
	Meriones unguiculatus		6,250	X			Europe

recession in the United States during the latter time period (GAO 2010), but is expected to recover following the end of the recession.

At least 3,900 species of nonindigenous amphibians, birds, mammals, and reptiles were imported to the United States in the years assessed during 1970–2010. In the 1970s, over 2,000 species were reported by the USFWS import data; this number nearly doubled by 2010. The cumulative number of species imported by 2010 ranged from 1.4 times (mammals) to 3.3 times (lizards) the number of species imported in the 1970s (Fig. 6–3). Birds accounted for half of the imported vertebrate species (Fig. 6–3). However, this number only represents 20% of their global species richness (Fig. 6–4). More turtle species were imported, relative to their global species richness, than any other taxonomic group (Fig. 6–4). Overall, the average number of species imported per year decreased from 1970–71 to 2008–2010. However, trends differed for each taxonomic group. The number of lizard and amphibian species imported per time period increased significantly ($P < .02$); turtle and snake species showed no significant change ($P < .10$); and bird and mammal species decreased significantly ($P < .04$).

The results described above reflect the surge of interest for amphibians and reptiles as pets, in the United States and abroad, in the last few decades of the 20th century (Hoover 1998; Auliya 2003; Tapley et al. 2011). Imports of these species have more than doubled since the early 1970s; this phenomenon is particularly apparent in the diversity of lizard species. A recent compilation of over 100 Internet dealers reported 779 amphibian and reptile taxa for sale online (Prestridge et al. 2011). Interest in a more diverse selection of amphibians and reptiles for pets has been accompanied by greater demand for specimens from these taxonomic groups. Many such specimens belong to just a few species that are popular pet animals, such as tree frogs (*Litoria caerulea*), green iguanas (*Iguana iguana*), and ball pythons (*Python regius*).

The increase in amphibian imports can be attributed to the food trade and education animals, as well as the pet trade. Pet amphibians such as fire-bellied toads and newts (*Bombina* sp. and *Cynops orientalis*) and African clawed frogs (*Hymenochirus* sp. and *Xenopus* sp.) accounted for 40% of imported specimens (Table 6–2). Most of the remaining 60% are from one species, the American bullfrog, *Lithobates catesbeianus* (Table 6–2). This species is native to the United States but is extensively farmed overseas in Asia and South America and then shipped back to the United States as food and education animals (Warkentin et al. 2009).

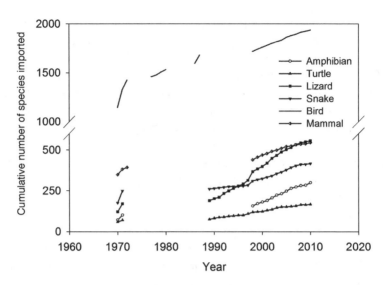

FIG. 6–3. Cumulative number of nonindigenous species imported to the United States over time (1970–2010) for amphibians, turtles, lizards, snakes, birds, and mammals.

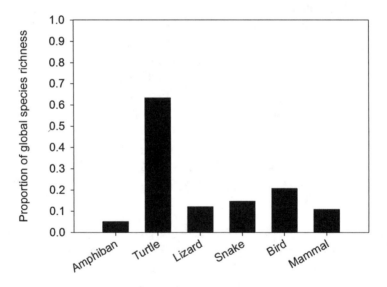

FIG. 6–4. Proportion of global species richness per taxonomic group imported to the United States from 1970 to 2010.

Turtles are particularly affected by both the pet and food trades (Thorbjarnarson et al. 2000). This taxonomic group has only about 300 species worldwide, and a majority of their global species richness is in trade. Although the number of turtle species imported is high given their global species diversity, the quantity of individuals has declined. This decline in US imports of turtles has been attributed to trade regulations enforced on foreign turtles, and increased availability of native turtles reared on ranching operations (Thorbjarnarson et al. 2000). However, even though US turtle imports have decreased, US exports have increased substantially. During 2002–2006, approximately 13 million turtles were exported annually, primarily to Asian markets (Romagosa 2009).

Some trends in international animal trade have remained consistent over the past 30 years, such as the wide diversity of bird species. Although a minimum 20% of their global diversity has been imported, birds made up 50% of all species imported in the 1970s, and about 30% in the 2000s. Birds are a speciose group (~10,000 species) that has been traded for centuries, and therefore the number of species in trade is not surprising, as more species should be accumulated over time (Fitzgerald 1989). With respect to the United States, the trade in wild birds existed as early as 1865, and more than 200 species were reported in 1906 alone (Oldys 1907). Despite their historical prominence, however, the diversity of bird species and the quantity of individuals in US imports decreased between the 1970s and the 2000s. This decline was likely due to the United States Wild Bird Conservation Act of 1992 (Engler and Parry-Jones 2007). Enforcement of this law resulted in Europe surpassing the United States as the most important foreign market for birds. A recent ban on the importation of wild birds enacted by the European Union in 2007 is expected to have the same effect on the number of birds imported to those countries as seen in the United States after 1992 (Carrete and Tella 2008). It has been suggested that the passage of the US Wild Bird Conservation Act (1992) contributed to the growth in the herpetological trade as an alternative to the bird trade (Hoover 1998).

The diversity of imported mammal species also has decreased over time, but the quantity of individuals has increased. In the early 1970s, 87% of mammal trade involved primates used as research animals (Banks 1976). Primate importation has decreased due to stricter regulations regarding trade in primates, as well as increased availability of individuals from captive breeding programs in the United States (Jorgenson and Jorgenson 1991). Despite the decrease in the primate trade, mammal imports increased. In the late 2000s, trade in primates was replaced by imports of small rodent species like desert hamsters (*Phodopus* sp., Table 6–2), captive bred in Europe. This trend may change again, as the majority of these hamsters were imported by the now-defunct US Global Exotics, a company raided by state and federal authorities in December 2009 (Brown 2009).

Geographic Patterns: Past and Emerging Trading Partners

All source countries that exported vertebrates ("Country of origin" on Form 3-177) to the United States were grouped into regions and subregions based on the United Nations Geoscheme (http://unstats.un.org /unsd/methods/m49/m49regin.htm). The regions used for comparison were Africa, Madagascar, Caribbean, Asia, Central and South America (including Mexico), North America (only considering Canada), Europe, Central Asia, Eastern Asia, Southeast Asia, Southern Asia, and Oceania. Data of all individuals imported per taxonomic group from each region for two 2-year time periods (1970–1971 and 2008–2009) were used to compare geographic patterns over time.

In 1970–1971 over 70% of amphibians and reptiles were imported from Central and South America (Fig. 6–5A-D). Eighty-percent of mammals and birds were imported from two and four regions, respectively (Fig. 6–5E-F). In 2008–2009, regions in Asia became more important trading partners for amphibians, lizards, and turtles, replacing Central and South American dominance in the early 1970s (Fig. 6–5A-D). The majority of imported snakes (74%, Fig. 6–5D) and birds (42%, Fig. 6–5D) now originate from Africa. Europe was the most dominant region of origin for mammals in 2008–2009 (Fig. 6–5F). Regional dominance can often be attributed to importation of just a few species. For example, in 2008–2009 half the amphibians imported from Eastern Asia were American bullfrogs, while desert hamsters (*Phodopus roborovskii* and *Phodopus sungorus*) composed 87% of mammals imported from Europe.

The primary cause for shifts in geographic patterns over time is country-specific restrictions in trade (Fitzgerald 1989). For example, in the 1970s, most animals originating from Central and South America were exported from Colombia. That country now restricts many of its wildlife exports, and as a result, animals are exported from other countries in South America or other regions in the world. Also related to this geographic shift is the advent of amphibian and reptile ranching operations in Asia, Africa, and

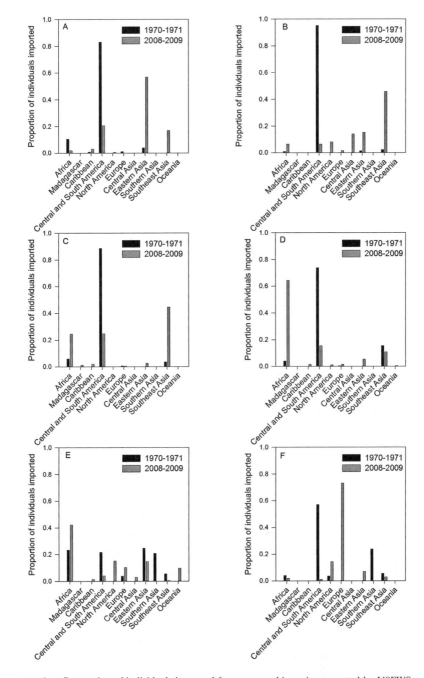

FIG. 6–5. Proportion of individuals imported from geographic region reported by USFWS importation data for 1970–1971 and 2008–2009 for amphibians (*A*), turtles (*B*), lizards (*C*), snakes (*D*), birds (*E*), and mammals (*F*).

South America. Ranching for American bullfrogs (East Asia and South America), iguanas (Central and South America), ball pythons (Africa), and red-eared sliders (*Trachemys scripta,* East and Southeast Asia) have created mass quantities of individuals for the export market; consequently the trade in these species has shifted to those regions.

Species Turnover

Understanding the dynamics of the live vertebrate trade can give insight to its long-term resilience to any perturbations, such as trade restrictions, in the global market. As an estimate for the elasticity of the live vertebrate trade, I calculated an approximation of turnover rate, which determines the presence of a species between the two time periods. An approximation of turnover rate was calculated between nonindigenous species imported from 1968–1972 and 2006–2010. The turnover rate is calculated as $TR=100(I_1+I_2)/(S_1+S_2)$; where I_1 is the number of species imported in the 1970s but not in the 1990s, and I_2 is number of species imported in 1990s but not in the 1970s. S_1 and S_2 are the number of species imported in the 1970s and 1990s, respectively.

There was substantial species turnover per time periods assessed (Fig. 6–6; 1968–1972, 2006–2010). The overall turnover rate of 63% means more than half the species imported in the 1970s were no longer imported in the late 2010s. Only 23% of all species were imported in both time periods. Again, this turnover rate varies among taxonomic groups (Fig. 6–6). Table 6–2 shows the top 5 species imported (by average number of individuals per time period) during the aforementioned two periods (1968–1972, 2006–2010). Roughly 94% of the top species in the 1970s were still found in trade in the late 2010s. This percentage suggests that popular species, once in trade, tend to remain there, albeit in lower numbers in some cases.

The high turnover rate in live vertebrate imports suggests that the demand for these species is largely elastic. The greater the elasticity in demand, the more likely different species, even from a different geographic region, can be substituted if supply of particular species is limited. For example, because of trade restrictions in Madagascan chameleons, different chameleon species from other regions of Africa were substituted in their global trade (Carpenter et al. 2004). Additionally, consumer demand can create shifts in the species, and quantity of individuals of those species, traded. This phenomenon can occur when a particular species portrays a character in a movie or television series, as exemplified by increased

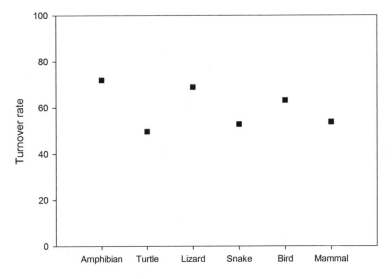

FIG. 6–6. Turnover rate between species imported in 1968–1972 and 2006–2010.

demand and trade in turtles ("Teenage Mutant Ninja Turtles"), clownfish ("Finding Nemo"), and owls ("Harry Potter" series) (Prosek 2010; Yong et al. 2011).

Summary

Import data extrapolated from USFWS Form 3-177 suggest that the US trade in live vertebrates, as typical of trade in most commodities, is dynamic. Each year's trade is subject to changes in supply and demand stemming from fluctuations in wild and captive populations, trade restrictions, and consumer preferences. These trade-related dynamics have led to changes in the species traded, the quantities of those species, and their geographic origin. The United States is the largest market in the international trade of live vertebrates, and its importation trends mirror several global patterns. First, the overall species diversity and quantity of individuals recorded by USFWS import data has increased substantially over time, a global trend corroborated by other research (Auliya 2003, Engler and Parry-Jones 2007, Carrete and Tella 2008). Within all the taxonomic groups, specific species are imported more frequently than others, suggesting that species used for trade are not randomly distributed among taxonomic groups (Romagosa et al. 2009).

Finally, any discussion about the wildlife trade requires at least brief comment on the effects on wild populations. The growth seen throughout all wildlife trade raises the question: Can all vertebrate species eventually be exploited for trade? Rarity and/or trade restrictions might prevent trade, but often these species become more desirable to collectors and the black market. This phenomenon, known as the "collector's vortex," creates increasing commercial value on a species as it becomes more rare, thereby encouraging illegal take from the wild (Wright et al. 2001; Courchamp et al. 2006; Rivalan et al. 2007). Several parrot and turtle species have been threatened or extirpated by this type of extinction vortex (Wright et al. 2001; Broad et al. 2003; Shepherd and Ibarrondo 2005). Additionally, depletion of one species encourages substitution of alternate species for trade (Roe et al. 2002; Broad et al. 2003), contributing to the accumulation of new species. If demand is for a commodity type, such as a lizard that can be kept easily in captivity, then a decline in trade in one species may correspond with an increase in another, not necessarily related species (Roe et al. 2002). These tendencies, given the current rate of trade in vertebrates, suggest that the rate of species accumulation in trade could maintain growth for some time.

Relationship to Invasion Process

The link between increasing international trade and the introduction of nonindigenous organisms is clear (Levine and D'Antonio 2003), and this trend appears no less apparent in the live vertebrate trade. With each new species that is imported, the risk of admitting a potentially invasive species and/or its companion pathogens and parasites also increases. To determine the contribution of the live vertebrate trade to biological invasions, I reevaluated the data and various analyses above within the framework of the invasion process. All species documented by the USFWS import data were categorized in four stages of the biological invasion process described in Fig. 6–1 (importation, introduction, establishment, and spread), following a review of the literature (e.g., Long 1981, 2003; Pranty 2004; Bomford et al. 2005; Kraus 2009). Unless otherwise indicated, only species introductions to the continental United States were considered. The number of species that have entered the biological invasion process through this pathway must be considered a minimum, as some species may not be captured by the USFWS data for the years analyzed, and records of failed introductions are rare (Kark and Sol 2005).

Quantitative and Taxonomic patterns

For each of the 6 taxonomic groups, I calculated a transition success rate, the percentage of nonindigenous species that successfully transitioned through each stage of the invasion process (also known as base rate, Smith et al. 1999). I also recorded the progression of the top species imported to the United States (Table 6–2) through the invasion process. A few species in Table 6–2 are indigenous to the United States; the introduction of these species to other regions within the continental United States was considered. Finally, the cumulative number of nonindigenous reptile species introduced to the continental United States over time was calculated for comparison with the accumulation of nonindigenous reptile species imported.

When USFWS data are framed within the invasion process, several trends become apparent. When all vertebrate groups are combined, 11% (range, 7%–16%) of species imported are introduced, 23% (range, 10%–42%) of those introduced transition successfully to the establishment stage, and 41% (range 33%–75%) of established species transition successfully to the spread stage. The proportion of species that transitioned successfully to the next stage varies among taxonomic groups. The most successful groups were lizards and mammals, with approximately 40% of all species introduced from these groups having established self-sustaining breeding populations. Least successful were turtle species, among which no nonindigenous species established breeding populations in the United States. Popularity in trade also appears to have an effect on transition success. Among the top imported species, 74% were recorded in the wild, 40% of those introduced established breeding populations, and 56% of established species spread from the original introduction site (Table 6–2). Introduction data by decade for reptiles indicate that the number of species introduced to the United States through human-use vectors (e.g., pet, food, and bait trades) is increasing rapidly over time. When these data are overlaid with importation data, results suggest that both are increasing at the same rate (Fig. 6–7). Lizards, which have the greatest increase in species imported over time, also have more species introduced and established than any other reptile group (Fig. 6–7).

The number of vertebrate species imported to the United States is increasing over time, with concomitant growth in the number of introductions (Temple 1992; Kraus 2009). In fact, most introductions of amphibian and reptiles worldwide have occurred in the past 30 years, a trend that reflects the increase in their trade during this period (Kraus 2009).

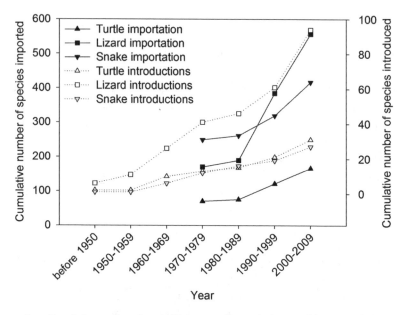

FIG. 6–7. Cumulative number of nonindigenous reptile species imported (1970–2009) and introduced (prior to 1950–2009) to the continental United States.

Additionally, as people immigrate to the United States from around the world, they bring with them food or medicinal animals used in their native cultures, as recently highlighted by the widely publicized introduction of the snakehead fish (*Channa* spp., Courtenay and Williams 2004). Probability of nonindigenous species establishment correlates strongly with the number of individuals released and/or the number of introduction attempts (propagule pressure, Kolar and Lodge 2001). This relationship is exemplified by the establishment worldwide of two heavily traded species, the American bullfrog and red-eared slider (Kraus 2009). Additionally, the transition success rates for the species in Table 6–2 suggest that frequently imported species are more likely to enter the invasion process.

Geographic Patterns: Past and Emerging Donor Regions

Finally, the native region for all nonindigenous species imported, introduced, established and spread in the United States was recorded to assess geographic patterns among species that enter the invasion process. Again, regions and subregions were based on the United Nations Geoscheme. For this comparison, an additional region, "Eurasia," was created because of the high number of species whose ranges overlapped both Europe and Asia. A transition success rate (all taxa combined) was calculated for each region: the number of species imported from a region that were introduced, the number of species introduced that successfully established, and the number of established species that spread from the original introduction site.

Central and South America is the most important donor region for species that enter the United States and progress through the second and third stages (introduction and establishment) of the invasion process (Fig. 6–8A-B). The fact that Central and South America contributes the most species to these stages (Table 6–3) should not be surprising, as more species are imported to the United States from this region than anywhere else. Africa and Southeast Asia are also important donor regions, each contributing over 60 species to the introduction stage, and over 15 species to the establishment stage (Fig. 6–8A-B, Table 6–3). For the final spread stage, Africa, Central and South America, Europe, and Southeast Asia all contribute an average of seven species (Fig. 6–8C). When all taxonomic groups and regions are combined, the success rate for the introduction-to-establishment transition is 22% (range, 8%–60%); the success rate for the establishment-to-spread transition is 42% (range 0%–100%, Table 6–3). For regions with at least five species introduced and established, species native to the Caribbean have the highest introduction-to-establishment success rate (60%), and species from Europe have the highest establishment-to-spread success rate (88%).

Transition success rates for the final two invasion stages for these two regions are high relative to those from other regions. The high success rate for animals imported from the Caribbean is likely due to that region's proximity and climatic similarity to southern Florida, where the species (all amphibians and lizards) from this region are established (Krysko et al. 2011). The history of European immigrants and their companion organisms (Jeschke and Strayer 2005), as well as climatic similarities (Tatem 2009), contributes to high transition success rates for European species. Due to lag times in introduction and establishment (Jeschke and Strayer 2005), the success rates for emerging trading partners, such as particular regions in Asia, may increase in the future (Bradley et al. 2012).

Summary

Which species introductions succeed or fail is a debated topic (Williamson and Fitter 1996), but success and failure alike undeniably begin with the

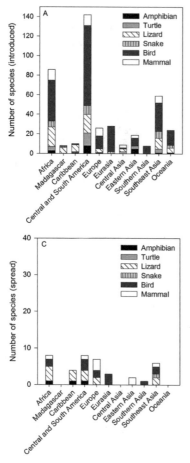

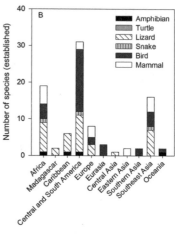

FIG. 6–8. Number of nonindigenous species imported to the United States and introduced (*A*), established (*B*), and spread from original introduction site (*C*) separated by native geographic region.

transport of species to a new range (Kolar and Lodge 2001; Duncan et al. 2003). Effects on native ecosystems cannot be known until after the arrival of new species, and post-establishment impacts often remain unknown for long periods owing to low detection and lag times (Crooks and Soulé 1999). Some species remain localized around the introduction site, such as the Veiled Chameleon (*Chamaeleo calyptratus*; Krysko et al. 2004). Others reproduce prolifically and expand their ranges like the Burmese python (*Python molurus bivittatus*). Traded since at least the early 1900s (Anonymous 1904), approximately 10,000 individuals were imported annually during 1997–2003. This species was first reported in Everglades Na-

TABLE 6.3 **Transition success rates for species per geographic region**

Region	Introduced	Established	Spread	Introduction-Establishment (%)	Establishment-Spread (%)
Africa	86	19	8	22	42
Madagascar	8	2	0	25	0
Caribbean	10	6	4	60	67
Central/ South America	142	31	8	22	26
Europe	26	8	7	31	88
Eurasia	28	3	3	11	100
Central Asia	9	1	0	11	0
Eastern Asia	19	2	2	11	100
Southern Asia	8	2	1	25	50
Southeast Asia	59	16	6	27	38
Oceania	24	2	0	8	0
Total	419	92	39	22	42

tional Park in 1979 (Snow et al. 2007), and the potential for detrimental cascading ecological effects has been documented only recently (Dorcas et al. 2012).

Additional concerns exist regarding health risk through the transmission of disease to native fauna and humans by vertebrate species in trade (Pavlin et al. 2009; Smith et al. 2009). Amphibians in the live animal trade are believed to be responsible for the spread of chytridiomycosis, an emerging fungal disease linked to the global population decline of amphibians (Weldon et al. 2004). The zoonotic disease transmission of monkeypox virus to humans from Gambian pouched rats (*Cricetomys gambianus*) and black-tailed prairie dogs (*Cynomys ludovicianus*), kept as pets in the United States, led to a complete ban in their trade after 2004 by the Centers for Disease Control and Prevention (Reed et al. 2004). The increased importation of live vertebrates from Asia, also an emerging center for infectious zoonotic disease, has created a call for concern (Coker et al. 2011). Additional introductions of nonindigenous animals and their associated parasites will continue if there is no effort to stem the flow of species moved globally through the live animal trade.

Reducing the risk of additional introductions from the live vertebrate trade pathway will require changes in the species used and the ways that the industries and their consumers acquire, keep, and dispose of those organisms (Keller and Lodge 2007). The ecological and economic costs associated with biological invasions have encouraged a more proactive approach to research on the biology of invasions, and created a need for a more predictive science (Kolar and Lodge 2001). A risk analysis approach enables creation of a predictive framework that can be utilized for regulation and management of vertebrate species imported into the United States. However, the challenge for researchers, managers, and government agencies is creating a system that can identify potentially harmful nonindigenous species, and prevent their entry, without prohibiting the importation of all species (Mack et al. 2000), and without creating undue stress for an already overwhelmed regulating federal agency.

Researchers have shown that, in certain scenarios, the economic benefits of pre-import screening would outweigh the costs. Keller et al. (2007) and Springborn et al. (2011) addressed this cost-benefit problem by applying bioeconomic models to the Australian ornamental plant industry and the US live reptile trade, respectively (see also chapter 8 of this volume). Both studies showed that screening of species proposed for import would produce net economic benefits compared with a policy of allowing all species. Because vertebrate species have high transition rates for the last two stages of the invasion process (Jeschke and Strayer 2005; this study), pre-import screening methods would be even more beneficial for this taxonomic group (Keller et al. 2007).

Even for preliminary screening methods, there is a need for global dissemination of invasive species information (Ricciardi et al. 2000). There is a shortage of scientific information on species that have transitioned through any of the stages of the invasion process and their subsequent management (Browne and De Poorter 2009). Such information is difficult to obtain because it requires searching through a wide variety of disciplinary journals or obscure sources and is often of variable quality (Ricciardi et al. 2000). In order to most efficiently and effectively implement risk assessment methods as outlined by Lodge et al. (2006), this information needs to be readily accessible.

The United States, as one of the largest global consumers of live vertebrates, should take a role in spearheading such an effort to incorporate import data within a global invasive species database. The utility of preserving these data in relation to biological invasions has not gone unnoticed. In June 2012, the Invasive Species Advisory Committee to the National Invasive Species Council recommended the incorporation of all species-specific import data associated with Form 3-177 into LEMIS or another accessible database. Currently, the USFWS LEMIS data, although invaluable, are not in an ideal format for a long-term dataset; therefore preparing and restructuring the data will take a considerable amount of time and work. However tedious the restructuring may be, the maintenance of these data within a global information network will secure their availability for future research on biodiversity conservation and biological invasions.

The Convention on Biological Diversity (CBD) has been working to coordinate international organization and agency cooperation to strengthen information services on invasive species that can be used in conducting risk assessments and developing early warning systems (CBD 2011). The development and maintenance of invasive species databases within a broadly accessible information network, such as the Global Invasive Species Information Network (GISIN), helps meet this need (Graham et al. 2008). Besides species-specific information on biological invasions, these databases should include data on global trade in these species, such as that derived from Form 3-177, in order to provide an estimate for potential propagule pressure as well as a long-term dataset on importation trends. These data can also contribute to an early warning system by identifying species whose trade has increased substantially, has progressed through additional stages of the invasion process, or has been reported as a disease vector. Strategies to identify and reduce propagule supply, whether through regulation or education, should help prevent future invasions through the live animal trade.

Policy Implications

Currently, there is no international regulatory framework for potentially invasive species in the live animal trade, but regulations are often implemented on a national basis. While certain countries have implemented bans or black/white lists, this process remains controversial for the potential trade restrictions and concern that these restrictions may create more underground trade that then cannot be monitored (Garner et al. 2009). Over the past 10 years, many international organizations (e.g., Convention on Biological Diversity, Convention on International Trade in Endangered

Species of Wild Fauna and Flora, International Union for the Conservation of Nature, World Trade Organization, World Organization of Animal Health) have discussed how to minimize biological invasions. Various suggestions have been made as to which international trade agreements are relevant to invasive species, but the coordination necessary to address international invasive species issues is complex (see Burgiel [chapter 13], C. Shine [chapter 14], and Brammeier and Cmar [chapter 16]).

One barrier to implementing a regulatory framework is a need for better data collection on the global trade in live animals. Unfortunately, few countries record importation/exportation data for species other than those regulated by the Convention on International Trade in Endangered Species (CITES; Schlaepfer et al. 2005). The CITES listing and permitting process has been suggested as a model to better track the trade in potentially invasive species, which would shift the focus of concern to the importing country rather than the exporting country (Office of Technology Assessment 1993). The same challenges that currently plague the effectiveness of CITES would apply (Blundell and Mascia 2005; Smith et al. 2011). For example, some countries lack the infrastructure for tracking species and monitoring their porous borders effectively. Regardless of these challenges, the CITES model is promising for development of an invasive species prevention program.

It is inevitable that some imported species will filter through the latter stages of the invasion process. Therefore, in order to have a complete and effective regulatory invasive species framework, an additional capacity for monitoring, control, and eradication is necessary. This requires reliable scientific information on nonindigenous species populations and their potential impacts. An insufficient understanding of the economic and ecological consequences of species introduced through trade often translates into a lack of political will and, ultimately, a lack of resources available to address issues caused by nonindigenous species at any stage along the invasion process.

Acknowledgments

This work was supported by the Center for Forest Sustainability at Auburn University, National Fish and Wildlife Federation Budweiser Conservation Scholarship, and Southern Regional Education Board–State Doctoral Scholars Program. I thank Craig Guyer, Mike Mitchell, the Guyer lab, the editors, and anonymous reviewers for suggestions that much improved this research and earlier drafts of this manuscript. I am grateful to Gary Townsend for his help and to Craig Hoover for discussions regarding USFWS LEMIS data. Finally, thanks to Reuben Keller for organizing the conference that inspired this volume, and to the editors for making the book a reality.

Literature Cited

Anonymous. 1904. Python waved over Grand St. sidewalk. *New York Times*, June 6.

Auliya, M. 2003. Hot trade in cool creatures: a review of the live reptile trade in the European Union in the 1990s with a focus on Germany. Brussels: TRAFFIC Europe.

Banks, R. C. 1970. Birds imported into the United States in 1968. Special scientific report–wildlife 136. Washington, DC: US Fish and Wildlife Service.

Banks, R.C. 1976. Wildlife importation into the United States, 1900–1972. Special scientific report–wildlife 200. Washington, DC: US Fish and Wildlife Service.

Banks, R. C., and R. B. Clapp. 1972. Birds imported into the United States in 1969. Special scientific report–wildlife 148. Washington, DC: US Fish and Wildlife Service.

Blundell, A. G., and M. B. Mascia. 2005. Discrepancies in reported levels of international wildlife trade. *Conservation Biology* 19:2020–2025.

Bomford, M., F. Kraus, M. Braysher, L. Walter, and L. Brown. 2005. Risk assessment model for the import and keeping of exotic reptiles and amphibians. Canberra, Australia: Bureau of Rural Sciences.

Bradley, B., D. M. Blumenthal, R. Early, E. D. Grosholz, J. J. Lawler, L. P. Miller, C. J.B Sorte, et al. 2012. Global change, global trade, and the next wave of plant invasions. *Frontiers in Ecology and the Environment* 10: 20–28.

Broad, S., T. Mulliken, and D. Roe. 2003. The nature and extent of legal and illegal trade in wildlife. *In The Trade in Wildlife: Regulation for Conservation*, edited by S. Oldfield, 3–22. London: Earthscan.

Brown, A. K. 2009. Dead rodents, skinny snakes found in Texas raid. *Associated Press*, December 16.

Browne, M., S. Pagad, and M. De Poorter. 2009. The crucial role of information exchange and research for effective responses to biological invasions. *Weed Research* 49:6–18.

Burgiel, S. 2014. From global to local: integrating policy frameworks for the prevention and management of invasive species. Chapter 13 of this volume.

Busack, S. D. 1974. Amphibians and reptiles imported into the United States. Wildlife Leaflet 506. Washington, DC: US Fish and Wildlife Service.

Carpenter, A. I., J. M. Rowcliffe, and A. R. Watkinson. 2004. The dynamics of the global trade in chameleons. *Biological Conservation* 120:291–301.

Carrete, M., and J. L. Tella. 2008. Wild-bird trade and exotic invasions: a new link of conservation concern? *Frontiers in Ecology and Evolution* 6:207–211.

Clapp, R. B. 1973. Mammals imported into the United States in 1971. Special scientific report–wildlife 171. Washington, DC: US Fish and Wildlife Service.

Clapp, R. B. 1975. Birds imported into the United States in 1972. Special scientific report–wildlife 193. Washington, DC: US Fish and Wildlife Service.

Clapp, R.B., and R. C. Banks. 1972. Birds imported into the United States in 1970. Special scientific report–wildlife 148. Washington, DC: US Fish and Wildlife Service.

Clapp, R. B., and R. C. Banks. 1973. Birds imported into the United States in 1971. Special scientific report–wildlife 170. Washington, DC: US Fish and Wildlife Service.

Coker, R. J., B. M. Hunter, J. W. Rudge, M. Liverani, and P. Hanvoravongchai. 2011. Emerging infectious diseases in Southeast Asia: regional challenges to control. *Lancet* 9765:599–609.

Convention on Biological Diversity, Secretariat. 2010. Pets, Aquarium, and Terrarium Species: Best Practices for Addressing Risks to Biodiversity. Technical Series No. 48. Montreal: SCBD.

Convention on Biological Diversity. 2011. Joint work programme to strengthen information services on invasive alien species as a contribution towards Aichi Biodiversity Target 9 (UNEP/CBD/SBSTTA/15/INF/14). http://www.cbd.int /doc/meetings/sbstta/sbstta-15/information/sbstta-15-inf-14-en.pdf.

Courchamp, F., E. Angulo, P. Rivalan, R. J. Hall, L. Signoret, L. Bull, and Y. Meinard. 2006. Rarity value and species extinction: the anthropogenic allee effect. *Public Library of Science Biology* 4. doi:10.1371/journal.pbio.0040415.

Courtenay, W. R., Jr., and J. D. Williams. 2004. Snakeheads (Pisces: Channidae)—a biological synopsis and risk assessment. United States Geological Survey Circular 1251.

Crooks, J. A., and M. E. Soulé. 1999. Lag times in population explosions of invasive species: causes and implications. In *Invasive Species and Biodiversity Management*, edited by O. T. Sandlund, P. J. Schei, and A. Viken, 103–125. Dordrecht: Kluwer Academic.

Defenders of Wildlife. 2007. Broken screens—the regulation of live animal imports in the United States. Report. Washington, DC: Defenders of Wildlife.

Dorcas, M.E. , J. D. Willson, R. N. Reed, R.W. Snow, M. R. Rochford, M. A. Miller, W. E. Meshaka Jr., et al. 2012. Severe mammal declines coincide with python proliferation in Everglades National Park. *Proceedings of the National Academy of Sciences USA* 109:2418–2422.

Duncan, R.P., T. M. Blackburn, D. Sol. 2003. The ecology of bird introductions. *Annual Review of Ecology and Evolutionary Systematics* 34:71–98.

Engler, M., and R. Parry-Jones. 2007. Opportunity or threat: the role of the European Union in global wildlife trade. Brussels: TRAFFIC Europe.

Fitzgerald, S. 1989. *International Wildlife Trade: Whose Business Is It?* Washington, DC: World Wildlife Fund.

Franke, J., and T. M. Telecky. 2001. *Reptiles as Pets: An Examination of the Trade in Live Reptiles in the United States.* Washington, DC: Humane Society of the United States.

GAO. 2010. Report to the Committee on Homeland Security and Governmental Affairs. Live animal imports: agencies need better collaboration to reduce the risk of animal-related diseases. GAO-11-9. Washington, DC: GAO. http://www.gao.gov/new.items/d119.pdf.

Garner, T.W.J, I. Stephen, E. Wombwell, and M. C. Fisher. 2009. The amphibian trade: bans or best practice? *EcoHealth* 6:148–151.

Graham, J., A. Simpson, A. Crall, C. Jarnevich, G. Newman, and T. J. Stohlgren. 2008. Vision of a cyberinfrastructure for nonnative, invasive species management. *Bioscience* 58:263–268.

Hanselmann, R., A. Rodriguez, M. Lampo, L. Fajardo-Ramos, A. A. Aguirre, A. M. Kilpatrick, J. P. Rodriguez, and P. Daszak. 2004. Presence of an emerging pathogen of amphibians in introduced bullfrogs *Rana catesbeiana* in Venezuela. *Biological Conservation* 120:115–119.

Hoover, C. 1998. The US role in the international live reptile trade: Amazon tree boas to Zululand dwarf chameleons. Washington, DC: TRAFFIC North America.

Hulme, P. E. 2009. Trade, transport and trouble: managing invasive species pathways in an era of globalization. *Journal of Applied Ecology* 46:10–18.

Jeschke, J. M., and D. L. Strayer. 2005. Invasion success of vertebrates in Europe and North America. *Proceedings of the National Academy of Sciences USA* 102:7198–7202.

Jorgenson, J. P., and A. B. Jorgenson. 1991. Imports of CITES-regulated mammals into the United States from Latin America: 1982–1984. In *Latin American Mammalogy*, edited by M. A. Mares and D. J. Schmidly, 322–335. Norman: University of Oklahoma Press.

Kark, S., and D. Sol. 2005. Establishment success across convergent Mediterranean ecosystems: an analysis of bird introductions. *Conservation Biology* 19:1519–1527.

Keller, R. P., and D. M. Lodge. 2007. Species invasions from commerce in live aquatic organisms: problems and possible solutions. *Bioscience* 57:428–436.

Keller, R. P., D. M. Lodge, and D. C. Finnoff. 2007. Risk assessment for invasive species produces net bioeconomic benefits. *Proceedings of the National Academy of Sciences USA* 104:203–207.

Kolar, C.S., and D.M. Lodge. 2001. Progress in invasion biology: predicting invaders. *Trends in Ecology and Evolution* 16: 199–204.

Kraus, F. 2009. *Alien Reptiles and Amphibians: A Scientific Compendium and Analysis.* Dordrecht: Springer-Verlag.

Krysko, K.L., J.P. Burgess, M.R. Rochford, C.R. Gillette, D. Cueva, K.M. Enge, L.A. Somma, et al. 2011. Verified non-indigenous amphibians and reptiles in Florida from 1863 through 2010: outlining the invasion process and identifying invasion pathways and stages. *Zootaxa* 3028:1–64.

Krysko, K.L., K. M. Enge, and F. W. King. 2004. The veiled chameleon, *Chamaeleo calyptratus*: a new exotic lizard species in Florida. *Florida Scientist* 67: 249–253.

Levine, J. M., and C. M. D'Antonio. 2003. Forecasting biological invasions with increasing international trade. *Conservation Biology* 17:322–326.

Lodge, D. M., S. Williams, H. J. MacIsaac, K. R. Hayes, B. Leung, S. Reichard, R. N. Mack, et al. 2006. Biological invasions: recommendations for US policy and management. *Ecological Applications* 16:2035–2054.

Long, J. L. 1981. Introduced birds of the world. New York: Universe.

Long, J. L. 2003. Introduced mammals of the world. Collingwood, Australia: CSIRO.

Mack, R. N., D. Simberloff, W. M. Lonsdale, H. Evans, M. Clout, and F. A. Bazzaz. 2000. Biotic invasions: causes, epidemiology, global consequences and control. *Ecological Applications* 10:689–710.

Miller, M. 2014. There ought to be a law! The peculiar absence of "organic" harmful nonindigenous species legislation. Chapter 15 of this volume.

Nijman, V. 2010. An overview of international wildlife trade from Southeast Asia. *Biodiversity and Conservation* 19:1101–1114.

Nilsson, G. 1977. *The Bird Business: A Study of the Commercial Bird Trade.* Washington, DC: Animal Welfare Institute.

Nilsson, G. 1990. *Importation of Birds into the United States 1986–1988.* Washington, DC: Animal Welfare Institute.

Office of Technology Assessment, United States Congress. 1993. *Harmful Nonindigenous Species in the United States.* OTA-F-565. Washington, DC: US Government Printing Office.

Oldys, H. 1907. Cage-bird traffic of the United States. In *Yearbook of the United States Department Agriculture, 1906,* 165–180. Washington, DC: United States Government Printing Office.

Paradiso, J. L., and R. D. Fisher. 1972. Mammals imported into the United States in 1970. Special scientific report–wildlife 161. Washington, DC: US Fish and Wildlife Service.

Pavlin, B.I., L.M. Schloegel, and P. Daszak. 2009. Risk of importing zoonotic diseases through wildlife trade, United States. *Emerging Infectious Diseases* 15:1721–1726.

Pranty, B. 2004. Florida's exotic avifauna: a preliminary checklist. *Birding* 36:362–372.

Prestridge, H.L., L. A. Fitzgerald, and T. J. Hibbitts. 2011. Trade in non-native amphibians and reptiles in Texas: lessons for better monitoring and implications for species. *Herpetological Conservation and Biology* 6:324–339.

Prosek, J. 2010. Beautiful friendship. *National Geographic Magazine* 217:120–131.

Reaser, J. K., and J. A. Waugh. 2007. *Denying Entry: Opportunities to Build Capacity to Prevent the Introduction of Invasive Species and Improve Biosecurity at US Ports*. Washington, DC: IUCN-World Conservation Union.

Reed, K. D., J. W. Melski, M. B. Graham, R. L. Regnery, M. J. Sotir, M. V. Wegner, J. J. Kazmierczak, et al. 2004. The detection of monkeypox in humans in the Western Hemisphere. *New England Journal of Medicine* 350:342–350.

Rhyne, A. L., M. F. Tlusty, P. J. Schofield, L. Kaufman, J. S. Morris, and A. Bruckner. 2012. Revealing the appetite of the marine aquarium fish trade: the volume and biodiversity of fish imported into the United States. *PLOS One*. doi:10.1371/journal.pone.0035808.

Ricciardi, A., W. W. M. Steiner, R. N. Mack, and D. Simberloff. 2000. Toward a global information system for invasive species. *BioScience* 50:239–244.

Rivalin, P., V. Delmas, E. Angulo, L. S. Bull, R. J. Hall, F. Courchamp, A. M. Rosser, and N. Leader-Williams. 2007: Can bans stimulate wildlife trade? *Nature* 447:529–530.

Roe, D., T. Mulliken, S. Milledge, J. Mremi, S. Mosha, and M. Grieg-Gran. 2002. Making a killing or making a living? Wildlife trade, trade controls and rural livelihoods. *Biodiversity and Livelihoods Issues No. 6*. London: IIED.

Romagosa, C. M. 2009. United States commerce in live vertebrates: patterns and contribution to biological invasions and homogenization. PhD diss., Auburn University.

Romagosa, C. M., C. Guyer, and M. C. Wooten. 2009. Contribution of the live vertebrate trade toward taxonomic homogenization. *Conservation Biology* 23:1001–1007.

Ruiz, G. M., and J. T. Carlton. 2003. Invasion vectors: a conceptual framework for management. In *Invasive Species: Vectors and Management Strategies*, edited by G. M. Ruiz and J. T. Carlton, 459–504. Washington, DC: Island Press.

Schlaepfer, M. A., C. Hoover, and C. K. Dodd. 2005. Challenges in evaluating the impact of the trade in amphibians and reptiles on wild populations. *BioScience* 55:254–264.

Shepherd, C. R., and B. Ibarrondo. 2005. The trade of the Roti Island snake-necked turtle *Chelodina mccordi*, Indonesia. Petaling Jaya: TRAFFIC Southeast Asia.

Shine, C. 2014. Developing invasive species policy for a major free trade bloc: challenges and progress in the European Union. Chapter 14 of this volume.

Smith, C. S., W. M. Lonsdale, and J. Fortune. 1999. When to ignore advice: invasion predictions and decision theory. *Biological Invasions* 1:89–96.

Smith, K., M. Behrens, L. Schloegel, N. Marano, S. Burgiel, and P. Daszak. 2009. Reducing the risks of the wildlife trade. *Science* 324:594.

Smith, M. J., H. Benítez-Díaz, M. A. Clemente-Muñoz, J. Donaldson, J. M. Hutton, H. N. McGough, R. A. Medellin, et al. 2011. Assessing the impacts of international trade on CITES-listed species: current practices and opportunities for scientific research. *Biological Conservation* 144:82–91.

Snow, R. W., K. L. Krysko, K. M. Enge, L. Oberhofer, A. Warren-Bradley, and L. Wilkins. 2007. Introduced populations of Boa constrictor (Boidae) and *Python molurus bivittatus* (Pythonidae) in southern Florida. In *Biology of the Boas and Pythons*, edited by R. W. Henderson and R. Powell, 416–438. Eagle Mountain, UT: Eagle Mountain Publishing.

Springborn, M. 2014. Reducing damaging introductions from international species trade through invasion risk assessment. Chapter 8 from this volume.

Springborn, M., C.M. Romagosa, and R. P. Keller. 2011. The value of non-indigenous species risk assessment in international trade. *Ecological Economics* 70:2145–2153.

Tapley, B., R. A. Griffiths, and I. Bride. 2011. Dynamics of the trade in reptiles and amphibians within the United Kingdom over a ten-year period. *Herpetological Journal* 21:27–34.

Tatem, A. J. 2009. The worldwide airline network and the dispersal of exotic species: 2007–2010. *Ecograph* 32:94–102.

Temple, S. A. 1992. Exotic birds: a growing problem with no easy solution. *Auk* 109:395–397.

Thorbjarnarson, J., C. J. Lagueux, D. Bolze, M. W. Klemens, and A. B. Meylan. 2000. Human use of turtles: a worldwide perspective. In *Turtle Conservation*, edited by M. W. Klemens, 33–85. Washington, DC: Smithsonian Institution Press.

US Fish and Wildlife Service. 2001. *Division of Law Enforcement Annual Report FY 2000*. Washington, DC: US Fish and Wildlife Service.

Warkentin, I. G., D. Bickford, N. Sodhi, and C. J. A. Bradshaw. 2009. Eating frogs to extinction. *Conservation Biology* 24:1056–1059.

Weldon, C., L. H. du Preez, A. D. Hyatt, R. Muller, and R. Speare. 2004. Origin of the amphibian chytrid fungus. *Emerging Infectious Diseases* 10:2100–2105.

Wilcove, D. S., D. Rothstein, J. Dubow, A. Phillips, and E. Losos. 1998. Quantifying threats to imperiled species in the United States. *Bioscience* 48:607–615.

Williamson, M., and A. Fitter. 1996. The varying success of invaders. *Ecology* 77:1661–1666.

Wright, T. F., C. A Toft, E. E. Enkerlin-Hoeflich, J. González-Elizondo, M. Albornoz, A. Rodríguez-Ferraro, F. Rojas-Suárez, et al. 2001. Nest poaching in neotropical parrots. *Conservation Biology* 15:710–720.

Yong, D. L., S. D. Fam, and S. Lum. 2011. Reel conservation: can big screen animations save tropical biodiversity? *Tropical Conservation Science* 4:244–253.

CHAPTER SEVEN

All in the Family: Relatedness and the Success of Introduced Species

Marc W. Cadotte and Lanna S. Jin

Introduction

Human activity has resulted in the increased movement of organisms around the world, which may have unintended or unexpected impacts for many ecosystems. Global demand for biologically based industries (e.g., biofuels) is increasing the movement of species, and it is therefore critical to adequately assess the potential impact of proposed species introductions (Ferdinands et al. 2011). To this end, a number of policy and management initiatives have been employed to reduce both the importation and impact from potentially harmful nonnative species (Inderjit 2005; Simberloff 2005). One approach that has been successfully developed is the assessment of the ecological and trait characteristics of potential introductions, where traits are scored on the basis of how likely they influence invasiveness. Species that possess many of the characteristics or traits that are known to be associated with invasiveness get high scores and are categorized as likely invaders. Such weed risk assessment (WRA) schemes developed for particular regions (and North America: Reichard and Hamilton 1997; Australia: Pheloung et al. 1999) have since been evaluated, modified, and employed elsewhere (Daehler et al. 2004; Gordon et al. 2008; Chong et al. 2011).

WRAs generally assign a score to species on the basis of ecological traits such as climatic breadth, dispersal capability, and method of reproduction, as well as information on the invasiveness of species elsewhere and that of close relatives (i.e., those species that have diverged from a shared common ancestor within the past several million years). While most of these traits seem to have a clear association with population growth and spread (e.g., seed dispersal), it is not immediately clear why the invasiveness of close relatives should matter for assessing risk. From the perspective of someone advocating for the economic benefits of a specific introduction, determining risk based on the evils of relatives may seem unfair and prejudicial.

It is generally recognized that closely related species share many of their basic ecological requirements and roles (Harvey and Pagel 1991; Futuyma 2010; Wiens et al. 2010). This basic assumption has been used to predict or explain the success of nonnative species in large observational studies (Strauss et al. 2006; Proches et al. 2008; Cadotte et al. 2010). Typically, WRAs score a species as being related to a problematic invader if they belong to the taxonomic grouping–such as in the same genus or subfamily (Pheloung et al. 1999, Krivanek and Pysek 2006). However, Reichard and Hamilton (1997) go so far as to recommend rejecting a species on the basis of membership in a family that includes problematic invaders. This may be an overly conservative requirement, as some families are extremely large, containing both economically important species and problematic invaders. For example, the grass family (Poaceae) consists of more than 10,000 species, including some of our most important agricultural crops (e.g., rice and wheat) and some very serious invaders (e.g., Japanese stillgrass [*Microstegium vimineum*] in the eastern United States and false brome [*Brachypodium sylvaticum*] in the northwestern United States). Therefore, it is of paramount importance to determine whether patterns of relatedness predict invasiveness. Also, understanding how the degree of relatedness predicts invasiveness is crucial for making sound management decisions. In this chapter we develop the rationale for why relatedness should inform the potential invasiveness of introduced species and we review studies that examine patterns of relatedness among invaders.

The Importance of Evolutionary History

The introduction of a nonnative species brings species that have been evolving separately for millions of years into contact. These species all have conserved traits (traits which have not changed for a long time) that were shaped by environments past. Closely related species, which share more of these conserved traits than distantly related species, tend to look and function in similar ways (Wiens and Graham 2005; Wiens et al. 2010). Despite the general similarity of close relatives, closely related species that evolve together are under acute pressure to diverge ecologically so that competition between species is reduced (Cox 2004; Vellend et al. 2007). For example, if two species share a food resource then each will potentially face less competition if they diverge to rely on different food sources. Such is the case with ecologically similar Galapagos finches, which when living on different islands have similar beak sizes, whereas when they occur together show differences in beak size and thus have specialized on different seed sizes (Schluter et al. 1985).

Here we present a conceptual model to explain why these evolutionary pressures and relationships matter for nonnative species success. We argue that closely related species should behave similarly when introduced to a new ecosystem, because their shared evolutionary history has left them with similar strategies for dealing with environmental conditions. Thus, closely related nonnative species should attain relatively similar range sizes and abundances. In contrast, we argue that closely related species that are native to the same ecosystem will have less similar patterns of abundance and distribution because they have faced evolutionary pressures to diverge, leading to different strategies. To illustrate this, we represent evolutionary relationships with a phylogenetic tree (Fig. 7–1) that connects species on the basis of their shared ancestry.

Closely related species that evolve separately tend to share key ecological traits because of the relatively limited power of adaptive change (i.e., evolution). The idea that phylogenetic relationships can be used to explain species abundances and co-occurrences arises from the conservation of key ecological traits and the relatively limited power of adaptive change to shift niche requirements (Futuyma 2010; Wiens et al. 2010). The opportunity for rapid evolutionary change depends upon available genetic variation and selective pressures (Lande 1979; Hansen and Houle 2008); moreover, it is limited by available niche opportunities and gene flow, both of which constrain population divergence (Futuyma 2010).

Regardless of the constraints, species interacting and evolving together over a long time period (referred to as *sympatric*) provide the main pressure for rapid species divergence (Case 1981; Schluter and McPhail 1992; Schluter 1994). While ecologically similar or closely related sympatric species are subject to strong pressure for ecological divergence, closely related species that live apart from one another (*allopatric*), while surely

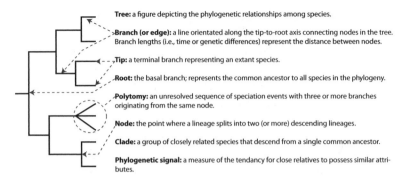

FIG. 7–1. A guide to phylogenetic terminology.

under their own evolutionary pressures, did not evolve based on the same set of evolutionary pressures occurring locally in the new habitat. Given this logic, we expect the abundance and distribution of an assortment of species assembled from different regions of the world to be influenced by relatedness. It is not that the current ecological niches of sympatric species are not informed by their evolutionary histories, rather that key changes have rapidly accumulated allowing these species to partition available resources, space, mutualistic partners, or enemy-free space. This rationale leads to the abundances and distribution of sympatric species being less informed by conserved traits than for allopatric species.

Two appropriate general models of evolutionary change can be used to further examine the dichotomy described above: random accumulated changes when species evolve apart and directional selection in sympatry. Randomly accumulated changes in traits can be modeled using a Brownian Motion (BM) model of evolution (Felsenstein 1985; Harvey and Pagel 1991). In the BM model, a trait changes from its ancestral state by drawing deviations from a normal distribution. Its current trait value is thus the product of a random walk (Box 7–1). On average, the longer the time since two species diverged, the more dissimilar their traits or ecological strategies are from one another (Box 7–1). It is often recognized that this model is overly simplistic, though analyses often find BM models of trait evolution can be useful to explain patterns of trait differences among large groups of species (Cooper and Purvis 2010; Harmon et al. 2010).

Where the BM model fails is when evolutionary change has been directional rather than random. This occurs when sympatric species are under selective pressure to diverge (e.g., as with Darwin's finches; Schluter et al.

BOX 7–1. **Modeling Trait Evolution**

Brownian motion (BM) evolution is the simplest model describing change in traits over time. In BM models, trait change occurs as a random walk from some initial trait value, with deviation sampled from a normal distribution (Felsenstein 1985; Harvey and Pagel 1991). Using Butler and King's (2004) formulation, the rate of change in a trait (X) is:

$$dX(t) = \sigma dB(t) \qquad \text{Eq1}$$

where, $dB(t)$ represents random values that are normally distributed with mean 0 and variance dt and σ scales the magnitude of fluctuations. For multiple lineages, total trait variation depends on the length of time (t) that the process operates, such that greater t results in greater trait dispersion (Fig. Box 7–1a). With BM evolution, it is important to note that the mean trait value stays constant, $\bar{X}_{t=1} = \bar{X}_{t=2}$ (Fig. Box 7–1a) and if two lineages start with identical trait values, $X_{1,t=0} = X_{2,t=0}$, then $\bar{X}_{1,t} = \bar{X}_{2,t}$ even if $\sigma_1 \neq \sigma_2$ (Fig. Box 7–1a). While BM models are overly simplistic, there are several plausible alternative, more complex models used to test for trait evolution (Cooper and Purvis 2010; Harmon et al. 2010). The most commonly used alternative is the Ornstein-Uhlenbeck (OU) model that introduces directional selective regimes on trait evolution (Hansen 1997; Butler and King 2004). OU models explicitly incorporate optimal trait values guided evolutionary trajectories. With the OU model, the rate of change in a trait is:

$$dX(t) = \alpha(\theta - X(t))dt + \sigma dB(t) \qquad \text{Eq2}$$

where, θ is the optimal trait value and α measures the strength of selection. The "pull" towards the optimal trait value, thus the rate of change, is greatest at the maximal value of $\theta - X(t)$. Fig. Box 7–1b shows OU evolution for the trait values of two lineages. If α is zero, then the OU model converts to a BM model.

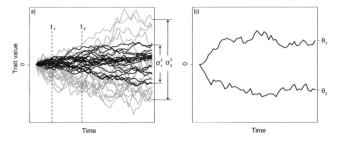

FIG. BOX 7–1. Two models of evolutionary change in a trait. The BM model (a) predicts that traits change according to random walk processes with greater divergence with greater time. The OU model (b) predicts rapid divergence to optima that in this case represent optima with reduced competition.

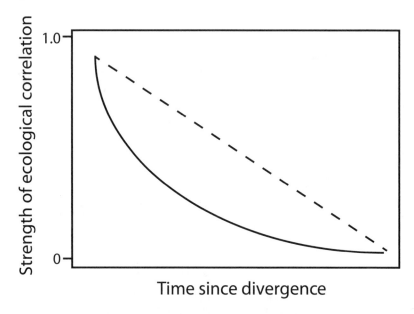

FIG. 7–2. The correlation between species traits or niches and the time since divergence predicted by the Brownian motion model (dashed line) and the Ornstein-Uhlenbeck model.

1985). A commonly used model for such divergence is an Ornstein-Uhlenbeck (OU) model (Hansen 1997; Butler and King 2004), which includes both trait optima and a measure of the strength of selection (Box 7–1). When dealing with multiple species, the optimal trait values can be based on some optimal differences that minimize the strength of competition. In computer simulations and mathematical models that allow species to evolve, these optimal distances emerge to minimize the negative effects of species' competitive interactions (Case 1981; Scheffer and van Nes 2006). Because this process is driven by competition, we refer to it as antagonistic OU evolution (as opposed to regular OU evolution, which specifies optimal trait values without considering competitive interactions among species).

Both BM and antagonistic OU models predict very closely related species to have similar trait values; conversely, very distantly related species are predicted to be dissimilar in trait values. However, where the models differ is that the OU model predicts rapid species divergence followed by stable trait values, whereas the BM model predicts that species will diverge at a constant rate over time (Fig. 7–2). Thus, since species rapidly diverge under the antagonistic OU model, we expect a weaker role for evolutionary history in explaining abundance and coexistence patterns because even relatively closely related species may be quite ecologically different from one another.

Why Evolutionary History Should Matter More for Nonnative Species than Natives

Assembling communities of nonnative species in new regions may, in effect, mimic random sampling from disparate allopatric assemblages. The result is an accumulation of species with idiosyncratic and provincial responses to evolutionary partners and antagonists. Whereas trait changes in one sympatric partner likely result in evolutionary responses in other species, the accumulation of multiple allopatric species essentially combines a random assortment of unique evolutionary responses.

Within sympatric associations, we expect species to be less similar to their close relatives. Conversely, randomly combining species from disparate habitats should result in assemblages that lack this niche separation among close relatives because there has not been pressure to diverge. Without this rapid divergence, coexistence should be based on existing conserved ecological traits. The ecological traits of species introduced from elsewhere likely conform to the BM model of trait divergence for any given trait, rather than an OU model with rapid divergence. Thus the phylogenetic relationships among nonnative species should be more important to explaining patterns of coexistence and abundance than for natives, which have coevolved together.

If the evolutionary "fit" for an introduced species to a new habitat is based more on conserved traits than recent evolutionary divergence between closely related species, then we expect phylogenetic relationships to more strongly correlate with patterns of abundance and distribution than for natives. Since nonnative species patterns may rely more on phylogenetic history, environmental filters should have a stronger effect on the success of nonnative species and we should see particular shared traits that "fit" the local environment and determine species' success (Mayfield et al. 2010). Thus closely related nonnative species should be more likely to occupy similar habitats and expand to similar range sizes than for two closely related natives, which are more likely to have evolved specialization.

Despite the predicted differences in nonnative and native phylogenetic community patterns, phylogenetic history should not be considered unimportant for natives. In native assemblages, phylogeny may explain some important community and ecosystem patterns (Cavender-Bares et al. 2009). If divergent selection has shaped traits important for local co-existence, disturbances or climatic events may still act on highly conserved traits (Prinzing et al. 2001, Helmus et al. 2010, Thuiller et al. 2011). Large-scale environmental influences on native diversity may be better predicted by phylogenetic patterns than traditional diversity measures (Chave et al. 2007), and phylogeny seems to be a good predictor in native species responses to climate shifts (Davis et al. 2010).

Review of Relatedness–Invasion Studies

Different evolutionary histories lead to different ecological patterns. Through a combination of historical events, vicariance, and environmental barriers, evolutionary history is unevenly distributed across the globe with various regions acting as centers of diversity for different groups of species. Species belonging to some taxonomic groups have been more successful at being transported by humans, and then becoming established and invasive (Stebbins 1965; Daehler 1998; Pysek 1998; Cadotte and Lovett-Doust 2001; Lockwood 2005). Two examples highlight the importance of evolution for the success of nonnative species. First are patterns of plant invasions in China. The cactus family (Cactaceae) evolved in Central America and was historically not present in Asia. Cactus species, especially prickly pear cacti (*Opuntia* sp.) and night-blooming cacti (or dragonfruits—*Hylocereus* sp.), have been disproportionately successful in invading arid regions in China (Wu et al. 2010). The diverse spurge family (Euphorbiaceae) is a family without many native species in China. Spurges have also been disproportionately successful invaders in arid regions of China (Wu et al. 2010).

Secondly, Californian grasslands were historically primarily composed of perennial grasses and generally lacking annual grasses (HilleRisLambers et al. 2010). Like other Mediterranean ecosystems, California chaparral has been heavily invaded by a number of different species, but exotic grasses have been especially successful, resulting in large ecosystem effects by altering fire regimes (D'Antonio and Vitousek 1992). These successful grasses have consisted of several groups of closely related grasses comprising mainly annual species (Cadotte et al. 2010).

Given the patterns described above, phylogenetic relationships between nonnative and native species have been used as a proxy for the potential of introduced species to naturalize and invade. For example, introduced fish tend to have higher impacts if they are phylogenetically more different from native species (Ricciardi and Atkinson 2004). In plants, nonnative species that are more distantly related to natives are also more invasive (Strauss et al. 2006; Diez et al. 2008; Davies et al. 2011). Some mechanisms that have been proposed to explain this pattern include niche differentiation (Davies et al. 2011), lack of pollen limitation (Morales and Traveset 2009; Burns et al. 2011), the response of soil biota and pathogens (Brandt et al. 2009), and lack of susceptibility to herbivores and pathogens that are found on natives (Gossner et al. 2009; Hill and Kotanen 2009, 2011). However, this pattern is not universally supported by existing data (Marco et al. 2010; Dostal and Paleckova 2011; Thomaz and Michelan 2011). For example, on Mediterranean islands, the presence of closely related native species has been shown to only marginally affect the ability of an introduced species to invade (Lambdon and Hulme 2006).

Phylogenetic patterns of species invasions seem to be quite dependent on the spatial scale of the observations. The above examples implicitly invoke mechanisms that operate at small scales (e.g., sharing disease, competing for resources). However, if we take a step back and examine the phylogenetic patterns of invasions, one feature that emerges is that nonnative species seem to be clustered into groups of closely related species at large scales, producing a lower than expected amount of total evolutionary time represented by species, despite increases in richness (Proches et al. 2008; Winter et al. 2009; Cadotte et al. 2010).

While the patterns of relatedness between nonnative and native species seem to be complex, our hypothesis is robust for linking the relatedness among nonnative species and their patterns of large-scale success. The hypothesis constructed here–that introduced species should show a stronger phylogenetic signal (see Fig. 7–1) in abundance and distribution than natives—needs two important pieces of information to test: detailed occupancy or abundance data and a good phylogeny for both nonnative and native species. Ideally, a phylogeny would be built from molecular data and perhaps be time calibrating, and the potential influences of phylogeny-building methods should be acknowledged (Cadotte et al. 2009). Further, testing of this hypothesis should be sensitive to both spatial and phylogenetic scales and care should be taken to delineate the scale of observation and what the potential species pool is for constructing the phylogeny (Swenson et al. 2006; Proches et al. 2008; Kembel 2009; Swenson 2009).

Several lines of evidence seem to support the hypothesis that successful introduced species will be nonrandom phylogenetically. By and large, the best predictor of introduced species success is how well the climate and habitat requirements are met in the new territory (Hayes and Barry 2008; Bomford et al. 2009), and close relatives tend to share climatic requirements and tolerances (Prinzing et al. 2001; Thuiller et al. 2011). This would explain why studies find that some families harbor disproportionately high numbers of nonnative species (Cadotte and Lovett-Doust 2001). These environmental effects are most likely present at larger scales, and so patterns of relatedness should be strong at large but not smaller scales. This geographical scale effect on the phylogenetic scale has been observed for New Zealand and Australian plant invaders, which show a phylogenetic signal in the number of sites occupied at large scales but not at smaller scales (Diez et al. 2008; Cadotte et al. 2009). When strong environmental pressures are created, such as in urban environments, species selected for are found in fewer clades (see Fig. 7–1) than expected by chance, resulting in low amounts of nonnative evolutionary history (Ricotta et al. 2009).

In a test comparing the patterns of relatedness for introduced and native species, Cadotte et al. (2010) examined plant community phylogenetic patterns across a large-scale richness gradient in California. They found that introduced species tended to be more phylogenetically clustered than natives, indicating that closely related introduced species tended to occur together within communities. Further, as one moves from one site to another, phylogenetic branches are lost and gained as composition changes (called turnover), and for introduced species turnover occurred primarily among the tips of the phylogenetic tree and not deeper branches. This indicates that introduced lineages are always present, even though specific species may or may not be present. This is in stark contrast to the natives, which were more phylogenetically evenly distributed within communities and showed higher phylogenetic turnover (Cadotte et al. 2010).

While there have not been a lot of tests on the hypothesis that successfully colonizing and invading introduced species should be closely related to other successful introduced species, evidence seems to confirm this. Even less evidence exists that this pattern is stronger in introduced compared with native species. However, there are two issues that potentially confound the nonnative phylogenetic pattern. Firstly, when the importation of nonnative species is strongly biased toward a few vectors that move many individuals from one site to another (Berg et al. 2002; Colautti et al. 2006), then random assembly cannot be assumed. Moving whole sympatric assemblages would result in coevolved niche evolution patterns remaining intact and the BM model of the phylogenetic trait distribution may be invalid. Secondly, if there are only a few axes on which selection operates, it is plausible that disparate communities may contain species where the selection pressures operated on the same set of attributes. This could be the case with traits such as the evolution of trade-offs in leaf shape, size, and chemistry (Wright et al. 2004).

The hypotheses and supporting evidence presented here confirm the WRA criterion that close relatives of problematic introduced species should not be released in a new habitat. However, there are a couple of shortcomings that need to be addressed. First, it may not be known whether close relatives are invasive because either close relatives have not had an opportunity to invade or a species does not have close relatives. It is possible that with a lack of information, species can get a pass in the risk assessment (Reichard and Hamilton 1997; Simberloff 2005). In this case, other factors such as reproductive strategy or climate matching should be seriously considered. Finally, there has not been enough research to know how closely related two species must be to be informative. At present, most research considers being related at the genus level to be sufficiently related.

Policy Implications

Approaches employing conceptual models can offer insights into policy options. The decision to allow the importation of nonnative organisms needs to be made, and often it is made on scant evidence. According to Romagosa (chapter 6 in this volume), the rate of organism importation has been increasing, with more than 4,000 vertebrate species legally imported into the United States over the past 40 years. The number of plant species purposely imported likely greatly eclipses this number. Policy makers are unlikely to have detailed natural history knowledge for all these species prior to the decision to import them. Policy makers need to weigh the possible risks of importation against the economic (and potentially cultural or aesthetic) benefits. While the benefits may be relatively straightforward to estimate or predict, the estimates of risk can be accompanied by substantial errors. Employing broad ecological understanding can aid in these decisions where evidence is lacking. Specifically, ecological theory can highlight those species that require more screening.

Weed risk assessments (WRAs) have used the invasiveness of close relatives as a major criterion for excluding a particular species. On the surface, this would seem to be an inherent bias that may judge a species according to its relatives and unduly limit the potential economic benefit from importing a particular species. However, using both a conceptual model and a review of the literature, we show that closely related species have similar invasion potentials and the ban on importing relatives of problematic invaders is well supported. Close relatives share similar ecologies and are limited by similar environmental conditions. Thus, closely related nonindigenous species appear to fill out similar ranges in their new territories.

While information may be lacking on exactly how closely related species should be to inform invasion potential, species in the same genera as successful invaders should not be imported. Regional and national exclusion lists should include all close relatives of any serious invader. Less clear is how closely an introduced species should be related to natives in order to inform importation decisions. Some evidence suggests that species closely related to natives will likely be controlled by diseases, predators, or herbivores of the native relatives.

Literature Cited

Berg, D. J., D. W. Garton, H. J. MacIsaac, V. E. Panov, and I. V. Telesh. 2002. Changes in genetic structure of North American *Bythotrephes* populations following invasion from Lake Ladoga, Russia. *Freshwater Biology* 47:275–282.

Bomford, M., R. O. Darbyshire, and L. Randall. 2009. Determinants of establishment success for introduced exotic mammals. *Wildlife Research* 36:192–202.

Brandt, A. J., E. W. Seabloom, and P. R. Hosseini. 2009. Phylogeny and provenance affect plant-soil feedbacks in invaded California grasslands. *Ecology* 90:1063–1072.

Burns, J. H., T.-L. Ashman, J. A. Steets, A. Harmon-Threatt, and T. M. Knight. 2011. A phylogenetically controlled analysis of the roles of reproductive traits in plant invasions. *Oecologia* 166:1009–1017.

Butler, M. A., and A. A. King. 2004. Phylogenetic comparative analysis: a modeling approach for adaptive evolution. *American Naturalist* 164:683–695.

Cadotte, M. W., E. T. Borer, E. W. Seabloom, J. Cavender-Bares, W. S. Harpole, E. Cleland, and K. F. Davies. 2010. Phylogenetic patterns differ for native and exotic plant communities across a richness gradient in Northern California. *Diversity and Distributions* 16:892–901.

Cadotte, M. W., M. A. Hamilton, and B. R. Murray. 2009. Phylogenetic relatedness and plant invader success across two spatial scales. *Diversity and Distributions* 15:481–488.

Cadotte, M. W., and J. Lovett-Doust. 2001. Ecological and taxonomic differences between native and introduced plants of southwestern Ontario. *Ecoscience* 8:230–238.

Case, T. J. 1981. Niche packing and coevolution in competition communities. *Proceedings of the National Academy of Science USA* 78:5021–5025.

Cavender-Bares, J., K. H. Kozak, P. V. A. Fine, and S. W. Kembel. 2009. The merging of community ecology and phylogenetic biology. *Ecology Letters* 12:693–715.

Chave, J., G. Chust, and C. Thebaud. 2007. The importance of phylogenetic structure in biodiversity studies. In *Scaling Biodiversity*, edited by D. Storch, P. M. Marquet, and J. H. Brown, 150–167. Santa Fe: Santa Fe Institute Editions.

Chong, K. Y., R. T. Corlett, D. C. J. Yeo, and H. T. W. Tan. 2011. Towards a global database of weed risk assessments: a test of transferability for the tropics. *Biological Invasions* 13:1571–1577.

Colautti, R. I., I. A. Grigorovich, and H. J. MacIsaac. 2006. Propagule pressure: a null model for biological invasions. *Biological Invasions* 8:1023–1037.

Cooper, N., and A. Purvis. 2010. Body size evolution in mammals: complexity in tempo and mode. *American Naturalist* 175:727–738.

Cox, G. W. 2004. *Alien Species and Evolution*. Washington, DC: Island.

Daehler, C. C. 1998. The taxonomic distribution of invasive angiosperm plants: ecological insights and comparison to agricultural weeds. *Biological Conservation* 84:167–180.

Daehler, C. C., J. S. Denslow, S. Ansari, and H. C. Kuo. 2004. A risk-assessment system for screening out invasive pest plants from Hawaii and other Pacific Islands. *Conservation Biology* 18:360–368.

D'Antonio, C. M., and P. M. Vitousek. 1992. Biological invasions by exotic grasses, the grass fire cycle, and global change. *Annual Review of Ecology and Systematics* 23:63–87.

Davies, K. F., J. Cavender-Bares, and N. Deacon. 2011. Native communities determine the identity of exotic invaders even at scales at which communities are unsaturated. *Diversity and Distributions* 17:35–42.

Davis, C. C., C. G. Willis, R. B. Primack, and A. J. Miller-Rushing. 2010. The importance of phylogeny to the study of phenological response to global climate change. *Philosophical Transactions B* 365:3201–3213.

Diez, J. M., J. J. Sullivan, P. E. Hulme, G. Edwards, and R. P. Duncan. 2008. Darwin's naturalization conundrum: dissecting taxonomic patterns of species invasions. *Ecology Letters* 11:674–681.

Dostal, P., and M. Paleckova. 2011. Does relatedness of natives used for soil conditioning influence plant-soil feedback of exotics? *Biological Invasions* 13:331–340.

Felsenstein, J. 1985. Phylogenies and the comparative method. *American Naturalist* 125:1–15.

Ferdinands, K., J. Virtue, S. B. Johnson, and S. A. Setterfield. 2011. Bio-insecurities: managing demand for potentially invasive plants in the bioeconomy. *Current Opinion in Environmental Sustainability* 3:43–49.

Futuyma, D. J. 2010. Evolutionary constraint and ecological consequences. *Evolution* 64:1865–1884.

Gordon, D. R., D. A. Onderdonk, A. M. Fox, and R. K. Stocker. 2008. Consistent accuracy of the Australian weed risk assessment system across varied geographies. *Diversity and Distributions* 14:234–242.

Gossner, M. M., A. Chao, R. I. Bailey, and A. Prinzing. 2009. Native fauna on exotic trees: phylogenetic conservatism and geographic contingency in two lineages of phytophages on two lineages of trees. *American Naturalist* 173:599–614.

Hansen, T. F. 1997. Stabilizing selection and the comparative analysis of adaptation. *Evolution* 51:1341–1351.

Hansen, T. F., and D. Houle. 2008. Measuring and comparing evolvability and constraint in multivariate characters. *Journal of Evolutionary Biology* 21:1201–1219.

Harmon, L. J., J. B. Losos, T. J. Davies, R. G. Gillespie, J. L. Gittleman, W. B. Jennings, K. H. Kozak, et al. 2010. Early bursts of body size and shape evolution are rare in comparative data. *Evolution* 64:2385–2396.

Harvey, P. H., and M. Pagel. 1991. *The Comparative Method in Evolutionary Biology*. Oxford: Oxford University Press.

Hayes, K. R., and S. C. Barry. 2008. Are there any consistent predictors of invasion success? *Biological Invasions* 10:483–506.

Helmus, M. R., W. Keller, M. J. Paterson, N. D. Yan, C. H. Cannon, and J. A. Rusak. 2010. Communities contain closely related species during ecosystem disturbance. *Ecology Letters* 13:162–174.

Hill, S. B., and P. M. Kotanen. 2009. Evidence that phylogenetically novel nonindigenous plants experience less herbivory. *Oecologia* 161:581–590.

Hill, S. B., and P. M. Kotanen. 2011. Phylogenetic structure predicts capitular damage to Asteraceae better than origin or phylogenetic distance to natives. *Oecologia* 166:843–851.

HilleRisLambers, J., S. G. Yelenik, B. P. Colman, and J. M. Levine. 2010. California annual grass invaders: the drivers or passengers of change? *Journal of Ecology* 98:1147–1156.

Inderjit, ed. 2005. *Invasive Plants: Ecological and Agricultural Aspects*. Basal, Switzerland: Birkhausser Verlag.

Kembel, S. W. 2009. Disentangling niche and neutral influences on community assembly: assessing the performance of community phylogenetic structure tests. *Ecology Letters* 12:949–960.

Krivanek, M., and P. Pysek. 2006. Predicting invasions by woody species in a temperate zone: a test of three risk assessment schemes in the Czech Republic (Central Europe). *Diversity and Distributions* 12:319–327.

Lambdon, P. W., and P. E. Hulme. 2006. How strongly do interactions with closely-related native species influence plant invasions? Darwin's naturalization hypothesis assessed on Mediterranean islands. *Journal of Biogeography* 33:1116–1125.

Lande, R. 1979. Quantitative genetic analysis of multivariate evolution, applied to brain-body size allometry. *Evolution* 33:402–416.

Lockwood, J. L. 2005. Predicting which species will become invasive: what's taxonomy got to do with it? In *Phylogeny and Conservation*, edited by A. Purvis, J. L. Gittleman, and T. Brooks, 365–384. Cambridge: Cambridge University Press.

Marco, A., S. Lavergne, T. Dutoit, and V. Bertaudiere-Montes. 2010. From the backyard to the backcountry: how ecological and biological traits explain the escape of garden plants into Mediterranean old fields. *Biological Invasions* 12:761–779.

Mayfield, M. M., S. P. Bonser, J. W. Morgan, I. Aubin, S. McNamara, and P. A. Vesk. 2010. What does species richness tell us about functional trait diversity? Predictions and evidence for responses of species and functional trait diversity to land-use change. *Global Ecology and Biogeography* 19:423–431.

Morales, C. L., and A. Traveset. 2009. A meta-analysis of impacts of alien vs. native plants on pollinator visitation and reproductive success of co-flowering native plants. *Ecology Letters* 12:716–728.

Pheloung, P. C., P. A. Williams, and S. R. Halloy. 1999. A weed risk assessment model for use as a biosecurity tool evaluating plant introductions. *Journal of Environmental Management* 57:239–251.

Prinzing, A., W. Durka, S. Klotz, and R. Brandl. 2001. The niche of higher plants: evidence for phylogenetic conservatism. *Proceedings of the Royal Society B* 268:2383–2389.

Proches, S., J. R. U. Wilson, D. M. Richardson, and M. Rejmánek. 2008. Searching for phylogenetic pattern in biological invasions. *Global Ecology and Biogeography* 17:5–10.

Pysek, P. 1998. Is there a taxonomic pattern to invasions? *Oikos* 82:282–294.

Reichard, S. H., and C. W. Hamilton. 1997. Predicting invasions of woody plants introduced into North America. *Conservation Biology* 11:193–203.

Ricciardi, A., and S. K. Atkinson. 2004. Distinctiveness magnifies the impact of biological invaders in aquatic ecosystems. *Ecology Letters* 7:781–784.

Ricotta, C., F. A. La Sorte, P. Pysek, G. L. Rapson, L. Celesti-Grapow, and K. Thompson. 2009. Phyloecology of urban alien floras. *Journal of Ecology* 97:1243–1251.

Scheffer, M., and E. H. van Nes. 2006. Self-organized similarity, the evolutionary emergence of groups of similar species. *Proceedings of the National Academy of Science USA* 103:6230–6235.

Schluter, D. 1994. Experimental evidence that competition promotes divergence in adaptive radiation. *Science* 266:798–801.

Schluter, D., and J. D. McPhail. 1992. Ecological character displacement and speciation in sticklebacks. *American Naturalist* 140:85–108.

Schluter, D., T. D. Price, and P. R. Grant. 1985. Ecological character displacement in Darwin finches. *Science* 227:1056–1059.

Simberloff, D. 2005. The politics of assessing risk for biological invasions: the USA as a case study. *Trends in Ecology and Evolution* 20:216–222.

Stebbins, G. L. 1965. Colonizing species of the native California flora. In *The Genetics of Colonizing Species*, edited by H. G. Baker and G. L. Stebbins, 173–191. New York: Academic.

Strauss, S. Y., C. O. Webb, and N. Salamin. 2006. Exotic taxa less related to native species are more invasive. *Proceedings of the National Academy of Sciences USA* 103:5841–5845.

Swenson, N. G. 2009. Phylogenetic resolution and quantifying the phylogenetic diversity and dispersion of communities. *Plos One* 4:e4390.

Swenson, N. G., B. J. Enquist, J. Pither, J. Thompson, and J. K. Zimmerman. 2006. The problem and promise of scale dependency in community phylogenetics. *Ecology* 87:2418–2424.

Thomaz, S. M., and T. S. Michelan. 2011. Associations between a highly invasive species and native macrophytes differ across spatial scales. *Biological Invasions* 13:1881–1891.

Thuiller, W., S. Lavergne, C. Roquet, I. Boulangeat, B. Lafourcade, and M. B. Araujo. 2011. Consequences of climate change on the tree of life in Europe. *Nature* 470:531–534.

Vellend, M., L. J. Harmon, J. L. Lockwood, M. M. Mayfield, A. R. Hughes, J. P. Wares, and D. F. Sax. 2007. Effects of exotic species on evolutionary diversification. *Trends in Ecology and Evolution* 22:481–488.

Wiens, J. J., D. D. Ackerly, A. P. Allen, B. L. Anacker, L. B. Buckley, H. V. Cornell, E. I. Damschen, et al. 2010. Niche conservatism as an emerging principle in ecology and conservation biology. *Ecology Letters* 13:1310–1324.

Wiens, J. J., and C. H. Graham. 2005. Niche conservatism: integrating evolution, ecology, and conservation biology. *Annual Review of Ecology, Evolution, and Systematics* 36:519–539.

Winter, M., O. Schweiger, S. Klotz, W. Nentwig, P. Andriopoulos, M. Arianoutsou, C. Basnou, et al. 2009. Plant extinctions and introductions lead to phylogenetic and taxonomic homogenization of the European flora. *Proceedings of the National Academy of Sciences USA* 106:21721–21725.

Wright, I. J., P. B. Reich, M. Westoby, D. D. Ackerly, Z. Baruch, F. Bongers, J. Cavender-Bares, et al. 2004. The worldwide leaf economics spectrum. *Nature* 428:821–827.

Wu, S. H., H. T. Sun, Y. C. Teng, M. Rejmánek, S. M. Chaw, T. Y. A. Yang, and C. F. Hsieh. 2010. Patterns of plant invasions in China: taxonomic, biogeographic, climatic approaches and anthropogenic effects. *Biological Invasions* 12:2179–2206.

CHAPTER EIGHT

Reducing Damaging Introductions from International Species Trade through Invasion Risk Assessment

Michael Springborn

Introduction

A series of recent investigations of invasive species risk from international flora and fauna trade identifies several important lessons for the design of screening tools used to determine which species are too risky for importation (Springborn et al. 2011; Schmidt et al. 2012; Lieli and Springborn 2013). From a welfare perspective, the choice of whether to accept or reject a nonnative species for importation depends on the anticipated benefits of importation, the level of anticipated losses if the species is revealed to be invasive, and finally the likelihood that the species will be invasive. This chapter outlines a risk assessment framework, shared by each of the articles cited above, which provides a rigorous structure for balancing trade benefits with the cost of potential invasion. Because the problem depends on a statistical model of species invasion threat and the challenge is to make a decision on whether to accept or reject each species based on weighing the benefits and costs, the components are brought together in what is referred to as a statistical-decision theoretic framework. Analysis of the framework demonstrates how costs and benefits combine to determine the optimal decision rule. The rule identifies a particular threshold of invasion threat beyond which the benefits of trade in a species no longer justify the potential damages from invasion. The optimal decision when invasion threat for a particular species surpasses the threshold is to not accept the species for importation.

This chapter begins with a formal description of the statistical decision problem of whether to accept or reject a proposed species for importation. While the construction of the model is at times technical, intuitively the decision problem turns on the balance between the likelihood of two different errors. The rate at which benign species are mistakenly designated as too risky and rejected for trade is the "false positive rate." The rate at which invasive species are mistakenly deemed benign and accepted for importation is the "false negative rate." As a decision maker becomes more conservative—lowering the threshold of an acceptable threat—benefits are enjoyed from lowering the rate at which invasives are mistakenly accepted—that is, the false negative rate. However, the fundamental trade-off is that this comes at the expense of increasing the rate of species mistakenly branded as invasive—that is, the false positive rate, which leads to higher costs of forgone trade. Previous analyses have made the simplifying assumption that the decision maker selects a threshold such that the false positive rate equals the false negative rate (Bomford et al. 2009; Keller et al. 2007). This simple rule of thumb for setting the threshold does not ensure that benefits justify the costs (Springborn et al. 2011).

The first rigorous decision-theoretic analysis for invasive species screening was that of Keller et al. (2007), which demonstrated positive net benefits from screening. Recent advances discussed in this chapter develop the intuition for the optimal decision in further detail. This analysis illuminates how setting the ideal threshold of acceptable invasion threat boils down to the ratio of import benefits to invasion losses. This structure communicates the essence of the decision problem in a simple fashion to practitioners in regulating agencies and biologists working on the topic.

Technical challenges and recent advances for the major modeling components are discussed here in turn. In general, estimating the threat posed by a particular species involves obtaining a "training data set" of existing species introductions, including information on attributes thought to be correlated with invasion risk (e.g., biogeographic variables and whether or not species have become invasive elsewhere). The threat model, which maps species attributes to an estimate of invasion threat, can be structured as a standard regression (e.g., Springborn et al. 2011) or regression tree (e.g., Schmidt et al. 2012). I highlight and discuss solutions for a pervasive but often overlooked nonrandom sampling problem in training data sets. This statistical issue arises where invasive species are purposely oversampled relative to the population of all species available for trade. Addressing this "endogenous stratified sampling" problem is essential for ensuring that model output—the estimate of species invasion threat—can be interpreted in probabilistic terms, rather than a unitless index.

In this chapter, I then discuss advances in determining values for the welfare elements of the decision framework, in particular for characterizing the welfare forgone when species are not accepted for importation. I outline coarse-grained valuation methods for use when only highly aggregated trade data are available (i.e., data for taxonomic groups but not individual species). I then turn to a recently developed method for valuing the loss of variety in imports explicitly when detailed trade data is obtainable—for example, at the species level. This section highlights data and research needs—for example, the value of expanding the maintenance of and access to trade data. I further argue that rigorous estimation of potential losses from invasion for a given taxon is the most pressing gap to fill for improving cost-benefit calculations.

Finally, I review results from three recent applications of the framework described—plant imports to the United States (Schmidt et al. 2012), plant imports to Australia (Lieli and Springborn 2013), and reptile and amphibian imports to the United States (Springborn et al. 2011). In each case the estimated net benefits from screening species for invasion risk are substantial. A novel method from one of the applications is described; it allows an analyst to collapse what is typically a two-step process (estimating species invasion threat and deciding whether or not to allow a species for importation) into a single step. The advantage to this integration is that threat estimation is no longer sensitive to misspecification of the form of the regression model (e.g., logistic). I conclude with recommendations for future research and interesting prospects for innovative methodologies.

The Statistical Decision Problem

Historically there has been some skepticism that predictive attributes of future invasion status could be identified (Enserink 1999; Williamson 1999). However, Kolar and Lodge (2001) and more recently Hayes and Barry (2008) review significant advances along this front. Hayes and Barry (2008) conducted a meta-analysis of 49 studies and identified variables found to be broadly associated with invasion success. Across various taxa they found that the two species attributes most strongly associated with establishment success were climate similarity (between existing and potential new ranges) and establishment success elsewhere. They also discuss a more extensive set of variables from species ecology, history, and biogeography that have been found to be predictive for certain taxa.

The model below describes how such data is used to inform the question of whether a particular species that has been proposed for import should be allowed entry. To formally describe the necessary data elements, let S_N represent the available training data set comprised of N observations (species). This includes the observed outcome of the binary variable y—which specifies the true nature of a particular species as either invasive ($y = 1$) or benign ($y = 0$). Data on a set of risk attributes that are potentially predictive of y is denoted by x. The full training data set is given by $S_N = \{(y_1, x_1), (y_2, x_2), \ldots, (y_N, x_N)\}$ for species $n = 1, \ldots, N$.

Starting with the model of invasion threat, let $p(x_n; \theta)$ represent the probability that a particular species n is invasive conditional on predictive attributes, x_n, and a vector of model parameters, θ: $p(x_n; \theta) = \Pr(y_n = 1 | x_n; \theta)$. The actions available to the decision maker for any species n, are to either reject the species ($a_n = 1$) or allow it to be imported ($a_n = 0$). Let $u_{a,y}$ represent the net benefits of taking screening action a given the true nature y of a proposed species (where the index n is suppressed to avoid clutter).

The objective driving the decision problem is to maximize expected net benefits per species through the binary screening decision specified by a:

(1.1) $$\max_a \left\{ p(x;\theta)u_{a,1} + [1 - p(x;\theta)u_{a,0}] \right\}.$$

Here the expected net benefits are given by the net benefits of taking action a when the species is invasive ($u_{a,1}$) and not invasive ($u_{a,0}$), with each case weighted by its probability. The optimal choice is to reject a species ($a = 1$) when the expected net benefits of doing so are greater than that of accepting ($a = 0$). Thus, rejecting is the optimal choice when the following inequality holds:

(1.2) $$p(x; \theta)u_{1,1} + [1 - p(x; \theta)]u_{1,0} > p(x; \theta)u_{0,1} + [1 - p(x; \theta)]u_{0,0}$$

Rearranging the terms in Equation (1.2), we arrive at the following decision rule: a proposal is optimally rejected when

(1.3) $$p(x;0) > \frac{(u_{0,0} - u_{1,0})}{(u_{1,1} - u_{0,1}) + (u_{0,0} - u_{1,0})} \equiv c.$$

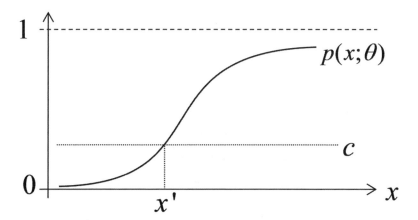

FIG. 8–1. The statistical decision problem.

Equation (1.3) shows that it is ideal to reject a species when the probability of establishment is greater than a cutoff defined as c. This relationship is depicted graphically in Fig. 8–1 for the case in which x is one-dimensional. The particular shape of p will be given by model estimation—here it is assumed to be s-shaped and increasing in x. It will be ideal to accept a species when the predictive attribute x is less than x'—that is, when the threat of invasion is less than the cutoff, $p(x, \theta) < c$. Otherwise the species should be rejected. Next I describe how the form of c can be simplified and estimated before turning to the statistical model for $p(x; \theta)$ in detail.

Because we are concerned with the *change* in expected net benefits relative to a preexisting baseline, a first step in specifying the utility components of Equation (1.3) is to characterize the status quo policy. For example, the status quo might be an open door policy where all species are allowed to be imported or a closed door policy where all species are banned. Here I take as a starting point an open door policy, which is essentially consistent with the current US approach (Lodge et al. 2006; Fowler et al. 2007). In this case, when a screening system is implemented and successfully excludes a true invasive, benefits take the form of avoided welfare losses from invasion. Let $V_I > 0$ represent the present discounted value of the long-run expected losses from an invasion in the group of species concerned. Let $V_T > 0$ represent the present discounted value of the stream of expected benefits from trade. When a species is rejected for importation, V_T is lost by the importing country, regardless of whether the species is invasive.

BOX 8–1. **The Australian Weed Risk Assessment Model**

The Australian Weed Risk Assessment (WRA) model, described in detail by Pheloung et al. (1999), is one of the longest running and most widely emulated invasive species screening systems. The WRA approach involves using information on species attributes that are correlated with the likelihood that a species will be invasive in order to make decisions on whether to allow proposed imports. Each proposed species receives a WRA score, a numerical index based on responses to 49 questions regarding attributes of a plant that are correlated with weediness. If the score exceeds a threshold set by experts, the proposed species is not allowed for importation. Three types of information are combined to generate the WRA score for a species. Biogeographic variables include observed distribution, climate preferences, and existing global weediness history of a plant. Undesirable traits are assessed—for example, to determine whether the species is noxious or parasitic. A final group of biological variables provides information for determining whether the species is likely to "reproduce, spread and persist" (Pheloung 1995, p. 11).

A training data set of 370 nonnative plant species was used to inform the construction of the original model and to determine the WRA score threshold above which species would no longer be accepted. This data set included 84 non-weed species and 286 weedy species, as determined by plant scientists. The WRA model is mathematically straightforward but it is not grounded on formal statistical or economic foundations (Caley et al. 2006). The WRA model differs from the framework discussed in this chapter in that the scoring and weighting of various predictive attributes is determined subjectively, as is the acceptable WRA score threshold. The WRA score serves as an index of invasion threat but is not formally a probability measure as developed here. Finally, the threshold beyond which the invasion threat is deemed unacceptable was determined subjectively by plant scientists in the WRA model. In this chapter the optimal threshold is derived in the description of the statistical decision problem and depends explicitly on estimates of the forgone benefits from restricted trade and the losses from invasion.

Turning to the specific utility components of Equation (1.3), the present value of expected net benefits of rejecting a species ($a = 1$) conditional on it truly being an invasive ($y = 1$) is given by $u_{1,1} = V_I - V_T$. Alternatively, when the species is rejected ($a = 1$) but not invasive ($y = 0$), the lost value of trade is the salient payoff: $u_{1,0} = -V_T$. Because net benefits are not affected when a decision does not change the status quo, in the open door case the change in utility when species are accepted is zero regardless of whether or not the species is truly invasive: $u_{0,0} = u_{0,1} = 0$. Given these payoffs under the four action-outcome possibilities, the cutoff from Equation (1.3) simplifies to

(1.4) $$c \equiv \frac{(u_{0,0} - u_{1,0})}{(u_{1,1} - u_{0,1}) + (u_{0,0} - u_{1,0})} = \frac{V_T}{V_I}.$$

This result shows that the optimal screening stringency—that is, the ideal upper bound on the acceptable invasion threat—depends only on the *ratio* of trade benefits to invasion losses. The same optimal decision rule also emerges if the status quo is a closed door policy[1]; however, the expected net benefits of the screening system will depend on the initial baseline policy. In the next section, I turn to estimation of the two parameters of concern from Equation (1.4), V_T and V_I.

Estimating Costs and Benefits

Welfare Benefits from Imports

An imperfect—but not rare—approach to characterizing the welfare impact of public policies that restrict economic activity is to use lost revenues as a proxy for the opportunity cost, or forgone benefits (Field and Field 2006). In the risk assessment context, Keller et al. (2007) make the simplifying assumption that the benefits from trade in a species are the total retail expenditures on those species. Although better data were not available to the authors at the time, this formulation of benefits can be improved. Such expenditures capture the aggregate amount paid by consumers, whereas a metric of the net benefits to consumers would capture the difference between the (maximum) willingness-to-pay and the expenditures, commonly referred to as the consumer surplus.

Next I describe a conceptual model of the welfare loss of forgoing imports; this model may be appropriate when general import data for a live plant or animal sector is available but species-specific data is not. This approach, which was implemented in Schmidt et al. (2012), is based on the framework for evaluating the opportunity cost of a simple quota restriction on imports from Feenstra (2004). Once an estimate of the lost surplus from restricting all trade in a sector is obtained, the amount is divided by the number of unique species in trade to arrive at an estimate of the average lost benefit per species.

The model is depicted in Fig. 8–2 where consumption (z) or production (y) of a good is represented on the horizontal axis and the price in dollars is represented on the vertical axis. If there were no international trade, the equilibrium outcome would be at the intersection of the domestic demand

(D) and supply (S) curves. However, given the availability of imports at a fixed world price (p^*) and without constraints on imports, the equilibrium occurs at the intersection of the demand curve and the horizontal line given by p^*. Here, consumption z_0 is comprised of domestic production y_0 and imports $m_0 = z_0 - y_0$. The import demand function (M) in the second panel is given by the horizontal distance between the domestic demand and supply curves in the first panel—that is, the amount of demand not satisfied by domestic production at a particular price. Domestically, producer surplus (profit) is given by L while consumer surplus is given by $(F + G + H + I + J + K)$.

Now consider the effect of an import quota set to $m_1 < m_0$. In the new equilibrium, consumption z_1 is comprised of domestic production y_1 and imports $m_1 = z_1 - y_1$. Now, producer surplus is given by $(L + F)$ while consumer surplus is given by $(J + K)$. Comparing domestic surplus before and after the advent of the quota, shows that the area given by $(G + H + I)$ is no longer included in either consumer or producer surplus. First, consider the area H, which is referred to as the "importing rents" or profit given the opportunity to purchase m_1 units of goods at price p^* on the world market and sell them domestically at price p_1. The allocation of this profit depends on how import quota licenses are distributed, which can take several forms (see Feenstra 2004, p. 256). Since we are interested in the value of all trade, then we will consider the case of a complete trade ban ($m_1 = 0$) in which instance H and K both disappear. The remaining lost surplus from the trade restriction is given by $(G + I)$, which appears in both panels.

In order to characterize the value of trade we need to calculate the area $(G + I)$, the lost domestic surplus given a trade ban. To facilitate this calculation, let the slope of the import demand curve M be given by dp/dm, let $R = p^* m_0$ represent the unrestricted trade expenditures, and let the price elasticity of import demand[2] (hereafter, "demand elasticity") be given by $\varepsilon_M = \dfrac{dm}{dp} \dfrac{p^*}{m_0}$. The calculation of lost surplus is then given by:

$$\textbf{(1.5)} \quad G + I = \left[\left(-\frac{dp}{dm} m_0\right) m_0\right] \bigg/ 2 = \left[\left(-\frac{dp}{dm} \frac{m_0}{p^*}\right) p^* m_0\right] \bigg/ 2 = R / (-2\varepsilon_M).$$

Since the slope dp/dm is negative, the elasticity parameter will also be negative and lost surplus as calculated from equation (1.5) will be expressed as a positive number.

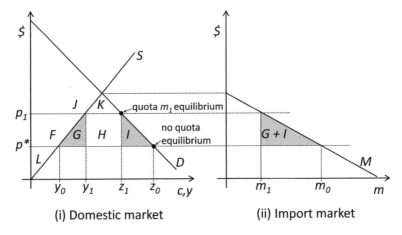

FIG. 8–2. Graphical model of the welfare opportunity cost $(G+I)$ of imposing an import quota level of m_l.

Under this model, import surplus for the sector as a whole is estimated from two inputs, import expenditures (R) and the demand elasticity (ε_M). Taking the approach outlined here, Schmidt et al. (2012) obtained import expenditures for plants for planting from the US Department of Commerce (USDOC 2010) via the Global Agricultural Trade System (GATS) online database. Demand elasticities for different US import sectors are available in the economic literature—for example, as recently estimated by Kee et al. (2008). Schmidt et al. divided the total surplus by the estimated number of unique species available for trade within the sector to generate an estimate of V_T.

The advantages of this approach are that only three aggregate parameter inputs are needed and the approach is simple enough that nonspecialists can conduct the calculations. One disadvantage is that elasticity parameter estimates for a given group of species may not be available if the group is not well differentiated within the harmonized system for categorizing international trade. Another key disadvantage is that imports of various species are lumped together in one homogenous pool, when arguably concerns over banning trade of a particular species are more about the loss of variety in species available for trade than about a small drop in undifferentiated trade. Next I describe a method for targeting the lost value of variety directly in estimating the opportunity cost of rejecting imports of a given species.

Benefits from Import Variety

Recently Blonigen and Soderbery (2010) advanced a microeconometric approach to parameterizing a utility model of product variety. This enabled Springborn et al. (2011) to generate sector-specific (herpetofaunal trade) estimates for a variety model parameter, which captures the degree to which consumers benefit from additional variety of species. The general method of moments estimation is quite technical, though example code is printed in Feenstra (2010, pp. 27–37). Both a limitation and strength of this approach is its dependence on a rich panel data set of import quantities and prices, collected over several years and differentiated by species.

While collection of import data disaggregated to the species level is not the norm, detailed import data for several taxa—including birds, mammals, reptiles, and amphibians—are obtainable from the US Fish and Wildlife Service (USFWS). Maintained in the Law Enforcement Management Information System (LEMIS) database, this information includes the annual quantity and stated customs value of various animal imports. Unfortunately these data are not as freely available as most sources of trade data—only a limited window of data is typically obtainable and one must use a Freedom of Information Act request to do so.

The econometric estimate of the variety model parameter is then used to calculate a closed form expression for the welfare impact of rejecting a species (Springborn et al. 2011, p. 2147). This welfare metric is derived using the same demand model as used in the econometric estimation. The approach, while data and computationally intensive, uses a consistent framework for both parameter estimation and welfare calculation.

Invasion Losses

Ideally analysts could arrive at objective estimates of both invasion losses (V_I) and trade benefits (V_T) to be used as clean inputs to the model. However, the development of screening systems is outpacing progress in estimation of V_I. Although the costs of species establishments are clearly large (Pimentel et al. 2005), there remains a lack of comprehensive data for most taxa. There is evidence that this information gap is starting to close—for example, Aukema et al. (2011) conducted a broad analysis of damages from nonnative forest insects in the United States. In the meantime, in the absence of such information, the next best available option is to consider informed scenarios to aid in decision making. For example, Springborn et al. (2011) anchor their analysis on a particular benchmark case where the level of V_I is chosen such that the (better understood) benefits of trade, V_T, are set equal to invasion losses, V_I, discounted by the probability that a randomly chosen species will be invasive.[3] This benchmark case is one in which the benefits of trade are equal to expected invasion losses—that is, a social decision maker would be indifferent to accepting all species or rejecting all species (at least in the absence of a functional screening system). If one believes that on average species in the particular group cause more damage than benefits, then V_I will be greater. Alternatively, if one believes that benefits on average outweigh the losses, then V_I will be smaller.

While not ideal, recommendations are being made in the literature and design decisions are being made in practice that ignore the implications of V_I and V_T in decision making altogether. For example, the heuristic for identifying a cutoff between accepted and rejected species suggested by Bomford et al. (2009) and the approach taken by Keller et al. (2007) are only optimal when a specific ratio V_I/V_T holds—one which will be true only by chance. The risk screening framework outlined here offers a structure for decomposing and for making explicit those components of the model that have previously received short shrift. It allows decision makers to understand how the stringency of a proposed cutoff maps into an implied ratio of the valuation of establishments versus trade and thus to a more informed and concrete debate.

Modeling Invasion Threat

The last component of the statistical decision problem depicted in Equation (1.3) is the parametric model for the threat of establishment, $p(x;\theta)$. The task here involves estimating the parameters, θ, of a model that maps predictive attributes, x, into an estimate of p. Two alternative approaches for estimating these parameters include standard regression and a regression tree.

Construction of a regression model begins with specifying an equation for the likelihood of invasiveness as a function of observed species attributes. Typically a generalized linear model is used, such as a binary logistic regression as in Bomford et al. (2009) and Lieli and Springborn (2013). A standard regression is a global model where a single predictive fit holds across the ranges of all variables. In contrast, regression trees feature a recursive partitioning of explanatory variables, which leads to a branching classification tree. In simplest form, at each node of the tree, the branch taken to a subsequent node depends on whether the level of a particular attribute for a species is above or below a certain threshold. Each branch of the tree eventually terminates in a particular classification—for example, invasive or not invasive if the classification is binary. An algorithm for solving the regression tree faces three tasks: (1) at each step deciding which variable to partition on and what the threshold should be, (2) how many partitions to make before finishing, and (3) how to predict the outcome variable (invasive or not invasive) for each x in a partition (Loh 2008). While traditional regression trees produce a single, "best" tree, Schmidt et al. (2012) use a "boosted" regression tree, which merges results from multiple models (trees) using a forward, stage-wise, numerical strategy to minimize the loss in predictive performance (Ridgeway 1999; Elith et al. 2008). As with standard regression, after the model is fitted, for any given species we can calculate an estimate of the probability that the true classification is invasive or not invasive, conditional on the predictive variables.

The Nonrandom Sampling Problem

Regardless of the approach to fitting the model of invasion threat, the fact that invasive species training data sets are typically not random samples presents an empirical challenge. As observed by Caley et al. (2006), when an effort is made to include a meaningful number of observations of the rare event of interest (invasion) the sample proportion of invasive species in the data is elevated above the population level. In this case the "population" is the pool of potential species for trade. When observations of a particular type are overrepresented, the sample is not random but rather "endogenously stratified." Despite the observations of Caley et al. (2006) this empirical problem is often overlooked (e.g., Bomford et al. 2009).

While this may seem like an esoteric technical point, the consequence of failing to correct for a stratified sample is that it is not appropriate to interpret fitted estimates of p as estimates of population probabilities. When invasives, which are rare in the population, are strongly overrepresented in the data fitted estimates of p will be shifted upwards, relative to the population. For example, while perhaps around 2% of plants for planting (Smith 1999) are invasive, approximately 23% of observations in the training data set for the Australian Weed Risk Assessment (Pheloung et al. 1999) are invasive. While fitted estimates of p from an uncorrected model still have meaning as a generic index of invasion threat, the decision rule in Equation (1.3) no longer holds. Thus, it is no longer appropriate to simply calculate the fixed cutoff rule given by V_T/V_I. It is still possible to identify an optimal cutoff, but this must be done numerically with the training data sample and thus is subject to sampling error.

To correct for the endogenously stratified sample, Caley et al. (2006) developed a numerical bootstrap approach in which the invasive (weed) observations were undersampled to achieve a ratio that matches estimates of the true proportion of invasives in the population. This approach is intuitive but has the downside of setting aside useful information (excluded weed observations) in each round of the bootstrap sample. Corrections for the stratified sampling problem have long been discussed in the social science literature (Manski and Lerman 1977; Cosslett 1993; Imbens and Lancaster 1996; King and Zeng 2001). These corrections do not rely on simulation methods (e.g., bootstrap) and involve reweighting rather than discarding data. Springborn et al. (2011) apply a rare events logistic regression routine that implements a correction from King and Zeng (2001), code for which is freely available for common statistical programs.

Estimating the Population Base Rate of Invasion

Correcting for an endogenously stratified sample where invasives are overrepresented relative to the population rate unsurprisingly requires an estimate of the true population proportion of invasive species, or the "base rate." For some groups of species it may simply be the case that information on the base rate is unavailable. For higher-profile groups of potential invasive species, such as plants for planting, estimates of the base rate may be found in the published literature—for example, Smith (1999) as used in Lieli and Springborn (2013). Springborn et al. (2011) demonstrate the value of archiving trade data with their approach to estimating the base rate for herpetofaunal species. Based on US Fish and Wildlife Service reports from 1970–1971 compiled by Romagosa et al. (2009), Springborn et al. estimate the base rate as the fraction of species identified in imports from 1970–1971 that are known to have established in the United States. This decades-old subset of trade data was selected to "minimize downward bias in the estimate that might stem from either a lag in establishment or a delay in recording establishment" (p. 2148).

Results and Future Directions

The Risk Assessment Payoff

Knowledge of the base rate not only enables correction for an endogenously stratified sample but also estimation of the anticipated net benefits from instituting a screening system. Given an open door baseline and a base rate of π, the per species expected net benefits (ENB) of instituting a screening system is given by

(1.6) $ENB = \underbrace{Pr(y=1)}_{\substack{\text{Probability that} \\ \text{a random} \\ \text{species} \\ \text{is an invader}}} \underbrace{Pr(\hat{y}=1|y=1)}_{TPR} \underbrace{[V_I - V_T]}_{\substack{\text{Net benefit} \\ \text{of excluding} \\ \text{an invader}}} + \underbrace{Pr(y=0)}_{\substack{\text{Probability that} \\ \text{a random species} \\ \text{is benign}}} \underbrace{Pr(\hat{y}=1|y=0)}_{FPR} \underbrace{[-V_T]}_{\substack{\text{Net benefit of} \\ \text{excluding a} \\ \text{benign species}}}$

$$= \pi \cdot TPR \cdot [V_I - V_T] + [1 - \pi] \cdot FPR[-V_T],$$

where Pr indicates probability, $\hat{y}$ is the decision maker's prediction of whether a species under scrutiny will be invasive ($\hat{y} = 1$) or not invasive ($\hat{y} = 0$), TPR is the true positive rate, and FPR is the false positive rate.

Recent estimates of ENB indicate substantial potential returns from screening. For US herpetofaunal imports Springborn et al. (2011) estimated ENB of $54K–$141K per species assessed, under central scenarios. Schmidt et al. (2012) estimated ENB per species for US plant imports of at least $80K–$140K, given conservative estimates of the potential losses from weeds. Lieli and Springborn (2013) arrived at much higher estimates of ENB per species for Australian plant imports of $3,800K–$4,200K. However, this Australian application—which actually predates Springborn et al. (2011) and Schmidt et al. (2012)—is based on cruder, "back of the envelope" estimates of lost gains from trade restriction and invasion losses taken from the literature. The main contributions of Lieli and Springborn (2013) consist of laying out the statistical-decision-making framework and developing the maximum utility methodology, rather than a rigorous estimate of ENB based on theoretically sound parameter inputs. Even so, the results across all three analyses consistently show economically significant returns to species screening.

These estimates do not include the cost of conducting the risk assessment for a particular species. A recent informal survey of practitioners across many countries found that the type of basic risk assessment evaluated here might cost in the range of $1,200–$12,000 per species (Jenkins 2011).

Future Directions

As described above, threat estimation and implementation of the optimal decision rule can be performed in a segregated two-step process. Under either a standard regression or regression tree approach, conditional probabilities of invasiveness are estimated in isolation before consequences of outcomes are evaluated in deciding whether to accept or reject a novel import.

It bears noting that regression involves optimization with respect to a generic objective that is not typically customized for the particular problem at hand. For example, in a simple linear regression, the model is fitted to minimize the sum of squared error, regardless of the direction of the error. For environmental risk in general and invasive screening in particular, we are in a position to eschew generic loss functions in favor of directly considering the specific costs of false positives and false negatives. The implication is that, instead of a segregated, two-step process, it may be more powerful to conduct model estimation and decision making in one unified step. Lieli and Springborn (2013) take this tack for a plant screening problem, using the "maximum utility" method (Elliot and Lieli 2013; Lieli and White 2010).

The maximum utility method exploits the idea that, for prediction of a binary variable (invasive/noninvasive), a global fit of the model is less important than a localized fit that partitions the space of the explanatory variables (x) into regions where proposals are either accepted or rejected in a way that maximizes net benefits. Returning to Fig. 8–1, the intuition for the maximum utility method is that for the decision problem at hand, the critical task is simply to identify the threshold x' at which $p(x;\theta)$ and c intersect—beyond this a more perfect global fit for $p(x;\theta)$ provides no extra benefit. As shown by Lieli and Springborn (2013), exploiting this feature allows the maximum utility approach to be less sensitive to specification error (specifically in the functional form of $p(x;\theta)$ and c, which is also allowed to vary over x).

Policy Implications

Overall, there is a strong case to be made that the United States—and countries similarly taking a reactive stance to trade in nonnative species—would be well served by shifting to a proactive approach where basic invasion risk assessment is conducted before imports begin—that is, before species have been unmasked as problematic. The analytical tools and data currently exist to implement screening for a number of different groups of live organisms in trade. Risk assessment will involve error—some benign species will be mistakenly excluded while some damaging species will slip through the screen. However, arguably a bigger error would be allowing an (unobtainable) error-free system to be the enemy of the good. Despite imperfections in species risk assessments, substantial benefits have been estimated across a number of live animal and plant imports.

One potential policy application of the risk assessment framework in this chapter would be in support of selecting plant species for a new "gray list" category known as "Not Approved Pending a Pest Risk Analysis" (NAPPRA) established in a 2011 amendment to the US Plant Protection Act (USDA-APHIS 2011). NAPPRA listings are restricted from importation but can be reevaluated through an extended pest risk assessment. A limited first round of additions to this category was largely composed of infrequently traded, low-value plants. Going forward, however, the USDA will be considering the addition of plants for which the overall desirability of trade restrictions is less immediately obvious. The approach outlined in this chapter provides a rigorous and transparent framework for making tough decisions about what level of risk is acceptable.

The risk assessment applications discussed in this chapter demonstrate the value of building and maintaining detailed data sets that can be used to better understand ecological dynamics (e.g., the baseline risk of invasion) and welfare impacts of restricting trade. Highly useful specialized trade data for some live animal imports are available from the US Fish and Wildlife Service, but acquisition is time consuming and long-term maintenance falls to third-party researchers. Given the public value of this data for better understanding the exchange of nonnative species and making better-informed decisions to regulate that exchange, the data should be made more easily accessible and its long-run maintenance should be ensured.

The decision framework presented here demonstrates the value of understanding the benefits of imports and the likely losses if species are invasive. While import demand models arguably capture the bulk of benefits to the importing country, future assessments should explore values that emerge further down the value chain in post-import markets. Finally, invasive species losses are clearly the most uncertain input variable. The multidisciplinary team assembled to assess nonnative forest insect damages in Aukema et al. (2011) provides a strong example of the kind of impact valuation effort that should be supported in future research to better inform invasive species management.

Notes

1. Under a closed door status quo, there is no change in net benefits when a species is rejected: $u_{1,0} = u_{1,1} = 0$. Net benefits from accepting a benign species is given by $u_{0,0} = V_T$, while mistakenly accepting an invasive generates payoffs given by $u_{0,1} = V_T - V_I$. Evaluating Equation (1.4) with these inputs leads to the same cutoff as in the open door case, V_T/V_I.

2. The price elasticity of import demand specifies the percentage change in import demand, m, given a percentage change in price, p.

3. To be more precise, Springborn et al. (2011) focus on the case of establishment rather than invasion.

Literature Cited

Aukema, J. E., B. Leung, K. Kovacs, C. Chivers, K. O. Britton, J. Englin, S. J. Frankel, et al. 2011. Economic impacts of non-native forest insects in the continental United States. *PLOS ONE* 6(9):e24587.

Blonigen, B.A., and A. Soderbery. 2010. Measuring the benefits of foreign product variety with an accurate variety set. *Journal of International Economics* 82(2):168–180.

Bomford, M., F. Kraus, S. C. Barry, and E. Lawrence. 2009. Predicting establishment success for alien reptiles and amphibians: a role for climate matching. *Biological Invasions* 11(3):713–724.

Caley, P., W. M. Lonsdale, and P. C. Pheloung. 2006. Quantifying uncertainty in predictions of invasiveness. *Biological Invasions* 8:277–286.

Cosslett, S. 1993. Estimation from endogenously stratified samples. In *Handbook of Statistics*, Vol. 11, edited by G. S. Maddala, C. R. Rao, and H. D. Vinod, 1–43. Amsterdam: North-Holland.

Elith, J., J. R. Leathwick, and T. Hastie. 2008. A working guide to boosted regression trees. *Journal of Animal Ecology* 77:802–813.

Elliot, G., and R. P. Lieli. 2013. Predicting binary outcomes. *Journal of Econometrics* 174(1):15–26.

Enserink, M. 1999. Predicting invasions: biological invaders sweep in. *Science* 285(5435):1834–1836.

Feenstra, R. C. 2004. *Advanced International Trade: Theory and Evidence*. Princeton: Princeton University Press.

Feenstra, R. C. 2010. *Product Variety and the Gains from International Trade*. Cambridge: MIT Press.

Field, B. C., and M. K. Field. 2006. *Environmental Economics: An Introduction*. Boston: McGraw-Hill/Irwin.

Fowler, A. J., D. M. Lodge, and J. F. Hsia. 2007. Failure of the Lacey Act to protect US ecosystems against animal invasions. *Frontiers in Ecology and Environment* 5:353–359.

Hayes, K. R., and S. C. Barry. 2008. Are there any consistent predictors of invasion success? *Biological Invasions* 10:483–506.

Imbens, G., and T. Lancaster. 1996. Efficient estimation and stratified sampling. *Journal of Econometrics* 74 (2):289–318.

Jenkins, P. 2011. Personal communication, August 11, 2011.

Kee, H. L., A. Nicita, and M. Olarreaga. 2008. Import demand elasticities and trade distortions. *Review of Economics and Statistics* 90:666–682.

Keller, R., D. M. Lodge, and D. C. Finnoff. 2007. Risk assessment for invasive species produces net bioeconomic benefits. *Proceedings of the National Academy of Sciences USA* 104:203–207.

King, G., and L. Zeng. 2001. Logistic regression in rare events data. *Political Analysis* 9:137–163.

Kolar, C., and D. Lodge. 2001. Progress in invasion biology: predicting invaders. *Trends in Ecology and Evolution* 16(4):199–204.

Lieli, R. P., and M. Springborn. 2013. Closing the gap between risk estimation and decision-making: efficient management of trade-related invasive species risk. *Review of Economics and Statistics* 95:632–645.

Lieli, R. P., and H. White. 2010. The construction of empirical credit scoring models based on maximization principles. *Journal of Econometrics* 157:110–119.

Lodge, D. M., S. Williams, H. J. MacIsaac, K. R. Hayes, B. Leung, S. Reichard, R. N. Mack, et al. 2006. Biological invasions: recommendations for US policy and management. *Ecological Applications* 16:2035–2054.

Loh, W. 2008. Classification and regression tree methods. In *Encyclopedia of Statistics in Quality and Reliability*, edited by F. Ruggeri, R. S. Kenett, and F.W. Faltin. New York: Wiley.

Manski, C. F., and S. R. Lerman. 1977. The estimation of choice probabilities from choice based samples. *Econometrica* 45(8):1977–1988.

Pheloung, P. C. 1995. Determining the weed potential of new plant introductions to Australia. Report to the Agriculture Protection Board, Western Australia.

Pheloung, P. C., P. A. Williams, and S. R. Halloy. 1999. A weed risk assessment model for use as a biosecurity tool evaluating plant introductions. *Journal of Environmental Management* 57:239–251.

Pimentel, D., R. Zuniga, and D. Morrison. 2005. Update on the environmental and economic costs associated with alien-invasive species in the United States. *Ecological Economics* 52:273–288.

Ridgeway, G. 1999. The state of boosting. *Computing Science and Statistics* 31:172–181.

Romagosa, C. M., C. Guyer, M. C. Wooten. 2009. Contribution of the live-vertebrate trade toward taxonomic homogenization. *Conservation Biology* 23:1001–1007.

Schmidt, J., M. Springborn, and J. Drake. 2012. Bioeconomic forecasting of invasive species by ecological syndrome. *Ecosphere* 3:46.

Smith, C.S. 1999. Studies on weed risk assessment. Master's thesis, University of Adelaide, South Australia.

Springborn, M., R. Keller, and C. Romagosa. 2011. The value of nonindigenous species risk assessment in international trade. *Ecological Economics* 70(11):2145–2153.

USDA-APHIS (US Department of Agriculture–Animal and Plant Health Inspection Service). 2011. Importation of plants for planting: establishment of category of plants for planting not authorized for importation pending a pest risk analysis. *Federal Register* 76(103), 31172–31210.

USDOC (US Department of Commerce, US Census Bureau). 2010. Foreign trade statistics. Accessed November 4, 2013. http://www.fas.usda.gov/gats/default.aspx.

Williamson, M. 1999. Invasions. *Ecography* 22:5–12.

Controlling the Bad: Reducing the Impacts of Established Invaders

Once an invasive species is established and causing harm, the goals for management and policy are to control its spread and to reduce its impacts. Because there are many different types of invaders, and because they invade many different ecosystems and have variable impacts, control programs need to be very circumstance specific. Ideally, these control programs will be based on detailed knowledge of the species in question, the invaded habitats, and some measurement (or estimate) of how much harm can be avoided by implementing different types and levels of control. With this information the optimal management and policy approaches can be determined.

This section includes four chapters that each deal with control of established invasive species. The first three chapters cover invasions by single species, beginning with the example of the emerald ash borer in chapter 9 by ecologist Jon Bossenbroek and his ecologist and economist colleagues. They combine ecological models of the spread of this forest pest with economic models to estimate how much harm could have been avoided through different management approaches. Next, ecologist Jim Kitchell and colleagues (chapter 10) consider how populations of the invasive sea lamprey in Lake Superior will respond to climate change, and what this will mean for fisheries. The last species-specific chapter is by civil engineer Robert Keller (chapter 11). He describes the design and utility of barriers for preventing the spread of invasive common carp in Australia.

The section ends with a chapter by ecologists Kirsten Prior and Jessica Hellmann (chapter 12). They investigate whether there is evidence to suggest that invasion success arises from the loss of natural parasites and pathogens when species are introduced. Their results are important for determining situations under which introducing natural enemies of invasive species (i.e., classical biological control) can be an effective way to reduce impacts.

PLATE I (FIG. I–I). A population of northern snakehead was discovered in a Maryland (USA) pond in 2002. This discovery generated a lot of media attention, much of which focused on the exaggerated ability of the fish to travel over land. This example of media is the front page of the New York *Daily News*, July 13, 2002. Used with permission.

PLATE 2 (FIG. 2–2). *Facing page*: An example of the use of cane toads in Australian popular culture (photograph by T. Shine).

PLATE 3 (FIG. 5–1). *This page, top*: *Cover of Sooper Yooper: Environmental Defender*, an illustrated children's book by Mark Newman and Mark Heckman. The book describes the efforts of the hero, depicted here on the cover, to protect the Great Lakes from invasive species. ©Mark Heckman. Used with permission.

PLATE 4 (FIG. 5–2). *This page, bottom*: Billy Cooper, the hero from Sooper Yooper: Environmental Defender, with a sea lamprey. ©Mark Heckman. Used with permission.

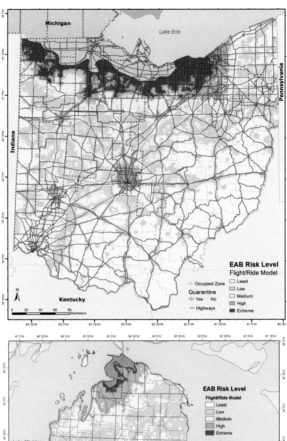

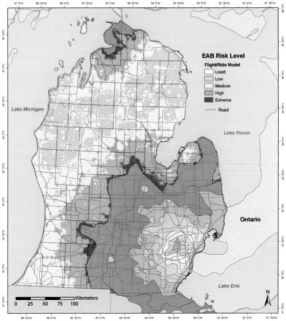

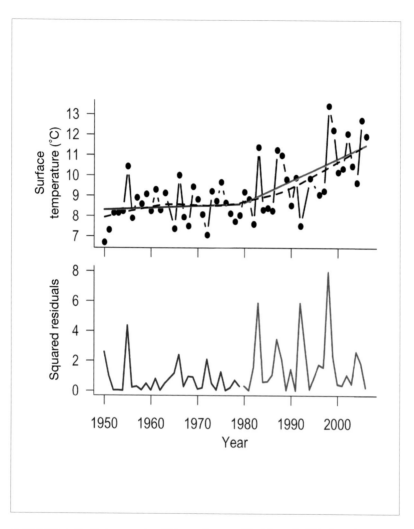

PLATE 5 (FIG. 9–3). *Facing page, top*: Risk map for emerald ash borer in Ohio (based on Prasad et. al 2010). Regions with highest risk are those most likely to be invaded.

PLATE 6 (FIG. 9–4). *Facing page, bottom*: Risk map for emerald ash borer in Michigan. Colors as for Plate 5.

PLATE 7 (FIG. 10–2). *This page*: Top panel: Changes in average annual surface temperatures for Lake Superior (redrawn from Austin and Coleman 2008) during the period 1950–2006. The dashed line represents a LOESS model. The two colored lines are linear regressions for the periods before and after 1980. Bottom panel: Squared residuals calculated from least squares regressions of the periods 1950–1979 and 1980–2006.

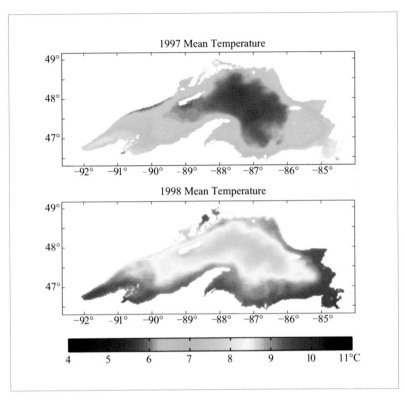

PLATE 8 (FIG. 10–3). *This page*: Average annual surface water temperatures for Lake Superior during 1997 (top panel) and 1998 (bottom panel) based on the model of Bennington et al. (2010). Water temperatures are represented in the color scale below. Units of latitude and longitude are represented in y-axis and x-axis, respectively.

PLATE 9 (FIG. 10–4). *Facing page*: Distribution of estimated days at preferred temperatures for Chinook salmon (top two panels) and white sucker (bottom two panels) as hosts for sea lampreys during 1997 and 1998 based on the model of Bennington et al. (2010). Chinook salmon are assumed to have a thermal preference of 13°C and white sucker thermal preference was assumed to be 18°C. Days per year are represented in the color scale. Units of latitude and longitude are represented in y-axis and x-axis, respectively.

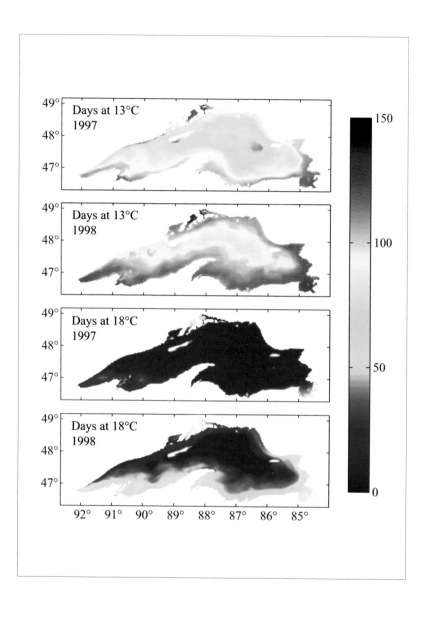

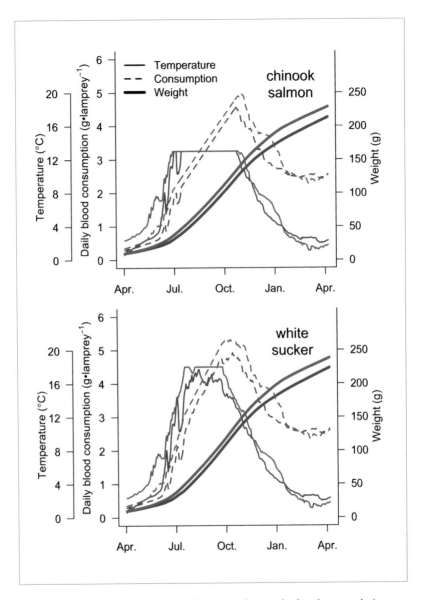

PLATE 10 (FIG. 10–5). Bioenergetics modeling results for growth of sea lampreys during 1997 and 1998 based on alternative hosts. Chinook salmon (top panel) were assumed to have a thermal preference of 13°C and white sucker (bottom panel) thermal preference was assumed to be 18°C. The blue lines are simulations for the cool year (1997) and the red lines simulations of the warm year (1998).

Evaluating the Economic Costs and Benefits of Slowing the Spread of Emerald Ash Borer

Jonathan Bossenbroek, Audra Croskey, David Finnoff,
Louis Iverson, Shana M. McDermott, Anantha Prasad,
Charles Sims, and Davis Sydnor

Introduction

The emerald ash borer (*Agrilus planipennis;* EAB) is poised to wipe out native ashes (*Fraxinus* spp.) in North America with expected catastrophic losses to ash tree forestry (MacFarlane and Meyer 2005). EAB was first discovered in Detroit in 2002. Most scientists hypothesize that it entered the United States through solid wood packing material transported in cargo ships and on planes. The beetles have continued their spread through firewood, wooden packing materials, and infested nursery trees. In only a few years, EAB has destroyed most ash trees within the Detroit Metropolitan area and spread through Michigan's lower peninsula and into Ohio, Indiana, and other states (Fig. 9–1). For the foreseeable future, EAB will continue to destroy urban and forested ash trees, causing substantial economic impacts.

Given this threat, immediate action is crucial for the preservation of ash trees. The US Department of Agriculture (USDA) and state and local agencies are already urgently attempting to prevent further spread of the EAB but are constrained by agency budgets, and little guidance exists about how best to allocate scarce funds to alternative methods of prevention and control. For government or private institutions, the allocation of

FIG. 9–1. Distribution of EAB as of May 1, 2012, based on data from the USDA and Canadian Food Inspection Agency (modified from http://www.emeraldashborer.info/files/Multi State_EABpos.pdf).

funding would address the level of investment in prevention provided the benefits outweigh the costs. Equally essential are analyses about which kinds of prevention methods—education and quarantine efforts, for example—are most cost-effective. Specifically, hard numbers are needed on the likely financial impacts of the emerald ash borer on urban areas and the ash industry, as well as benefits of possible alternative prevention and control options.

Our project has two basic objectives: first, to provide estimates of the regional economic impact emerald ash borer will potentially inflict upon the ash forestry in Ohio and Michigan; and second, to provide policy makers with quantitative guidance for cost-effective alternative strategies to control, prevent, or slow the spread of emerald ash borer.

Methods and Results

1) Current and Potential Distribution of Ash Trees and Emerald Ash Borer

The only known limiting factor in the potential distribution of EAB is the presence of ash trees. The insect is host specific to the genus *Fraxinus* and infests all North American ash species, though different species have different susceptibilities to becoming infested (Poland and McCullough 2006). Thus, to understand the potential distribution of the emerald ash borer requires knowing the distribution of ash at local and regional scales. We estimated the distribution of ash trees in Ohio and Michigan in three ways—a coarse-level analysis of the eastern United States, a fine-scale analysis of Ohio and Michigan, and an assessment of ash in urban areas.

We first used Forest Inventory and Analysis data (USDA Forest Service, Miles et al. 2001) to map and quantify the four species of ash that account for the vast majority of ash in rural settings in the eastern states (Fig. 9–2). The four species are *Fraxinus americana* (white ash), *F. pennsylvanica* (green ash), *F. nigra* (black ash), and *F. quadrangulata* (blue ash). A detailed estimate of ash resource availability in Ohio and Michigan was developed by combining estimates of ash basal area per Forest Inventory and Analysis (FIA) plot with a Landsat Thematic Mapper (TM)-based classification of forest types.

We also assessed the abundance of ash in private and public areas within cities, where it is a common landscaping tree. Sydnor et al. (2007) surveyed 200 communities in Ohio and found, for every 1,000 residents, an average of 20.5 ash street trees, 38.3 park trees, and 320.9 ash trees on personal property. Thus, EAB has had and will continue to have a large impact, not only on native forests but also on urban and suburban communities.

2) Risk of the Spread of Emerald Ash Borer

As ash trees occupy much of the eastern United States, the second step of our study was to predict how quickly EAB spreads in both forests and cities. Modeling the insect's spread requires the integration of several models and data layers because EAB disperses by multiple mechanisms. Flight is their natural mode of dispersal, but they also spread by way of human transport of goods and services (BenDor et al. 2006), especially firewood and wood products.

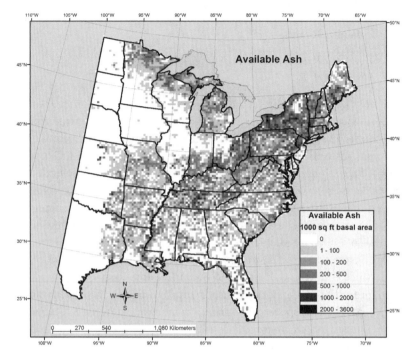

FIG. 9–2. Amount of ash available to the EAB. It is the product of the basal area of ash and percent forest (from Iverson et al. 2010).

Our model of movement comprised a "flight" model and an "insect-ride" model (Prasad et al. 2010). The flight component was based on the SHIFT model, a spatially explicit cell-based model designed to estimate the potential migration of trees under current fragmented landscapes, and including the northward climatic pressure shown to exist now and increasingly so into the future (Iverson et al. 1999, 2004; Schwartz et al. 2001). This model was adapted to match the "front" of the spread of EAB, based on the known front location in 2006, the abundance of EAB behind the front, and the quantity of ash ahead of the front (based on step 1). The flight component was based on state and federal data documenting the spread of EAB during 1998–2006, along with several assumptions (see Prasad et al. 2010).

The flight model simulated local-scale movements, which includes the flight capabilities of EAB and human-assisted local movements. The movement of firewood and wood products are typically thought of as long-distance vectors; however, they also play a role in spread at local scales. To examine the importance of human-mediated dispersal at the local scale, we developed a diffusion model to fit the spread of the "wave front" through Lucas County, Ohio. Two components drive the velocity of a diffusion model: r, the intrinsic rate of increase of a population, and D, a distance coefficient. Because specific population growth parameters are unknown for EAB, the intrinsic growth rate was tested over several orders of magnitude. To test the hypothesis that natural dispersal is not a major factor of the invasion, we adjusted the model parameter values to generate dispersal patterns similar to those observed in Lucas County. These parameter values were then assessed for their validity, namely whether these values were consistent with known life history traits of EAB or other similar species. On the basis of the diffusion model, we estimated mean values for the intrinsic growth rate (r) of 76 and a distance coefficient (D) of 803 km. The value of r estimated by our model for EAB is substantially higher than for other insects, including long-horned beetle (0.02; Akbulut et al. 2007) and Mexican rice borer (0.11; Sétamou et al. 2003). Also, the furthest recorded flight of an individual borer is 20 km (Taylor et al. 2006), which is much lower than the rate estimated by our model. Thus, our analysis suggests that local dispersal, even at a small scale, is driven primarily by human-mediated movement rather than natural diffusion.

Though the flight model was based on the empirical pattern of spread, the insect-ride model incorporated the mechanisms of long-distance dispersal that are known to move EAB. We modified the SHIFT model by weighting factors related to potential human-assisted movements of infested wood or just hitchhiking insects: traffic on roads, urban areas, various wood products industries (including nurseries), population density, and campgrounds. To register the increased probability of insects invading areas adjacent to highways by somehow attaching to vehicles moving down the road, we assigned higher weights to two widths of major road corridors and weighted the risk on the basis of data obtained from Annual Average Daily Traffic volumes for Michigan and Ohio (National Highway Planning Network). Wood products industries were responsible for some outlier EAB invasions, so a scheme was developed to weight buffers around individual businesses dealing in wood products (data from the listing of Standard Industrial Classification [SIC] codes from Dunn and Bradstreet).

Finally, campgrounds are likely destinations of human-assisted EAB transport (Muirhead et al. 2006), primarily through the (mostly illegal) movement of firewood. First, we used a gravity model to identify the relative risk of different campgrounds becoming infested with the insect. Gravity models use distance and attraction of a destination to determine where people choose to travel (Bossenbroek et al. 2001). Thus, the relative risk of a campground becoming infested is based on campground attractiveness, which was estimated on the basis of the number of campsites and distance to the current distribution of EAB. Our results provided a relative rate of propagule pressure for each campground within the study area. Campgrounds with high rates of propagule pressure were predicted to exist throughout the study area and not just close to the current range of EAB. By combining the insect-ride and insect-flight models, we could predict the relative risk of EAB introduction in Ohio and Michigan (Fig. 9–3; Fig. 9– 4).

To verify our model, we compared the output of our model to confirmed EAB observations as of 2007. The EAB data we used for verification was from a survey by the Ohio Department of Agriculture during 2003–2007 wherein they used girdled trees to monitor for EAB presence. Trees in the survey that were found to have EAB are called "detection trees." As of 2007, 255 detection trees were outside the occupied zone (Fig. 9–3). Of these locations, 32% fell in our highest-risk class, which primarily captured zones very near the core with high risk from both ride and flight models. Among the other detection trees, 30% fell in the high-risk class, 35% in the medium classes, 3% in the low classes, and 0 in the least class. This comparison suggests that our modeling efforts are capturing key aspects of the spread of EAB.

The detection tree data also provided insights about the importance of the different anthropogenic dispersal vectors in explaining the observed pattern of spread. Detection trees within Ohio were not randomly placed, so we compared the portion of total detection trees with the proportion of positive trees within particular distances of roads, campgrounds, and wood products industries. Most likely for convenience of sampling, ~50% (depending on year) of detection trees were within 2 km of a highway. However, ~75% of the positive detection trees were within 2 km of a highway, suggesting that highways are an important vector for the spread of EAB. We also found that roads with positive trees are generally nearer campgrounds and especially wood product industries. Thus, we conclude that roads, more than other anthropogenic factors, are the best predictors of long-distance dispersal by EAB.

Our model, by combining the insect's flight characteristics and human-facilitated movement, results in a map of spread that we believe estimated

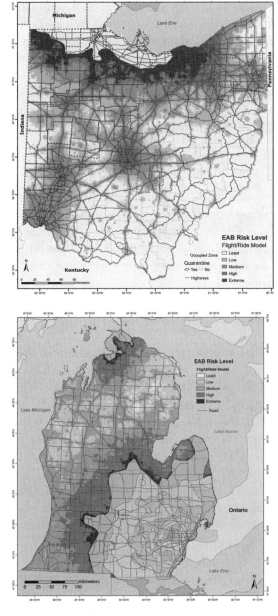

FIG. 9–3. Risk map for EAB in Ohio (based on Prasad et. al 2010). See also color plate.

FIG. 9–4. Risk map for EAB in Michigan. See also color plate.

risk areas for the next 2–4 years with much better accuracy than simple imputed statistical maps. We are able to outline degrees of risk in our maps that agree reasonably well with positive EAB locations. Our mapping effort should help managers better anticipate future risk from EAB, despite uncertain information, by locating areas of higher risk, thereby allowing managers to focus where infestations are most likely to occur. It may also help state and county agencies in the placement of traps or detection trees, or in sample plot design for researchers.

3) Spatially Explicit Estimation of Ash Tree Value

Estimating the value of ash trees in a spatially explicit manner required an inventory of ash stocks from across Ohio, both on forested land and in urban settings, as well as an estimate of losses to the overall economy. We focused on two primary elements of the economic impact of EAB. First, we assessed the value and potential costs of replacing ash trees in urban settings, including households, parks, and streets. Second, we developed a computable general equilibrium (CGE) model to estimate overall losses to the state economies of Ohio and Michigan in the case of a complete loss of ash stock.

To examine the local impact of an EAB invasion on urban and suburban communities Sydnor et al. (2007) used a survey of 200 Ohio Tree Cities. The costs assessed by these surveys included loss of landscape value, stump removal, and tree replacement. Total costs per 1,000 residents ranged from $157,600 to $664,800. When these values were extrapolated to the entire state of Ohio, Sydnor et al. (2007) estimated that the potential costs or losses from EAB-killed ash could range from $1.8 billion to $7.6 billion.

We recognize that assessing "costs" to individuals does not accurately convey how the EAB invasion impacts the larger statewide economy. To determine the annual regional economic consequences for Ohio and Michigan, a CGE model for the region was developed. CGE models are appropriate for determining the economic consequences of EAB outbreaks because they account for the fact that ash trees are inputs into many consumer goods. Seemingly local impacts of the invasion create ripple effects throughout the statewide economy owing to input interactions between industries. Our multisector, general equilibrium method for determining these complex regional economic consequences involves inter-industry linkages, production inputs (including labor and capital), households, government receipts and expenditures, and trade. Impacts to households occur when the spreading beetle causes prices and incomes to adjust to changing market conditions, and are calculated by measuring how much consumers are hurt by the resulting price changes.

The loss of ash tree harvests affects the Ohio and Michigan economies in four ways (Fig. 9–5). First, ash-dependent industries (e.g., logging operations, sawmills, processed wood firms) experience an input shortage along the supply chain, where a loss of ash trees in the logging sector trickles down to sawmills and processed wood firms. For this study, the loss of ash as an input was modeled by reducing the amount of final goods that could be produced by each industry. Second, parks, and consequently recreation activities, are impacted by a loss of ash trees, modeled as an increase in operating costs due to the removal and replacement of dead ash trees. Third, the demise of ash trees in residential yards means households must spend money to remove the trees. To incorporate these fees, household income was reduced by the average cost of ash removal. In addition, household expenditures on the removal of ash trees by landscaping businesses that remove and replace ash trees were taken into account. Fourth, state and local governments must remove ash trees, mainly along streets, which diverts funds from other public services. Government expenditures also lead to an additional increase in demand for landscaping businesses that provide ash tree replacement and removal. Given limited time and money, removal impacts affecting households, parks and recreation activities, and the state government would persist for 10 years until 100% of the ash were replaced, assuming that 10% of the trees would be removed annually. In the event that all ash trees would be destroyed, the impact on the logging sector would persist for upwards of 40 years.[1]

IMPLAN (Impact Analysis for Planning; MIG) data from 2003 provided detailed economic information for the CGE model. In addition, we aggregated 509 industry sectors into 14 categories. The logging, sawmill, and processed wood sectors were designated following Hushak (2005). The 11 remaining sectors (finished wood products, building, business services, transportation and storage, furniture manufacturing, consumer wood goods, recreation, paper products manufacturing, garden supply stores, parks, and miscellaneous) were grouped according to Iverson and Sydnor's research, which sorted through Ohio's industries and ranked each company's chance of using ash products from 1 (lowest) to 6 (highest), on the basis of the company's Standard Industrial Classification (SIC) number provided by the US Department of Commerce. Finally, we allowed

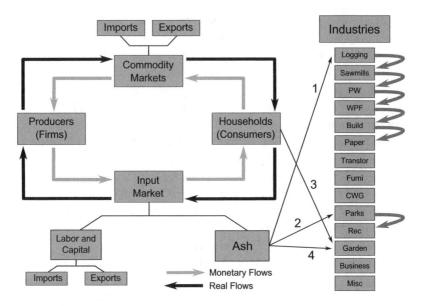

FIG. 9–5. The Ohio economy in terms of the sectors expected to be impacted by EAB. Arrow 1 represents the vertically integrated impacts from a loss of ash beginning with the logging sector. Arrow 2 represents the cost impact to parks and recreation sectors. Arrow 3 portrays the removal fee and consequent increase in demand of the garden sector. Arrow 4 represents the state governments' replacement fee and increase in demand for the garden sector. CWG, consumer wood goods; PW, processed wood; Transtor, transportation and storage; WPF, wood products finished. (Image from McDermott 2011.)

industries the ability to substitute other wood (e.g., maple) for ash. As ease of substitution for ash decreased, loss of productivity increased. The minimum impact occurred when the logging sector was affected, while the maximum impact occurred when all industries were unable to substitute for ash. In total, the model accounts for 7,776 combinations of substitution possibilities.

On the basis of the CGE model, the median impact from a complete loss of ash trees would result in an annual loss of $58.20 million in Ohio and $57.96 million in Michigan (Table 9–1). The maximum impact would result in an annual loss of $59.25 million in Ohio and $59.39 million in Michigan. On average, the states of Ohio and Michigan would experience an annual loss of 0.01% and 0.02% of their 2003 GDP, respectively. The sectors in which prices are projected to increase in the future are logging, sawmills, processed wood, finished wood products, paper, parks, and recreation. All other sectors' domestic prices would decrease.

TABLE 9–1. **Summary of annual median welfare impact from complete loss of ash harvest.**

Mode of Imapct	Ohio Impact	Michigan Impact
Vertically integrated production (excluding parks and recreation sectors)	−$2.85 million	−$3.81 million
Parks and recreation cost impacts	−$2.95 million	−$3.70 million
Household income reduction	−$51.92 million	−$49.93 million
Garden sector demand increase (household)	$5,924.00	$6,665.00
State cost impact	−$492,363.00	−$537,459.00
Garden sector demand increase (state)	$847.00	$1,032.00
Total average impact	−$58.20 million	−$57.96 million

The loss of ash affects households differently. For both Ohio and Michigan, the household income bracket of $25–$35K per year was the most heavily impacted. The results for the two states were similar, so we provide details on Ohio only here. On average, a household in the $25–$35K income bracket lost $14.32 annually (Table 9–2). Households in this income bracket yield the largest percentage of their income to removal of dead ash trees, which made the biggest impact on welfare in this model. Since the <$10K and $10–$15K household brackets own the lowest shares of labor and capital, they are impacted slightly differently from other household brackets. The <$10K bracket decreased consumption from all sectors with increased prices, and increased its consumption in all other sectors except for logging and sawmills, of which it purchased nothing. The $10–$15K bracket decreased consumption in sectors for which the price increased and decreased consumption in the building and consumer wood goods sectors. Household income brackets above $25K decreased their consumption of all commodities but decreased their consumption proportionally less in sectors whose prices decreased. On average, commodity prices for logging, processed wood, finished wood products, paper, recreation, and parks all increased. The IMPLAN data show that higher-income households consumed the most recreation and parks services and lost a higher amount of income to the removal of ash trees. These results imply that the mid- and higher-income brackets are the most impacted by EAB outbreaks.

The group with the largest annual welfare impact ($51.92 million in Ohio) is households faced with the removal of ash trees. In contrast, the average industry and parks/recreation impacts from an EAB invasion in Ohio result in an annual loss of $2.85 million and $2.95 million, respectively.

TABLE 9–2. **Annual Ohio impacts per individual households.**

Household Income	Minimum Impact ($\varepsilon_{LOG} = 1$)	Maximum Impact (*all ε's = 1*)	Average Impact (over 7,776 scenarios)
<$10K	$1.13	$0.69	$0.91
$10-$15K	$0.31	$0.20	$0.26
$15-$25K	–$0.42	–$0.37	–$0.40
$25-$35K	–$14.31	–$14.32	–$14.32
$35-$50K	–$15.11	–$15.32	–$15.21
$50-$75K	–$16.81	–$17.36	–$17.08
$75-$100K	–$17.34	–$18.28	–$17.81
$100-$150K	–$19.19	–$20.49	–$19.84
>$150K	–$21.48	–$23.22	–$22.35

4) *Cost and Effectiveness of Different Prevention and Control Strategies*

Slowing or stopping the spread of an invasive species, such as EAB, is often a goal of local, state and federal managers. Different strategies can be utilized, such as EAB eradication at the edge of the expanding population, increasing outreach and education efforts to reduce human-mediated spread, or preventing the establishment of new long-distance/outlier populations. To evaluate the appropriate investment in prevention strategies, we assessed the cost and effectiveness of postinvasion control approaches in two ways. First, we collected data on the costs of removing outlier populations. Second, we modeled the effectiveness of eradicating outlier populations to slow the spread of EAB in Ohio (Croskey 2009).

Early in the invasion of EAB, when an outlier population was detected, the typical response in both Michigan and Ohio was to quarantine the area and attempt to eradicate the population through tree removal. Invaded trees and all host trees within a minimum half-mile buffer zone were removed. Removal was subsequently followed by herbicide treatment of stumps in woodlots and grinding of landscape stumps. For thirteen outlier sites in Michigan, removal cost nearly $6 million for more than 216,000 trees. However, despite the amount of money spent on these efforts, there was very little assessment of their effectiveness. Over time the eradication efforts have proved largely ineffectual, as new populations of the borer emerged. Likewise, managers in Ontario, Canada, cut down all ash trees within a 10-km radius of an infestation in 2003, which also proved to be ineffective, as additional populations were discovered just outside the ash-free zone in 2004 (Muirhead et al. 2006).

To assess the ability of eradication programs of outlier populations to slow the spread of EAB, we developed a spread model based on the risk map of Ohio from Prasad et al. (2010; Fig. 9–3). The spread model predicted the infested status of a location on the basis of two factors: the distance to a known source of EAB and the risk level from the aforementioned map. In order to run multiple trials of our stochastic model in a timely manner, we increased the cell size of the Prasad et al. (2010) model by a factor of ten for the long-distance risk map, resulting in cells measuring 2,700 m by 2,700 m. In addition, we randomly pre-seeded long-distance infestations into the map; this reflected the long-distance observations observed in Ohio the same year the wave front reached the state (i.e., the first "year" of this model). The local spread component of the model was developed to mimic the wave front of the invasion known thus far, which occurred at about 20 km per year between 1998 and 2006 (Prasad et al. 2010).

To test the hypothesis that eradication would significantly slow the invasion spread and therefore lower annual costs, eradication events were built into the model. To simulate an eradication program, a certain percentage of infested cells were randomly returned to uninfested after each year of the model. Several levels of eradication were analyzed—5%, 10%, 15%, 20%, 25%, 50%, 75%, and 95%. Model simulations for each level of eradication were run for 30 years, 20 times. The locations of infested cells were plotted on a census map, allowing us to calculate the number of humans impacted and therefore potential costs incurred by the infestation each year. The census values obtained were averaged for each year across all trials.

The statewide spread model without any eradication events predicted complete infestation of the state within 26 years. However, the last seven years of the infestation under this strategy show little increase in the number of people affected—99% of the state of Ohio is infested at year 19. The various eradication strategies postponed EAB infestation by 2–4 years. With 95% eradication occurring each year, the state was completely infested in 26 years. The halfway point of the infestation in Ohio, 5.5 million people, occurred after eight years under the no-eradication strategy, and eleven years into the infestation under the 95%-eradication strategy. In sum, while eradication programs can slow the spread of EAB, whether such efforts are economically beneficial is questionable.

5) Optimizing Resources by Linking Distribution and Spread Models with Estimates of Potential Damages

To link models of spread, impact, and effectiveness of control efforts we used Stochastic Dynamic Programming (SDP) to optimize inputs from the various models. In the absence of well-defined dose-response relationships over strategies of prevention and control we narrowed the focus to an implementation of public policy early in the invasion, when the beetle was first spreading from Detroit, MI. We also focused the SDP framework on the optimal timing and stringency (or aggressiveness) of strategies to control the spread. In our model, optimal timing depends on the extent of the spread, current damages, expected future damages, cost of the strategy, impact of the strategy, and stochastic dynamic spread of EAB.

Policies to control an invasive species like EAB are investments of scarce resources in the face of uncertainty. Natural variability in factors like weather and biological processes of the species (growth, mortality, movement) as well as uncertainty arising from human interactions can randomly alter the dynamics of spread. This variability in spread dynamics translates into uncertain costs and benefits of control actions. Controlling EAB can also be postponed and requires the consideration of two kinds of irreversibilities that work in opposition and determine the effect of uncertainty on the optimal control decision. First, environmental damage due to EAB can be partially or totally irreversible. On the basis of a precautionary principle argument, this irreversibility calls for more immediate control actions than suggested by traditional cost-benefit analysis. Second, policies aimed at controlling the spread of the beetle impose sunk costs on society that may prove to be undesirable as more information on the invasion is revealed over time, causing decision makers to be more cautious about allocating scarce resources to control (an economic precautionary principle). This "fear of regret" delays control action beyond what would be suggested by traditional cost-benefit analysis. There is also a tradeoff between the timing and stringency of EAB policy strategies in terms of whether spread is slowed, stopped, or reversed. More stringent policies commit decision makers to larger sunk costs, which cause more cautious behavior by decision makers. Thus more stringent EAB policies will be delayed further into the future.

An alternative to cost-benefit analysis that incorporates irreversibility and uncertainty is the real options examined by Dixit and Pindyck (1994). Real options analysis specifies a stochastic process for the asset of interest (here, uninvaded area) and solves for the total value of the investment opportunity in order to determine an optimal threshold below which investment is optimal and above which investment should be postponed. However, our analysis extends the existing literature in three important ways. First, the costs and benefits of EAB control imply the existence of an optimal degree of control in addition to an optimal time to control that are linked and must be considered jointly. Second, the cost of stopping EAB spread is likely to increase over time as the invasion progresses, which differs from the traditional assumption of a fixed investment cost. Finally, we incorporate upper bounds on the stochastic process that correspond to physical barriers (Great Lakes) to EAB spread. Since decision makers may consider benefits of control that accrue within their jurisdiction only, we also allow for the possibility of perceived barriers (state and federal boundaries). When faced with these barriers, decision makers will optimally engage in a significantly lower amount of control.

CONTROLLING A BOUNDED EAB INVASION.   The model assumes EAB was introduced into Detroit, MI, in 1998, became established in a new habitat, is spreading from the point of introduction, and is causing damage in the areas it is spreading. We assume that the rate of spread can be permanently reduced from the current rate to some lower rate through a control policy at a particular cost. A risk-neutral social planner must optimally select 1) the rate to which invasive species spread should be reduced, and 2) the socially optimal time to do so, in order to minimize the expected present value of EAB damages and control costs.

An optimization problem such as this must be solved in two steps (Saphores and Carr 2000): 1) the optimal stringency of the policy to be adopted based on current known damage and 2) whether a policy with that level of stringency should be adopted immediately. At each instant in time optimal stringency minimizes expected damage and cost from that point forward by lowering the spread rate at a sunk cost (control cost that cannot be recovered). Immediate policy adoption lowers the growth in expected damage but incurs the sunk cost and forfeits the option value (a conditional value of information regarding the characteristics of the invasion). Otherwise policy adoption is postponed until the next instant in time where the decision maker determines a new spread rate reduction on the basis of current damage and is faced with the same binary choice concerning policy adoption. Thus possible outcomes for the EAB control policy are: 1) immediate slowing of spread, 2) delayed slowing of spread, 3) immediate reversal of spread, or 4) delayed reversal of spread.

For our model, management was conducted in ten different zones (Fig. 9–6). As EAB spreads outward from the introduction point, damages are incurred by ecosystems and by industries that rely on these ecosystems. The optimal timing of EAB control relies on comparing termination and continuation values. The termination value is the minimum of the precontrol or postcontrol damage position and represents the payoff received by controlling immediately. The precontrol damage position is equal to the value of all precontrol future damages. The postcontrol damage position is equal to the sum of the postcontrol future damages and the control cost. The continuation value represents what one would receive by abstaining from invasive species control and the value of being able to postpone the control decision. The value of being able to postpone control efforts in order to gain more information about EAB spread is known as the option or time value. This option value is an additional cost that must be overcome in order to trigger an investment in EAB control and arises as a result of the uncertainty inherent in the spread of the beetle.

For the dispersal of EAB through the management zones we parameterized a geometric Brownian motion (GBM) process with the known dispersal patterns from 1998 and 2006. The spread data provided in step 2 is used to calculate maximum likelihood estimates of the drift and volatility parameters for the GBM spread process. According to Sharov and Liebhold (1998), the cost of eradicating a similar invasive forest insect, gypsy moth, was $31,000 per square kilometer. Given the estimated 175-km length of the EAB population front in 2002, and assuming all control activities take place in a barrier zone along the population front that is 1 kilometer wide, a plausible upper bound for the cost of stopping EAB spread in 2002 would be $31,000 * 175 = $5 billion, which implies that the total cost of stopping EAB spread in 2002 would have been approximately $3.6 billion. Actual values will vary depending on the control method (tree removal, pesticide application, biological control, quarantines) and could be adjusted as needed.

Since most invasive species policies are formulated at the state and federal level, control policies are presented from the perspective of the federal government as well as individual states. In our scenario, the barrier in several management zones is marked by a natural feature like a lake, while in other cases barrier location depends on who makes the decision to invest in EAB control. If the decision is made at the state level, state boundaries mark the upper absorbing barrier, whereas national boundaries delineate that barrier in decisions made by federal policy makers. This implies that our results will be most applicable to state and federal governments developing isolated EAB policies, but could easily be extended to allow for cooperation among government entities. Any decision to control spread at the federal level instead of the state level would sig-

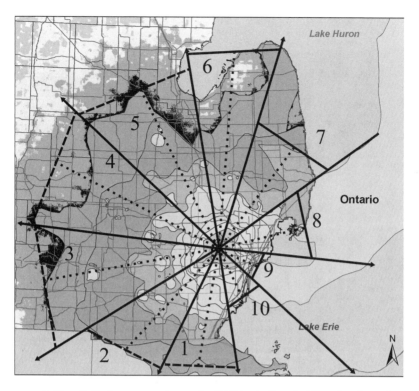

FIG. 9–6. Hypothetical management zones for EAB in the Detroit, MI, region (from Sims 2009).

nificantly increase the upper absorbing barrier in zones 1, 2, and 3. Spread in these zones will largely be responsible for invasion of the rest of the country.

OPTIMAL EAB CONTROL STRATEGY. The optimal EAB control strategy as predicted by our real options model depends on whether the decision making is based on a federal or state perspective. Table 9–3 presents characteristics of the optimal policy to control EAB at the federal level and the state level for Michigan. The total expected present value of EAB damages within the United States under this optimal control policy is $647 million and $726 million for the federal and state scenarios respectively. The optimal investment in EAB control in the state of Michigan by the federal government is over $2 billion, while it would only be $1.1 billion

TABLE 9-3. **Optimal control of EAB by the federal government or state of Michigan.**

	Manager	Management zone						
		1	2	3	4	5	6	7
Upper barrier (km)	Federal	1600	2100	3200	273	627	167	122
	State	80	108	278	273	627	167	122
Time to reach upper barrier (yrs)	Federal	334.55	294.54	219.76	48.18	129.97	3.57	2.15
	State	1.04	3.90	62.23	48.18	129.97	3.57	2.15
% reduction in spread rate	Federal	95.5%	91.7%	90.6%	86.40%	92.1%	0.4%	0.03%
	State	0.00%	0.30%	85.8%	86.4%	92.1%	0.40%	0.02%
Optimal spread rate (%/yr)	Federal	0.02	0.02	0.03	0.06	0.04	0.49	0.48
	State	0.46	0.23	0.04	0.06	0.04	0.49	0.48
Present Value cost (millions $)	Federal	672.60	214.74	247.96	390.22	509.54	0.14	0.00
	State	0.00	0.05	228.40	390.56	509.54	0.14	0.00
Expected Present Value damages (millions $)	Federal	68.23	47.19	33.15	74.43	50.96	227.04	145.87
	State	73.64	108.42	46.13	74.09	50.96	227.04	145.87

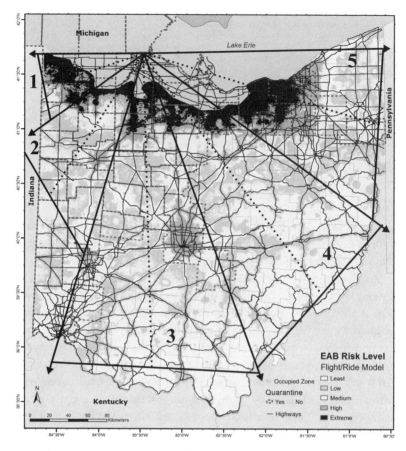

FIG. 9–7. Hypothetical management zones for EAB in Ohio (from Sims 2009).

by the state of Michigan. Because of the lower level of control, expected damages within Michigan under state-level control are greater than under the federal policy.

Given the optimal response to EAB by the state of Michigan in 2002, Ohio would become the next state to consider the option to control the borer. Specifically, the EAB population front would have reached Ohio in 2004 somewhere near the city of Toledo. Because the control problem facing Ohio is fundamentally different than the problem facing Michigan, five new management zones are created that encompass Ohio (Fig. 9–7). Characteristics of the optimal policy to control the invasion by Ohio are presented in Table 9–4. Given the parameters selected to represent EAB

TABLE 9-4. **Optimal control of EAB by the state of Ohio.**

	Management zone				
	1	2	3	4	5
Upper barrier (km)	112.59	200.16	341.94	321.09	258.54
Time to reach upper barrier (yrs)	11.92	34.78	58.75	64.39	44.56
% reduction in spread rate	10.50%	71.00%	81.50%	72.10%	78.50%
Optimal spread rate (%/yr)	0.27	0.12	0.08	0.07	0.10
Present Value cost (millions $)	4.02	109.18	160.19	55.22	166.90
Expected Present Value damages (millions $)	54.45	28.64	17.72	9.79	25.91

spread in Ohio, state officials would find it optimal to spend nearly $496 million on EAB control within the state. As a result EAB damages in Ohio would total approximately $147 million.

A full comparison of EAB control policies formulated at the state and federal levels would require that the control option be evaluated in each potentially invaded state starting when that state is expected to become invaded. However, data limitations become an even bigger issue as an invasion progresses. For that reason we limited our analysis to the control decisions of two states: Michigan and Ohio. Even with this limited analysis EAB will clearly spread much faster under state-level control compared with federal EAB policies. If the federal government had optimally responded to the EAB invasion in 2002, $2 billion would have been spent on control and EAB might still be contained within the state of Michigan. However, if EAB control had been relegated to individual states, over $1.5 billion would have been optimally spent by Michigan and Ohio to control EAB and the invaded area would be much larger.

Conclusions

Our analyses suggest it would have been optimal to spend over $1 billion at the beginning of the EAB invasion to slow or stop its progress. Considering we addressed only Ohio and Michigan, however, we cannot say to what extent it is worth investing in eradication events now or in the future without further analysis. Nonetheless, each step of our project contributed important results that will benefit managers of EAB and invasive species generally:

Step 1: Currently there is no known limit to the range of EAB except the distribution of ash trees themselves. Our efforts, however, showed that for an economic analysis trees grown for harvest and for recreation/land-scaping are important to consider.

Step 2: EAB dispersal is a result of both human-mediated and natural spread. Even at the local scale, human-mediated dispersal is important. At the state scale, roads are the primary human factor associated with EAB spread compared with campgrounds or the wood products industry.

Step 3: The financial impact of EAB invasion will be substantial to individual members of a community, involving the costs of removal and replacement of dead trees. Using a CGE, we estimate that the welfare loss to the states of Ohio and Michigan will exceed $110 million annually. These losses will affect mostly individuals earning over $25,000.

Step 4: Eradication efforts for EAB were expensive and rarely successful, though formal evaluation of the effectiveness of eradication programs has been minimal. Eradication of long-distance outbreaks in Ohio would have slowed the spread by several years, yet the entire state is expected to be infested within 20 years.

Step 5: Slowing the spread of EAB at the beginning of its invasion would have been worth more than $1 billion. The total amount it is worth for management actions is also dependent on the agency (federal vs. state in this case) making the decision.

Emerald ash borer will continue to spread regardless of human intervention. However, we have demonstrated that slowing its spread can reduce the rate of impact and eventual welfare loss.

Policy Implications

Our analysis of the emerald ash borer invasion addresses several complex issues in generating invasive species policy, such as coordinating uncertainty in ecological processes and uncertainty in the timing and magnitude of control investment or effort. Addressing each of these issues highlights several implications for policy makers. First, policy makers should encourage investment in models of long-distance spread. The spread of invasive species is often a result of multiple vectors, including natural and anthropogenic mechanisms as well as an understanding of potential habitat. The anthropogenic forces that move invasive species frequently result in long-distance jumps, ultimately causing the overall rate of spread to rapidly increase. These vectors are often similar for multiple species, so any understanding of how one pathway, such as movement of firewood, affects the spread of a species will aid in predictions of other related species that have yet to be introduced into nonnative areas.

Second, understanding how resources that are impacted by invasive species are integrated into larger regional economies can influence policy decisions. Our analyses highlight that the damages of invasive species are probably not restricted to specific localities or limited demographic groups. Rather, because our economy is a complex network of sectors, the impact to one sector of the economy is likely to influence many other sectors, with some sectors perhaps benefitting even though the economy has an overall loss in welfare. Likewise, households with disparate income categories may be impacted differently. Thus, policy makers may need to balance their response because of how these different constituencies are impacted.

Third, our economic analyses highlight the importance for policy makers to recognize the temporal dynamics of an invasion. The real options framework employed in this work emphasizes the importance of timing for slowing or stopping the spread of invasive species. Policy decisions and their subsequent implementation occur at specific times and are subject to uncertainty and changes in damages. By responding quickly to an invasion, policy makers sacrifice the option of gaining more knowledge about a species and how to control it, which may reduce the cost of control. On the other hand if implementation is delayed, the damages incurred due to the invasive species are likely to increase.

Finally, prevention of invasive species in the first place should be a primary policy goal. We predict EAB will cause an annual loss of welfare of several million dollars for Michigan and Ohio. Although a focus on prevention and early eradication is not a new idea (see Simberloff 2003), our findings about the long-term economic impacts of EAB will hopefully help convince policy makers that prevention, monitoring, and early detection are well worth the cost (Leung et al. 2002).

Acknowledgments

This work was funded in part by the PREISM Program of the USDA (awarded to Bossenbroek, Iverson, and Finnoff) and a Sigma Xi Grant-in-Aid of Research (awarded to Croskey). This is publication No. 2014–001 from the University of Toledo Lake Erie Center.

Notes

1. Using an average growth rate for ash trees as 22.5 inches a year and an average maturity height of 885 inches leads to a rotation period of 40 years (Arbor Day 2009; Ohioline 2009; Treehelp.com 2009).

Literature Cited

Akbulut, S., A. Keten, I. Baysal, and B. Yüksel. 2007. Effect of seasonality on the reproductive potential of *Monochamus galloprovincialis* (Coleoptera: Cerambycidae) reared in black pine logs under lab conditions. *Phytoparasitica* 36:187–198.

BenDor, T. K., S. S. Metcalf, L. E. Fontenot, B. Sangunett, and B. Hannon. 2006. Modeling the spread of the emerald ash borer. *Ecological Modeling* 197:221–236.

Bossenbroek, J. M., C. E. Kraft, and J. C. Nekola. 2001. Prediction of long-distance dispersal using gravity models: zebra mussel invasion of inland lakes. *Ecological Applications* 11:1778–1788.

Croskey, A. K. 2009. Evaluating the costs of the emerald ash borer invasion in Ohio. Master's thesis, University of Toledo.

Dixit, A. K., and R. S. Pindyck. 1994. *Investment under Uncertainty*. Princeton, NJ: Princeton University Press.

Hushak, L. 2005. Economics of the forest products industry in Ohio. Unpublished manuscript.

Iverson, L. R., A. Prasad, J. Bossenbroek, D. Sydnor, and M. W. Schwartz. 2010. Modeling potential movements of the emerald ash borer: the model framework. In *Advances in Threat Assessment and Their Application to Forest and Rangeland Management*, edited by J. M. R. Pye, H. Michael, Y. Sands, D. C. Lee, and J. S. Beatty, 581–597. Portland, OR: US Department of Agriculture, Forest Service, Pacific Northwest and Southern Research Stations.

Iverson, L. R., A. Prasad, and M. W. Schwartz. 1999. Modeling potential future individual tree-species distributions in the eastern United States under a climate change scenario: a case study with *Pinus virginiana*. *Ecological Modelling* 115:77–93.

Iverson, L. R., M. W. Schwartz, and A. M. Prasad. 2004. How fast and far might tree species migrate in the eastern United States due to climate change? *Global Ecology and Biogeography* 13:209–219.

Leung B., D. M. Lodge, D. Finnoff, J. F. Shogren, M. A. Lewis, G. Lamberti. 2002. An ounce of prevention or a pound of cure: bioeconomic risk analysis of invasive species. *Proceedings of the Royal Society B* 269:2407–2413.

MacFarlane, D. W., and S. P. Meyer. 2005. Characteristics and distribution of potential ash tree hosts for emerald ash borer. *Forest Ecology and Management* 213:15–24.

McDermott, S. M. 2011. Economics with a view: invasive species ecosystem exter-
nalities. PhD diss., University of Wyoming.

Miles, P. D., G. J. Brand, C. L. Alerich, S. W. Woudenberg, J. F. Glover, and E. N.
Ezzell. 2001. The forest inventory and analysis database: database description
and users manual, version 1.0. GTR NC-218. St. Paul, MN: North Central Research Station, USDA Forest Service.

Muirhead, J. R., B. Leung, C. Overdijk, D. W. Kelly, K. Nandakumar, K. R. March-
ant, and H. J. MacIsaac. 2006. Modeling local and long-distance dispersal of
invasive emerald ash borer *Agrilus planipennis* (Coleoptera) in North America.
Diversity and Distributions 12:71–79.

Poland, T. M., and D. G. McCullough. 2006. Emerald ash borer: invasion of the
urban forest and the threat to north Americas ash resource. *Journal of Forestry*
104:118–124.

Prasad, A. M., L. Iverson, M. Peters, J. Bossenbroek, S. N. Matthews, D. Sydnor,
and M. Schwartz. 2010. Modeling the invasive emerald ash borer risk of spread
using a spatially explicit cellular model. *Landscape Ecology* 25:353–369.

Saphores, J. D., and P. Carr. 2000. Real options and the timing of implementation
of emissions limits under ecological uncertainty. In *Project Flexibility, Agency,
and Competition: New Developments in the Theory and Application of Real Options*, edited by M. J. Brennan and L. Trigeorgis. New York: Oxford University
Press.

Schwartz, M. W., L. R. Iverson, and A. M. Prasad. 2001. Predicting the potential future distribution of four tree species in Ohio, USA, using current habitat
availability and climatic forcing. *Ecosystems* 4:568–581.

Sétamou, M., J. S. Bernal, T. E. Mirkov, and J. C. Legaspi. 2003. Effects of snow-
drop lectin on Mexican rice borer (Lepidoptera: Pyralidae) life history parameters. *Journal of Economic Entomology* 96:950–956.

Sharov, A. A., and A. M. Liebhold. 1998. Bioeconomics of managing the spread of
exotic pest species with barrier zones. *Ecological Applications* 8:833–845.

Simberloff, D. 2003. How much information on population biology is needed to
manage introduced species? *Conservation Biology* 17:83–92.

Sims, C. 2009. Essays on the bioeconomic control of invasive species and forest
pests. PhD diss., University of Wyoming.

Sydnor, T. D., M. Bumgardner, and A. Todd. 2007. The potential economic impacts
of emerald ash borer (*Agrilus planipennis*) on Ohio, U.S., communities. *Arbori-
culture and Urban Forestry* 33:48–54.

Taylor, R. A., T. M. Poland, L. S. Bauer, K. N. Windell, and J. L. Kautz. 2006.
Emerald ash borer flight estimates revised [abstract]. In Emerald Ash Borer
and Asian Longhorned Beetle Research and Development Review Meeting,
October 29–November 2, 2006; Cincinnati, OH, FHTET 2007–04, edited by
V. Mastro, D. Lance, R. Reardon, and G. Parra, 10–12. Morgantown, WV: US
Forest Service, Forest Health Technology Enterprise Team.

CHAPTER TEN

Climate Change Challenges in the Management of Invasive Sea Lamprey in Lake Superior

James F. Kitchell, Timothy Cline, Val Bennington, and
Galen A. McKinley

Introduction

The invasive and parasitic sea lamprey (*Petromyzon marinus*) has contributed to major declines of many native fish populations in the
Great Lakes. Control programs now maintain lamprey populations at low
levels. As a result, many fish populations have increased, thereby restoring
the socioeconomic benefits of their fisheries. In Lake Superior, recovery of
host species during the period 1960–1980 also produced a major increase
in growth responses for surviving lamprey. Superior is the most rapidly
warming lake on the planet. Since 1980, climate warming has increased sea
lamprey growth rates with concomitant increase in fecundity.

Sea lamprey effects on native lake trout have been the dominant focus of research and management. This analysis expands attention to the
interaction of climate change and lamprey effects for two other host species, one introduced and the other native. We focus on Chinook salmon
(*Oncorhyncus tshawytscha*), an introduced species that supports a major
recreational fishery, and white sucker (*Catostomus commersoni*), an abundant native species. We couple bioenergetics models with a general circulation model of the lake to provide spatial analyses of lamprey growth
response and impacts. Combined, these provide regional management
guidance that can increase the effectiveness of lamprey control programs
in the face of warming Lake Superior temperatures.

Many readers of this diverse volume will be relatively unfamiliar with the history of Great Lakes issues and the biology, management, and trophic interactions involving the sea lamprey. Accordingly, a glossary is provided for terminology used in this context. This chapter offers an evaluation of challenges created for resource managers given a recent increase in average temperatures and in the interannual variability of temperature conditions, and their association with increase in the observed size of adult sea lampreys.

First, we review the history of changes in the Lake Superior ecosystem due to historical effects of commercial fisheries exploitation, invasion by sea lamprey, responses of selected fish species vulnerable to sea lamprey, and the recent increase in water temperature owing to climate change. Second, we apply modeling tools to analyze sea lamprey effects on selected host species as associated with variability in time and space of water temperature changes observed since 1980. We use strong contrast in temperature to evaluate sea lamprey growth responses in a relatively recent cool year (1997) during this overall period of increased temperature, and 1998, the hot El Niño year. Our results demonstrate that climate effects amplify the feeding rates and increase host mortality owing to larger lampreys. Lastly, we discuss how warming waters and increasing interannual variability may pose several challenges to lamprey management in the future.

Background

Sea Lamprey in the Great Lakes

Symposium volumes recount profound changes in the Laurentian Great Lakes owing to sea lamprey invasion (Smith 1980; Smith and Tibbles 1980; Christie and Goddard 2003). The sea lamprey is an ancient cartilaginous fish native to the North Atlantic Ocean. It has an anadromous, semelparous (see glossary) life history that includes parasitic feeding on hosts followed by migrations into freshwater streams or rivers where spawning occurs and death of adults follows. The larval stage (ammocete) burrows into soft sediments where it feeds for 3–17 years by filtering suspended microorganisms. Growth rates and abundance of ammocetes depend on stream productivity, spawning lamprey size and abundance, and population dynamics that respond to droughts and floods. At a size of 120–160 mm and weight of 3–10 grams, ammocetes undergo metamorphosis and develop a suctorial oral disc encircled with teeth. The resulting "transformer" life-history stage then migrates downstream to lakes or the ocean, attaches to host fishes, rasps a hole in the host's skin and feeds parasitically on blood and soft tissues. The adult lamprey stage persists for 12–24 months, then it ceases feeding and completes its spawning migration. The 12-month length of the adult stage is viewed as the most common duration (Moody et al. 2011).

Sea lamprey first invaded the upper Laurentian Great Lakes in the early 1800s, soon after completion of the Erie and Welland Canals. That allowed anadromous adult lampreys to circumvent the obstacle of Niagara Falls. In the century that followed, sea lamprey spread to each of the Great Lakes and was first reported in Lake Superior in 1949 (Scott and Crossman 1973). Continuing rise in lamprey abundance coincided with declining host abundances owing to both commercial fisheries and lamprey effects on host populations. That was especially documented through population declines of fisheries for lake trout, a native demersal or bottom-dwelling fish, and among the large fishes that are preferred hosts of sea lamprey. This combination of new mortality agents caused extirpation of the lake trout populations in Lake Ontario, Lake Erie, Lake Michigan, plus all but the northern waters of Lake Huron and the western waters of Lake Superior. This catastrophic outcome caused economic collapse for hundreds of coastal fishing communities (Kitchell and Sass 2008) and evoked a US-Canada treaty that formed the Great Lakes Fishery Commission (GLFC 2012) in 1955.

The GLFC was charged with two major responsibilities, namely control of sea lamprey populations and restoration of native fish communities. By 1958, use of a lampricide that effectively kills the ammocete life stage began in Lake Superior natal streams and gradually expanded to spawning streams of the eastern lakes over the following two decades. The lampricide must be applied at regular intervals because treatments do not result in total removal of ammocetes. Surviving ammocetes still produce adult lampreys now estimated at 5%–10% of former maximum abundances before use of lampricides. The lamprey control program continues to seek alternative methods such as barrier dams that inhibit upstream migration by adults, release of sterile males, and chemical attractants and repellants (Christie and Goddard 2003), but the use of lampricides remains the main control practice. The recovery of native lake trout in Lake Superior began during the 1960s after the successful management of lamprey and has been very dramatic (Bronte et al. 2003), although less successful in other lakes.

Sullivan and Adair (2010) provide an extensive review of lamprey control work for each of the Great Lakes. Practices in both US and Canadian waters include surveys of ammocete abundance in scores of streams, rivers, and lentic areas (i.e., still waters such as lakes, ponds, and swamps) for hundreds of habitats followed by lampricide treatment of those habitats where sizes of ammocetes suggest imminent emergence and transformation to adult feeding morphology. Subsequent surveys evaluate the efficiency of treatment in removing ammocetes, including use of mark-recapture methods to estimate abundance of spawning adults in selected waters. For Lake Superior, adult lamprey abundances are commonly monitored in a subset of rivers and streams where spawning occurs each year. The 12–20 monitored rivers are distributed across the east-west coastal waters of the United States. Although some rivers are regularly monitored, efforts differ among years in response to new research survey objectives and budget constraints. Estimated abundances of adult sea lampreys since 1980 are presented in Fig. 10–1.

Review of Lamprey Studies

There is a substantial body of experimental evidence (Farmer and Beamish 1973; Farmer et al. 1977; Swink 1990), field surveys (Christie and Goddard 2003), and theoretical literature about lamprey-host interactions (Bence et al. 2003). After emerging from natal streams, small lampreys attach to a diversity of host species and function as parasites. Multiple attacks are common. Some host mortality (about 25%) owes to infection of lamprey-induced wounds (Swink and Hansen 1989; Swink 1990). Large lampreys prefer larger hosts and can feed at rates that exceed host blood renewal rate. Host death is a consequence and the lamprey effects are those of a predator. If feeding rates exceed blood renewal rates of the host and the parasite switches to an alternative host, larger lampreys may also engage in more frequent attacks and can produce higher wounding rates (Madenjian et al. 2008).

Thus, the importance of lampreys as agents of mortality depends on two ratios—relative abundance of hosts per lamprey and lamprey size relative to that of the host. In general, about half of lamprey attacks result in death of the host species and an individual sea lamprey kills approximately 18 kilograms (40 pounds) of host fishes per year. That estimate is generally based on lamprey growth, yet substantial variability in death due to attacks owes to attachment sites and secondary infections (Madenjian et al. 2008). A study conducted in Lake Ontario revealed that sea lamprey attacks accounted for approximately half of total mortality rates for adult lake trout (Bergstedt and Schneider 1988).

Climate Change

The growing body of climate change evidence (Hansen et al. 2006) includes declining ice cover from long-term records of ice phenology as a climate proxy (Magnuson et al. 2000) and increases in water temperatures in large, deep lakes (Hampton et al. 2008; Dobiesz and Lester 2009). Lakes can serve as integrated indicators of accelerated rates of change in annual cycles when the extent and duration of ice cover alter albedo (i.e., reflective power) and local weather effects (Austin and Colman 2008; Desai et al. 2009). Austin and Colman (2008) document water temperature trends in Lake Superior that represent some of the most dramatic, recent changes. In the period of 1900–1980, lake-wide average surface water temperatures were in the range of 8°–9°C. Beginning in about 1980, warming has accelerated, resulting in current average water temperatures in the range of 11°–12°C. This increase is much greater than the rate of change in regional air temperature and is directly associated with reduced ice cover (Austin and Colman 2007). Over the past three decades, ice cover has been reduced by about half and wind velocities have increased (Desai et al. 2009). Although highly variable, annual duration of positive thermal stratification (see glossary) has generally increased.

By comparison, average annual rates of warming due to climate change in many lakes are in the range of only 0.1°–0.3°C per decade (Coats et al. 2006). Since 1985, however, rates of warming for Lake Superior are significant in that it is the world's largest lake by area and is warming at 1.21 (±0.68)°C per decade—the most rapid warming rates reported to date (Desai et al. 2009). Climate change effects can alter species distributions, interactions, and ecosystem functions (Carpenter et al. 2006). The observed rapid warming of Lake Superior provides opportunity for tests of climate-altered species interactions.

The average size of adult sea lamprey has steadily increased from about 150 grams in 1960 to greater than 200 grams in recent records (Jorgensen and Kitchell 2005b). This increase accompanied the observed warming since 1980 and a several-fold increase in abundance of native lake trout subspecies. Both the Chinook salmon and lean lake trout represent primary hosts (Bronte et al. 2003) and have experienced an increase of wounding and mortality effects by the larger sea lamprey (Moody et al. 2011).

Fisheries Management

Stocking of juvenile salmonids increased during the 1960s in an effort to restore commercial and recreational fisheries. Pacific salmon species (Chinook, coho, and steelhead) and brown trout were a primary focus. In addition, lake trout were stocked as part of the general management goal of restoring native fish populations. Stocking rates rose for Lake Superior until the mid-1980s, then declined as natural reproduction by native lake trout increased and was deemed sufficient to sustain restoration (Bronte et al. 2003). Monitoring programs were widely developed as a means for assessing lamprey effects on lake trout. The stocking of salmon and lake trout continues in other lakes but natural reproduction by salmon has increased in all lakes. Salmon abundance is monitored through recreational fisheries, but lamprey wounding data are not regularly collected.

On the basis of catch rates of commercial fisheries in Lake Superior, relative abundance of lake trout increased approximately ten-fold during the period of 1960–1985 (Bronte et al. 2003). Two lake trout subspecies or morphotypes are common—the nearshore or "lean" form and the deepwater or siscowet form. Siscowet abundance now substantially exceeds that of the lean form and the latter shows lessened rate of increase in recent years (Bronte et al. 2003).

The lake trout morphotypes differ in two ecologically important ways. Lean lake trout are the primary target of recreational and commercial fisheries because their nearshore habitat preference makes them more vulnerable to fisheries. During thermal stratification, their preferred habitat includes a preferred temperature of about $10°$ C while that for siscowet corresponds with the colder temperatures ($4°–5°C$) of deep water habitats (Moody et al. 2011). Both forms of lake trout are monitored for lamprey attack frequencies. At comparable sizes, siscowet exhibit a much higher wounding frequency than lean lake trout. As known from laboratory studies (Farmer et al. 1977) and the bioenergetics models based on laboratory results (Kitchell and Breck 1980), lamprey at $10°C$ would feed more than twice as rapidly as those at $4°C$. This suggests a greater mortality rate per attack for lean lake trout owing to higher lamprey feeding rates in their preferred habitat where temperatures are warmer. Bergstedt and Schneider (1988) found the greatest abundance of lamprey-induced mortality for lake trout carcasses at depths corresponding to preferred temperatures of lean lake trout.

Recent Changes in Lamprey Abundance and Size

As stated above, methods for estimating adult lamprey abundance derive from mark-recapture and catch per unit effort in the subset of rivers or streams where monitoring programs occur during the spring runs of spawning-phase adults. Owing to budgetary constraints, monitoring is not possible for many of the known spawning streams and rivers. About one-third of major spawning sites are monitored and the program changes periodically (Hansen and Jones 2008). Targeted systems include sustained efforts both on a small subset of rivers, sites where evidence of outbreaks or where new ammocete populations may have developed, and on those where new treatment methods have been developed (e.g., sterile male releases, attractant or repellant applications, and newly constructed barrier dams). Adult lamprey abundance in Lake Superior has been variable since 1980 but generally averages about 75,000 spawners per year (Mullett et al. 2003). Abundance data are presented for the period of 1980–2009 (Fig. 10–1). Estimates range from less than 30,000 to more than 300,000 adult lamprey. Many estimates of density are in excess of the management goals represented by the target and confidence intervals derived from monitoring wounding rates on lean lake trout. Interannual differences in adult lamprey populations are highly variable and remain relatively unexplained except for the prospect that variable recruitment of transformers derives from the conditions that regulate ammocete populations. In addition, there is no apparent correspondence with similar variability in other Great Lakes. In other words, mysteries remain about processes that regulate adult lamprey abundances (Hansen and Jones 2008). Nevertheless, larger adult lampreys have higher fecundity and the general changes in abundance increase in proportion.

As the control program expanded, adult lamprey size increased by about 20%, which corresponds with increased abundance of host species, and then leveled off during 1980–1985 (Bronte et al. 2003). A second phase of increased lamprey size is apparent since 1985 up until the most recent data, a period of variable but consistently high relative abundance of hosts and an additional increase of 10%–15% in adult lamprey sizes. This increase in growth rates corresponds with rapid increase in water temperatures (Fig. 10–2). In general, host abundance is now relatively high and lamprey abundance is much lower owing to the control program, yet adult

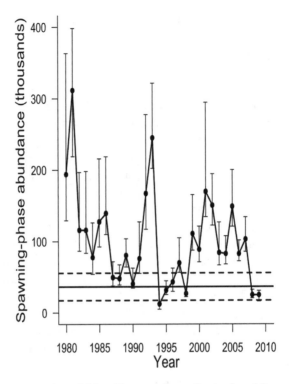

FIG. 10–1. Annual lake-wide population estimates for adult sea lamprey in Lake Superior for 1980–2009 based on mark-recapture methods (redrawn from Sullivan and Adair 2010). Error bars represent 95% confidence intervals. The horizontal solid line indicates the management target abundance with 95% confidence intervals (dashed lines).

lamprey size continues to increase and wounding rates on large lake trout have increased (Ebener 2007). We propose that this increase in host abundance combined with the effects of temperature change is responsible for increasing lamprey feeding and observed growth. An independent way to evaluate that prospect is to test for effects on other host species associated with the changes in temperature conditions.

Methods

As a complement to the work on lake trout as hosts (Moody et al. 2011), this analysis uses Chinook salmon (*Oncorhyncus tshawytscha*) as repre-

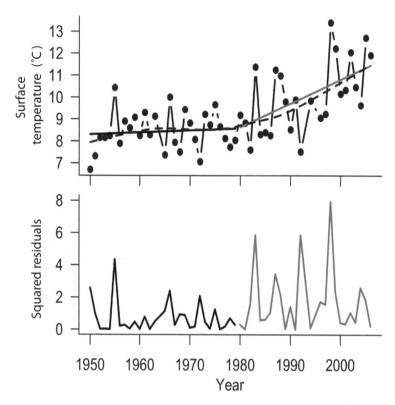

FIG. 10–2. Top panel: Changes in average annual surface temperatures for Lake Superior (re-drawn from Austin and Coleman 2008) during the period 1950–2006. The dashed line represents a LOESS model. The two lines are linear regressions for the periods before (darker line) and after (lighter line) 1980. Bottom panel: Squared residuals calculated from least squares regressions of the periods 1950–1979 and 1980–2006, line shading as for top panel. See also color plate.

sentative of pelagic hosts and white sucker (*Catostomus commersoni*) as representative of a demersal, nearshore host. Both species have higher thermal preferences than lake trout and, therefore, may be more indicative of climate change effects.

Our analyses of climate change effects are based on daily records of Lake Superior surface temperature data from the National Oceanic and Atmospheric Administration (NOAA) mid-lake buoy and a water temperature modeling approach developed by Bennington et al. (2010) that expands the temperature estimates across space and time for the entire Lake Superior system. Our analyses of temperature effects on sea lamprey growth focus on the contrasts of thermal conditions for 1997, a relatively

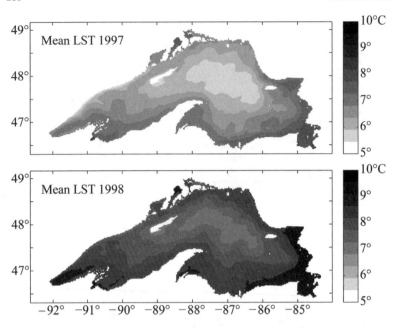

FIG. 10–3. Average annual surface water temperatures for Lake Superior during 1997 (top panel) and 1998 (bottom panel) based on the model of Bennington et al. (2010). Water temperatures are represented in the color scale below. Units of latitude and longitude are represented in y-axis and x-axis, respectively. See also color plate.

"cool" year, and 1998, a hot year (Fig. 10–3). These two years illustrate the strong and increasing interannual variability in Lake Superior.

Interannual temperature variability is substantial in Lake Superior. High temperature extremes have the potential to produce highly fecund lampreys and increase future lamprey populations. Mean annual surface temperatures taken from Austin and Coleman (2008) are presented in Fig. 10–2 for the periods that preceded and then followed the beginnings of climate warming. We tested for increased interannual variability by calculating squared residuals from least squares regressions of the average annual surface temperatures from the prewarming and postwarming periods (Fig. 10–2). While the mean annual temperatures have increased by more than 3°C, the magnitude of variability has doubled since about 1980 and represents the increase in variability generally expressed in current climate change models.

We developed estimates of lamprey feeding and growth rates based on the "Wisconsin" bioenergetics modeling tools (Hanson et al. 1997; Ney

1990) that have been widely employed (Hansen et al. 1993; Hartman and Kitchell 2008). The sea lamprey versions of this modeling approach are described in a series of sources and applications (Kitchell and Breck 1980; Kitchell 1990; Madenjian et al. 2003; Jorgensen and Kitchell 2005a, b; and Moody et al. 2011).

The bioenergetics model uses initial weight, final weight, known diet composition, physiological parameters, and annual thermal history to calculate changes in feeding rate required to balance energy budgets. The result is expressed in a value of "P" which scales from 0 to 1.0 as the proportion of maximum possible feeding required to accomplish the observed weight of spawning adult lamprey. This "P" value is comparable with that used by Kitchell and Breck (1980) and is quite different from that commonly used in estimating probabilities for statistical analyses. For example, a P of 0.8 indicates that an individual lamprey fed at 80% of maximum possible feeding rate in order to grow from the initial weight of a transformer to final adult weight over a specific period of time. Values of P can be specified to estimate achieved growth under contrasting conditions or estimated from observations of known growth. Parameters for the lamprey model are those used by Moody et al. (2011). Initial weight was set at 10 g for transformers on the basis of recent monitoring results from Lake Superior tributaries. Final weights were from data sets maintained by the US Fish and Wildlife Service as part of the GLFC monitoring programs.

Chinook salmon was chosen as a representative of pelagic habitats. Chinook salmon also occupy warmer waters than lake trout during periods of thermal stratification and have been widely used in other studies based on bioenergetics modeling (Stewart and Ibarra 1991; Negus et al. 2008). As a representative of a nearshore, demersal host, we chose white sucker for three reasons: 1) The preferred temperature of white sucker contrasts with that of the pelagic Chinook salmon and demersal lake trout. 2) As a nearshore species, white sucker is among the potential host species immediately available as transformers emerge from natal streams. Host choice after emergence increases the prospect that lamprey will subsequently be associated with members of that host species as thermal habitat partitioning develops during summer months. 3) White sucker was the focus of pioneering laboratory studies on temperature effects and host mortality owing to sea lamprey (Farmer and Beamish 1973).

Bioenergetics models commonly use an estimate of optimum temperature, "Topt," for growth (Hanson et al. 1997) based on laboratory studies, temperatures recorded by archival tags attached to individual fish (Moody et al. 2011), or depth distribution based on hydroacoustics used in field studies. For Chinook salmon, Stewart and Ibarra (1991) assumed Topt in Lake Michigan as 15°C, while Madenjian et al. (2004) report archival tag temperatures of 13°–17°C in Lake Huron, and Negus et al. (2008) used maximum temperatures of 8°–12°C for salmon in different ecoregions of Lake Superior. We chose 13°C as the most appropriate representation of Topt for Chinook salmon in Lake Superior. We chose a preferred temperature of 18°C for white sucker based on laboratory studies in thermal gradients (Reynolds and Casterlin 1978). Parameters for energetics models of both species derive from the compilations presented in Hanson et al. (1997). Kitchell and Breck (1980) estimated optimal temperatures for lamprey feeding and growth at 18°C. Thus, neither host provided thermal histories that exceeded the energetics constraints of lamprey growth.

Results and Discussion

The Lake Superior temperature model (Bennington et al. 2010) provides an estimate of days at preferred temperatures for both host species in cold year (1997) and warm year (1998) as presented in Fig. 10–4 for the paired comparisons of contrasting interannual temperature effects and host species. In both years, warming initially develops in nearshore habitats and expands into the pelagic habitats. Owing to latitudinal and bathymetric variations, warmer surface waters accumulate on southerly shores. Sea lampreys attached to white sucker hosts in nearshore habitats will consequently have a very different annual thermal history than those attached to the more pelagic Chinook salmon.

Two parsimonious approaches are available for evaluating climate change effects on fish growth rates. One would be based on observed changes in lamprey growth and adult sizes. Unfortunately, available lamprey growth data do not allow distinction of hosts chosen. Accordingly, we chose the alternative of assuming comparable feeding and growth rates for lampreys, known preferred temperatures, and assumed but contrasting thermal histories for both host species (Magnuson 2010). That assumption was used in testing temperature effects by fixing a P value of 0.8 for lamprey feeding on both host species. This P value represents an overall bioenergetics model estimate for sea lamprey growth to adult sizes over a 12-month parasitic feeding period (Moody et al. 2011). Lamprey growth, feeding rates, and annual thermal histories are presented for both years and the alternative host species in Fig. 10–5.

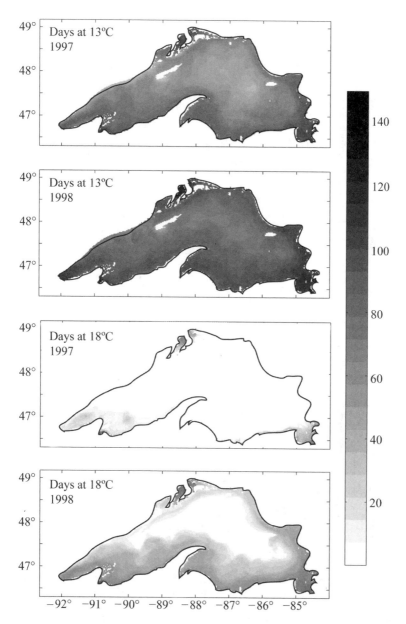

FIG. 10-4. Distribution of estimated days at preferred temperatures for Chinook salmon (top two panels) and white sucker (bottom two panels) as hosts for sea lampreys during 1997 and 1998 based on the model of Bennington et al. (2010). Chinook salmon are assumed to have a thermal preference of 13°C and white sucker thermal preference was assumed to be 18°C. Days per year are represented in the color scale. Units of latitude and longitude or represented in y-axis and x-axis, respectively. See also color plate.

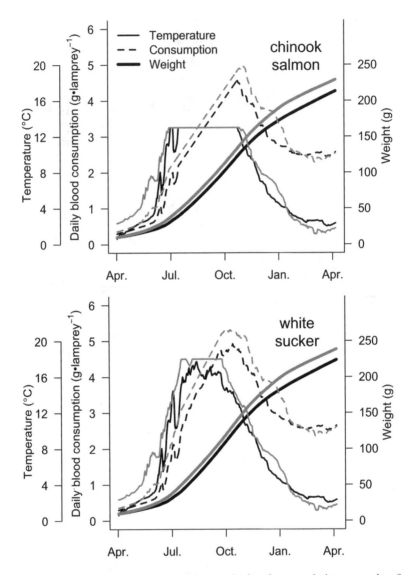

FIG. 10–5. Bioenergetics modeling results for growth of sea lampreys during 1997 and 1998 based on alternative hosts. Chinook salmon (top panel) were assumed to have a thermal preference of 13°C and white sucker (bottom panel) thermal preference was assumed to be 18°C. The dark lines are simulations for the cool year (1997) and the light lines simulations of the warm year (1998). See also color plate.

Adult lamprey sizes and feeding rates have increased in concert with the effects of interannual differences in thermal conditions. The general outcome is a 10 % increase in adult lamprey sizes owing to the contrast of thermal histories for 1997 and 1998. For both Chinook salmon and white sucker as host species, these results indicate the general effects of warming owing to climate change apparent since 1980. Results of simulated adult lamprey sizes are within the range of those observed (Fig. 10–5).

Regional and habitat differences in temperature effects are apparent in Fig. 10–3 and Fig. 10–4, but we do not know how host abundance and/or host selection may contribute to the observed differences in adult lamprey size. We tested for regional effects of host abundance by evaluating differences in adult size based on lamprey size data in the spring spawning runs that followed the growth year. Data were provided for the Middle, Misery, and Bad rivers by the US Fish and Wildlife Service. They included sizes of 507 adult lamprey in 1998 and 734 adults in 1999 from systems distributed across the range of spawning rivers along the southern reaches of Lake Superior. The range of adult sizes was 144–195 grams in 1998 and 181–227 grams in 1999. Average increase in adult size for the three rivers was 15%.

Lampreys do not exhibit homing behavior. Instead, adults migrate into stream and river systems based on chemical cues from ammocetes and seasonal water temperature conditions at the mouths of streams and rivers (Sorensen and Vrieze 2003). As an independent test of the assumed universal $P = 0.8$ for lake-wide lamprey growth, we assume that those sampled in monitoring programs derive from feeding in waters proximate to the system where adult lampreys have been captured. Percent differences in adult sizes included a 14% increase from the Middle River in southwestern Minnesota, an 11% increase from the Bad River that flows into Wisconsin's Chequamegon Bay, and a 30% increase from the Misery River located at the base along the southern shore of Michigan's Keweenaw Peninsula. Using the differences in recorded adult sizes for these rivers with large sample sizes in both years and the thermal histories for alternative hosts, we estimated differences in feeding rates (P values) required to accomplish differences in observed adult sizes for Chinook salmon as the host species. Simulations were conducted with annual thermal histories representative of the waters near each respective stream. Differences in outcomes owe largely to regional differences in temperature conditions and/or host abundances. Results are summarized in Table 10–1. Note that P values are more similar than differences in adult sizes and feeding rates,

TABLE 10–1 **Bioenergetics model results for sizes, total feeding rates, and P values of adult sea lampreys in each of three rivers. The P values represent proportions of maximum growth rate (scaled as 0 – 1) achieved in order to achieve observed weight (Hanson et al. 1997). We assumed a 12-month growth period for April 1997 through April 1998 or April 1998 through April 1999. Sizes are based on spawning adults captured in 1998 and 1999. Sample sizes are in parentheses.**

River	Adult size (g)		Blood Consumption (g)		P Value	
	1998	1999	1998	1999	1998	1999
Middle (Minn.)	181 (150)	206 (244)	728	876	0.80	0.79
Bad (Wisc.)	169 (78)	188 (46)	714	825	0.72	0.72
Misery (Mich.)	175 (241)	227 (136)	737	994	0.73	0.80

which provides additional evidence of important temperature effects at local and regional scales.

Adult lamprey sizes increased by 14% and total blood consumption increased by 20% for lamprey from the Middle River. Changes for the Bad River were an 11% increase in size and 15% increase in feeding rates. Those for the Misery River were a 30% increase in size and a 35% increase in feeding rates. The Misery River is located in southern shore regions where warming effects are most apparent. Unfortunately, we know too little about host abundances proximate to these rivers and the seasonal dynamics of distribution for lampreys attached to alternative hosts.

Armstrong and Schindler (2011) conducted a meta-analysis of P values for predatory fishes. For 649 entries, modes of frequency distributions were at about P values of 0.3 and 0.6. Values of 0.7–0.8 were very rare. Consequently, lamprey feeding and growth rates in Lake Superior are at very high levels. In general, the P values for Lake Superior lampreys reflect both the continuing high abundance of host species and the increase in thermal conditions that provided greater growth rates. In other words, our assumption of P = 0.8 for feeding rates on Chinook salmon or white sucker hosts is probably conservative.

The most recent report on sea lamprey abundance and management practices (Sullivan and Adair 2010) states that about 50% of lamprey attacks result in death of host fishes. Based on the results presented in Fig. 10–5, temperature differences between 1997 and 1998 can account for about 10%–12% of interannual differences in adult lamprey size and, therefore, at least an equivalent effect on host species. On the basis of sizes of returning adults in specific river systems there is even stronger evidence of differences in temperature and host effects. Unfortunately, those cannot be directly distinguished beyond the constraints of assumptions made above. Monitoring and temperature-tagging efforts applied to juvenile lampreys can help resolve these confounding effects (Bergstedt and Swink 1995) and can guide selection of monitoring and treatment efforts for river systems where larger adult lamprey are consistently apparent.

Climate Change and Host Abundance Interactions

Temperature changes in Lake Superior indicate a substantial and continuing warming trend. In addition, interannual variability in thermal conditions is increasing. Larger lampreys result from increasing thermal conditions, and that has potentially profound effects on predatory effects by lamprey and, therefore, on food web interactions that derive from greater host mortality rates. Because many host populations have generally recovered from previous effects of overfishing and sea lamprey invasion, host abundances are relatively high. Climate change effects increase water temperatures and combine with greater host abundances to create environmental conditions where lamprey growth rates are very high. Recent reports of density-dependent reductions in adult Chinook salmon growth rates (Negus et al. 2008) suggest increasing competition among piscivorous fishes and decline in forage abundance. Consequent changes in host:lamprey size ratios increase the prospect of host mortality caused by attacks of larger lampreys and, therefore, an overall increase in the effects of lampreys as apex predators in these systems.

Our modeling results offer a basis for advice to management and research efforts. Larger lampreys can have three important effects on populations of host species. First, vulnerability to direct mortality increases as the host-lamprey size ratio declines and lamprey feeding exceeds blood renewal rates for the hosts (Kitchell 1990). Second, attack rates should increase if more hosts are killed per attack and/or new hosts are sought. Indirect mortality owing to infection and wounding rates should increase. Those expectations are confirmed by recent reports from Lake Superior where wounding rates are increasing (Ebener 2007). Third, larger lampreys will have higher fecundity. The result will be an increase in egg production, the consequent increase in ammocete abundance per spawning female, and the positive feedback that would ultimately increase adult lamprey populations. Interannual variability in thermal conditions is increasing. Given that the parasitic phase of lamprey life history typically lasts only one year, increase in interannual variability in adult lamprey size and the consequent effects on host populations will elevate the challenges for lamprey management.

There is an important contradiction in outcomes wherein greater host mortality rates may not be equally reflected in the lake trout monitoring programs. Monitoring of host wounding rates is focused on live fish captured as an indicator of lamprey attacks. But, bigger lampreys kill more hosts and dead fish tend to sink, which makes them less likely to appear in monitoring efforts (Bergstedt and Schneider 1988). That could produce a decline in observed wounding rates while mortality rates are increasing. Testing for that possibility is among the challenges for research and management.

As evidenced by the historic collapse of fish communities in the Great Lakes, sea lamprey has functioned as an apex predator of these ecosystems. While abundance is generally controlled by the GLFC lamprey treatment programs, substantial interannual differences in adult abundance derive from a highly variable residual of surviving ammocetes that continues to fuel the population of adult lamprey. Climate warming allows those individuals to feed and grow at more rapid rates during their parasitic phase. As a result, the number of host fishes wounded or killed increases. As larger lampreys exert more predator-like effects, we should expect growing evidence for a trophic cascade that passes down through the Lake Superior food web (Kitchell et al. 2000). Challenges persist for the future of estimating lamprey effects. Again, positive feedback with a negative slope is developing in that more hosts fuel lamprey growth rates and larger lampreys have greater effects on population dynamics of host species and abundance of adult lampreys.

Policy Implications

There is an interesting irony and a conflict of objectives at the larger scales of lamprey management. Lampreys are the focus of ongoing conservation efforts in the river systems of the northwest US states (e.g., Oregon and Washington) where dam construction on major rivers, such as the Columbia River system, has impeded access to natal streams for anadromous adult lamprey species. Native Americans in this region view traditional spring harvest of lampreys as a cultural resource (Aquatic Community 2011). Reduction in adult lamprey abundance fuels debate about treaty rights. In this region, lamprey populations are not a focus of marine fisheries management. But, the reduced abundance of native lampreys is viewed as a violation of the cultural icon and an endangered natural resource taken into the courts for resolution.

In many rivers of western Europe, dams constructed in the previous century impede upstream access for migrating adult lampreys. In addition, urbanization and agricultural practices have caused decline in water quality to levels that severely limit habitat for ammocete production. Conservation initiatives include focus on restoration of lamprey populations as an indicator of success in restoration of river systems. Again, native lamprey populations are the target of endangered resource management on the basis of their role both as environmental indicators and as a critical resource in support of cultural traditions (Burton 2010).

The Great Lakes region of the United States and Canada offers very different challenges to sea lamprey management policies. There are growing efforts in support of stream and river habitat restoration through removal of antiquated dams in the Great Lakes watersheds (Hart and Poff 2003). Hundreds of small, century-old, and dysfunctional mill dams are deemed as safety hazards and as impediments to habitat connectivity for native species. Removal of dams is charged to natural resource management agencies and often funded by conservation groups promoting and supporting habitat restoration. These efforts open the prospects for development of new spawning and natal habitat for the invasive sea lamprey plus, of course, the associated costs required for survey efforts and sustained lamprey control.

In general, sea lamprey management in the Great Lakes has been an outstanding success. Development of lamprey population control and recovery of native fish populations are the major and sustained successes. Decline of adult sea lamprey abundance in Lake Superior (Fig. 10–1) is representative of those successes. Yet, the program faces a combination of new challenges. Among these challenges are climate change effects on thermal conditions in Lake Superior that have produced extended periods of warmer water (Cline et al. 2013). Growth rates of adult lampreys have increased owing to both more abundant hosts and the ongoing climate change effects that extend both the duration and extent of optimal growth during the parasitic phase of lamprey life history. Greater wounding and mortality rates for hosts follow. The larger size of adult lampreys also produces a greater reproductive output. Ongoing research and management actions offer promise for more effective lamprey control through development of chemical attractants or repellants, release of sterile males, and

specially constructed barrier dams that restrict access to lamprey spawning habitats without constraining recovery of other anadromous native
fish species. On the other hand, the recent increase in the broader goals
of dam removal projects focuses on century-old, dysfunctional dams and
may expand the potential spawning habitat for adult lamprey. These combine in the general challenge seeking explanation for control of variable
recruitment success of adult lampreys in the face of climate change effects
that can increase both droughts and floods in natal streams.

Recent economic conditions impose fiscal constraints for both US
and Canadian waters managed by federal and state agencies. Budget
limitations restrict both the research and management programs that can
respond to effects of sea lamprey in the Great Lakes and continue the
recoveries of multibillion dollar fisheries resources. Mediation of climate
change effects will take a substantial and continuing international effort.
Improvements in sea lamprey control for the Great Lakes are more trac-
table, proximate, and sustainable (Hansen and Jones 2008). Clearly, this
is one battle we can continue to win in the ongoing struggle of managing
invasive species.

Acknowledgments
This project was supported by grants to James Kitchell from the University of Wisconsin Sea Grant Institute and to Galen A. McKinley from
the National Science Foundation. We are deeply grateful to Jean Adams
(United States Geological Survey) and Jessica Barber (USFWS) for providing access to sea lamprey data. We thank Jeff Jorgensen, Greg Sass, and
Sture Hansson for help in assembling and conducting preliminary data
analyses. John Magnuson has provided continuing insights and advice in
our perception of climate change effects on Great Lakes fishes.

Glossary
In general, the terminology that follows pertains to fishes, their life histories, and their habitats and behaviors.

Ammocete: The larval life history stage of lampreys that live burrowed in
sediments of streams and rivers as filter feeders.

Anadromous: Life histories that involve migration of sexually mature
adults from marine environments or large lakes into rivers or streams
where spawning takes place.

Demersal: Fishes that live near bottom habitats, typically at greater
depths than in the nearshore environment.

Fecundity: Estimates of the number of eggs produced by mature female fishes when spawning.

Positive thermal stratification: Aquatic scientists specializing in physical processes refer to positive thermal stratification when the warm waters of lower density (>4°C) rise to surface depths of the water column. Commonly, this is simply referred to as "stratification"—that is, warmer waters on top of colder waters. When water temperatures drop to <4°C, colder water rises to the surface. This is termed *negative thermal stratification* and precedes ice formation.

Semelparous: Life histories that involve a single reproductive event usually followed by death of the spawning adults.

Transformer: The life history stage of lampreys wherein ammocetes change their morphology from filter-feeding then develop the oral morphology and digestive system that allows feeding as parasites.

Literature Cited

Aquatic Community. 2011. www.aquaticcommunity.com/news/.

Armstrong, J. B., and D. E. Schindler. 2011. Excess digestive capacity in predators reflects a life of feast and famine. *Nature* 476:84–88.

Austin, J. A., and S. M. Colman. 2007. Lake Superior summer water temperatures are increasing more rapidly than regional air temperatures: a positive ice-albedo feedback. *Geophysical Research Letters* 34(6):1–5.

Austin, J. A., and S. M. Colman. 2008. A century of temperature variability in Lake Superior. *Limnology and Oceanography* 53:2724–2730.

Bence, J. R., R. A. Bergstedt, G. C. Christie, P. A. Cochran, M. P. Ebener, J. F. Koonce, M. A. Rutter, and W. D. Swink. 2003. Sea lamprey (*Petromyzon marinus*) parasite-host interactions in the Great Lakes. *Journal of Great Lakes Research* 29 (Suppl 1):253–282.

Bennington, V., G. A. McKinley, N. Kimura, and C. H. Wu. 2010. General circulation of Lake Superior: mean, variability, and trends from 1979 to 2006. *Journal of Geophysical Research* 115:C12015.

Bergstedt, R. A., and C. P. Schneider. 1988. Assessment of sea lamprey (*Petromyzon marinus*) predation by recovery of dead lake trout (*Salvelinus namaycush*) from Lake Ontario, 1982–85. *Canadian Journal of Fisheries and Aquatic Science* 45:1406–1410.

Bergstedt, R. A., and W. D. Swink. 1995. Seasonal growth and duration of parasitic life stage on the landlocked sea lamprey (*Petromyzon marinus*). *Canadian Journal of Fisheries and Aquatic Science* 52:1257–1264.

Bronte, C. R., M. P. Ebener, D. R. Schreiner, D. S. DeVault, M. M. Petzold, D. A. Jensen, C. Richards, and S. J. Lozano. 2003. Fish community change in Lake Superior, 1970–2000. *Canadian Journal of Fisheries and Aquatic Science* 60:1552–1574.

Burton, A. 2010. Let's pray for the lampreys. *Frontiers in Ecology and the Environment* 8:392. www.frontiersinecology.org.

Carpenter, S. R., R. DeFries, T. Dietz, H. A. Mooney, S. Polasky, W. V. Reid, and R. J. Scholes. 2006. Millennium ecosystem assessment: research needs. *Science* 314:257–258.

Christie, G. C., and C. I. Goddard. 2003. Sea Lamprey International Symposium (SLIS II): advances in the integrated management of sea lamprey in the Great Lakes. *Journal of Great Lakes Research* 29 (Suppl 1):1–14.

Cline, T. J., J. F. Kitchell, and V. Bennington. 2013. Climate change expands the spatial extent and duration of preferred thermal habitat for Lake Superior fishes. *PLOS ONE* 8(4):e62279.

Coats R., J. Perez-Losada, G. Schladow, R. Richards, and C. Goldman. 2006. The warming of Lake Tahoe. *Climatic Change* 76:121–148.

Desai, A. R., J. A. Austin, V. Bennington, and G. A. McKinley. 2009. Stronger winds over a large lake in response to weakening air-to-lake temperature gradient. *Nature-Geoscience* 2:855–858.

Dobiesz, N. E., and N. P. Lester. 2009. Changes in mid-summer water temperature and clarity across the Great Lakes between 1968 and 2002. *Journal of Great Lakes Research* 35:371–384.

Ebener, M. P., ed. 2007. The state of Lake Superior in 2000. Great Lakes Fishery Commission Special Publication 07–02.

Farmer, G. J., and W. H. Beamish. 1973. Sea lamprey (*Petromyzon marinus*) predation on freshwater teleosts. *Journal of Fisheries Research Board Canada* 30:601–605.

Farmer, G. J., F. W. H. Beamish, and P. F. Lett. 1977. Influence of water temperature on the growth rate of landlocked sea lamprey (*Petromyzon marinus*) and the associated rate of host mortality. *Journal of Fisheries Research Board Canada* 34:1373–1378.

Great Lakes Fisheries Commission (GLFC). 2012. Protecting our fishery. http://www.glfc.org/.

Hampton, S. E., L. R. Izmesteva, M. V. Moore, S. L. Katz, B. Dennis, and E.A. Silow. 2008. Sixty years of environmental change in the world's largest freshwater lake—Lake Baikal, Siberia. *Global Change Biology* 14:1947–1958.

Hansen, G. J. A., and M. L. Jones. 2008. A rapid assessment approach to prioritizing streams for control of Great Lakes sea lampreys (*Petromyzon marinus*): a case study in adaptive management. *Canadian Journal of Fisheries and Aquatic Science* 65:2471–2484.

Hansen, J., M. Sato, R. Ruedy, K. Lo, D. Lea, and M. Medina-Elizade. 2006. Global temperature change. *Proceedings of the National Academy of Sciences USA* 103:14288–14293.

Hansen, M. J., D. Boisclair, S. B. Brandt, S. W. Hewett, J. F. Kitchell, M. C. Lucas, and J. J. Ney. 1993. Applications of bioenergetics models to fish ecology and management: where do we go from here? *Transactions of the American Fisheries Society* 122:1019–1030.

Hanson, P. C., T. B. Johnson, D. E. Schindler, and J. F. Kitchell. 1997. *Fish Bioenergetics 3.0.* Sea Grant Technical Report, University of Wisconsin Sea Grant Institute, Madison, WI. 119 pages.

Hart, D. D., and N. L. Poff. 2003. A special section on dam removal and river restoration. *BioScience* 52(8):652–655.

Hartman, K. J., and J. F. Kitchell. 2008. Bioenergetics modeling progress since the 1992 symposium. *Transactions of the American Fisheries Society* 137:216–223.

Jorgensen, J. C., and J. F. Kitchell. 2005a. Growth potential and host mortality of the parasitic phase of the sea lamprey (*Petromyzon marinus*) in Lake Superior. *Canadian Journal of Fisheries and Aquatic Science* 62:2343–2353.

Jorgensen, J. C., and J. F. Kitchell. 2005b. Sea lamprey (*Petromyzon marinus*) size trends in a salmonid stocking context in Lake Superior. *Canadian Journal of Fisheries and Aquatic Science* 62:2354–2361.

Kitchell, J. F. l990. The scope for mortality caused by sea lamprey. *Transactions of the American Fisheries Society* 119:642–648.

Kitchell, J. F., and J. E. Breck. 1980. A bioenergetics model and foraging hypothesis for sea lamprey. *Canadian Journal of Fisheries and Aquatic Science* 37:2159–2168.

Kitchell, J. F., S. P. Cox, C. J. Harvey, T. B. Johnson, D. M. Mason, K. K. Schoen, K. Aydin, et al. 2000. Sustainability of the Lake Superior fish community: interactions in a food web context. *Ecosystems* 3:545–560.

Kitchell, J. F., and G. G. Sass. 2008. Great Lakes ecosystems: invasions, food web dynamics and the challenge of ecological restoration. In *Ecological History of Wisconsin*, edited by D. Waller and T. Rooney, 157–170. Chicago: University of Chicago Press.

Madenjian, C. P, B. D. Chipman, and J. E. Marsden. 2008. New estimates of lethality of sea lamprey (*Petromyzon marinus*) attacks on lake trout (*Salvelinus namaycush*): implications for fisheries management. *Canadian Journal of Fisheries and Aquatic Science* 65:535–642.

Madenjian, C. P., P. A. Cochran, and R. A. Bergstedt. 2003. Seasonal patterns in growth, blood consumption, and effects on hosts by parasitic-phase sea lampreys in the Great Lakes: an individual-based model approach. *Journal of Great Lakes Research* 29 (Suppl 1):332–346.

Madenjian, C. P., D. V. O'Connor, S. M. Chernyak, R. R. Rediske, and J. P. O'Keefe. 2004. Evaluation of a chinook salmon (*Oncorhynchus tshawytscha*) bioenergetics model. *Water Resources* 635:627–635.

Magnuson, J. J. 2010. History and heroes: the thermal niches of fishes and long-term lake dynamics. *J Fish Biol* 77:1731–1744.

Magnuson, J. J., D. M. Robertson, B. J. Benson, R. H. Wynne, D. M. Livingstone, T. Arai, R. A. Assel, et al. 2000. Historical trends in lake and river ice cover in the northern hemisphere. *Science* 289:1743–1746.

Moody, E. K., B. C. Weidel, T. D. Ahrenstorff, W. P. Mattes, and J. F. Kitchell. 2011. Evaluating the growth potential of sea lampreys (*Petromyzon marinus*) feeding on siscowet lake trout (*Salvelinus namaycush siscowet*) in Lake Superior. *Journal of Great Lakes Research* 37:343–348.

Mullett, K. M., J. W. Heinrich, J. V. Adams, R. J. Young, M. P. Henson, R. B. McDonald, and M. F. Fodale. 2003. Estimating lake-wide abundance of spawning-phase sea lampreys (*Petromyzon marinus*) in the Great Lakes: extrapolation from sampled streams using regression models. *Journal of Great Lakes Research* 29 (Suppl 1):240–252.

Negus, M. T., D. R. Schreiner, T. N. Halpern, S. T. Schram, M. J. Seider, and D. M. Pratt. 2008. Bioenergetics evaluation of the fish community in the western arm of Lake Superior in 2004. *North American Journal of Fisheries Management* 28:1649–1667.

Ney, J. J. 1990. Trophic economics in fisheries: assessment of demand-supply relationships between predators and prey. *Reviews in Aquatic Sciences* 2:55–81.

Reynolds, W. W., and R. E. Casterlin. 1978. Behavioral thermoregulation and diel activity in white sucker (*Catostomus commersoni*). *Comparative Biochemistry and Physiology* 59A:261–262.

Scott, W. B, and E. J. Crossman. 1973. *Freshwater Fishes of Canada*. Bulletin 184, Fisheries Research Board of Canada, Ottawa, Canada. 966 pages.

Smith, B. R. 1980. Introduction to the proceedings of the 1979 Sea Lamprey International Symposium (SLIS). *Canadian Journal of Fisheries and Aquatic Science* 27:1585–1587.

Smith, B. R., and J. J. Tibbles. 1980. Sea lamprey (*Petromyzon marinus*) in Lakes Huron, Michigan and Superior: history of invasion and control, 1936–1978. *Canadian Journal of Fisheries and Aquatic Science* 27:1780–1801.

Sorensen, P. W., and L. A. Vrieze. 2003. The chemical ecology and potential application of the sea lamprey migratory pheromone. *Journal of Great Lakes Research* 29 (Suppl 1):66–84.

Stewart, D., and M. Ibarra. 1991. Predation and production by Salmonine fishes in Lake Michigan, 1978–1988. *Canadian Journal of Fisheries and Aquatic Science* 48:909–922.

Sullivan, P., and R. Adair. 2010. *Integrated Management of Sea Lamprey in the Great Lakes*. 2009. Ann Arbor, MI: Great Lakes Fishery Commission.

Swink, W. D. 1990. Effect of lake trout size on survival after a single sea lamprey attack. *Transactions of the American Fisheries Society* 119:996–1002.

Swink, W. D., and L. H. Hansen. 1989. Survival of rainbow trout and lake trout after sea lamprey attack. *North American Journal of Fisheries and Aquatic Science* 9:35–40.

CHAPTER ELEVEN

Ecological Separation without Hydraulic Separation: Engineering Solutions to Control Invasive Common Carp in Australian Rivers

Robert Keller

Introduction

The wild ancestor of the common carp (*Cyprinus carpio*) originated in the Caspian Sea, then migrated naturally to the Black and Aral seas, dispersing east into Siberia and China and west as far as the Danube River (Balon 1995). Through human activity, carp spread throughout Asia and Europe as an ornamental and aquaculture species. Carp were released into the wild in Australia on a number of occasions in the nineteenth and twentieth centuries but did not become widespread until after a 1964 release from a fish farm into the Murray River near Mildura, close to the Victorian-South Australian border. The spread of carp throughout the Murray-Darling Basin coincided with, and was largely caused by, widespread flooding in the early 1970s.

The Murray-Darling Basin covers approximately 1 million square kilometers, about one-seventh of Australia's land area. The basin yields 50% of Australia's food production, and is home to about 180 indigenous fish species and about 25 exotic fish species.

Common carp is now the most abundant large freshwater fish in the Murray-Darling Basin and the dominant species in many fish communities in southeastern Australia. Recent river surveys in New South Wales

(Harris and Gehrke 1997; Koehn et al. 2000) found that carp represent more than 90% of fish biomass in some rivers and have reached densities of one fish per square meter of water surface. Carp also occur in Western Australia and Tasmania and have the potential to spread through many more of Australia's water systems. They can migrate any time of year, with some individuals moving as far as 230 km.

Nevertheless, despite the large proportion of carp in fish biomass, many of the claims regarding the environmental effects of carp are difficult to confirm because of the lack of information on waterway health before their introduction. For many waterways the decline in quality indicators—such as turbidity, excessive nutrients, and chemical pollution—took place before carp were introduced, through activities such as catchment clearing, removal of bankside vegetation, stream channelization, pesticide use, and overfishing of native species.

Carp feed by sucking soft sediment into their mouths. Food items are separated and retained, while the sediments are ejected back into the water. This habit (known as roiling) muddies the water. The suspended sediment causes a number of problems for native fish species, including clogging of their gills and inhibited visual feeding (Koehn et al. 2000).

Carp pose an economic threat by affecting industries that depend on pristine water quality and aquatic habitats. Such industries include domestic and irrigation water suppliers, agriculture, tourism, and commercial and recreational fisheries. Carp cause significant damage to aquatic plants and increase water turbidity, negatively impacting native aquatic fauna, habitat, and ecosystems. It is less clear what the impacts of carp are on native fish populations—many of which were in decline before carp became widespread. Carp carry a number of disease organisms. Some of these, such as the Asian fish tapeworm (*Bothriocephalus acheilognathi*), now occur in Australia, and may pose a serious risk to native fish.

The biology and ecology of carp are two major reasons why the species is such an important and successful vertebrate pest in Australia. Carp exhibit broad environmental tolerances and thrive in habitats modified by humans. In Australia and elsewhere, they have benefited from modification of river systems, including construction of dams and other barriers to fish movement, reduced river flows, and inundation of floodplains, changes that conversely have had major detrimental impacts on native fish.

Carp in Australia generally live 15–20 years and can grow as large as 1.2 m and 60 kg. Specimens weighing 10 kg are relatively common in southeastern Australia. Carp breed at 2–4 years of age, and usually spawn in late spring or early summer. The eggs develop rapidly, hatching within 2–6 days. A female carp may spawn several times in one season, producing up to 1.5 million eggs per year, depending on the size of the fish. Fertilization is close to 100%, but mortality of eggs and larvae is commonly high.

To date, carp control has mainly consisted of commercial harvesting or poisoning. Whilst these options may reduce carp numbers, and poisoning may occasionally eradicate them from isolated areas, other options are being explored for more widespread control. Biological approaches include the so-called "daughterless technology" in which the sex-determination gene is modified, resulting in the production of male offspring only (Grewe 1997; Thresher and Bax 2003). This technology is still under development and tests to date have shown that the gene is only partially inherited.

However, even if eventually successful, many years would elapse before large numbers of carp containing the modified gene were ready to be released into rivers. Furthermore, because carp have a very long life cycle, the approach would need to operate over 40–50 years, and be used in conjunction with other short-term control techniques, to have a lasting effect.

Thus, although biological control techniques may eventually provide a solution, there exists an urgent immediate need to control the movement of carp within and between river systems to prevent their further expansion.

This chapter examines the use of structural means to control and prevent the spread of carp. Two approaches are considered—the first relying on behavioral differences between carp and indigenous fish species, and the second relying on indiscriminate trapping of all species. In the following, each is discussed separately with reference to a case study. Generalized discussion and conclusions complete the chapter.

The Williams Carp Cage

The so-called Williams Cage owes its name to Allan Williams, an employee of Goulburn Murray Water, who developed the design after observing that carp jump and Australian native fish do not. The development of the cage and its current status have been described by Stuart et al. (2006a) and a brief description, drawn from this reference, is presented in the following.

A schematic of the cage is shown in Fig. 11–1. The cage is constructed from 40 mm galvanized steel covered with 25 mm square wire mesh. The downstream compartment (fish entrance) incorporates a cone-trap to reduce escapement of trapped fish, and spans the full width of the fishway channel. The second compartment, to hold fish that had jumped, is 1.2 m long, spans the full width of the fishway and usually operates at approximately 0.6 m water depth.

The height of the jumping baffle, separating the two compartments, was set to 0.15 m above the water surface. Observations of native fish behavior indicated that this was the minimum height required to prevent native fish negotiating the baffle by jumping. Observations of adult carp indicated that some fish were able to jump at least 0.5 m high.

Details of the native fish exit gate are shown in Fig. 11–2.

The results of field trials undertaken during 2002–2006 are summarized in Table 11–1. They show that the Williams cage separated 91% of adult carp into a confinement area. In contrast, of 19,641 native fish that entered the cage, more than 99.9% passed through it and exited the fishway.

The results in Table 11–1 indicate that the Williams cage is successful in separating carp from native species in a fishway setting. However, trapping 91% of adult carp in the field test implies that 9% passed through. Given the fecundity and hardiness of carp this 9% still represents a sufficiently large percentage to ensure viable spawning populations upstream.

The cage has also undergone limited testing in a riverine application (Stuart et al. 2006a). For this test, the cage was located within the river but below a weir. Carp were observed to be abundant below the weir with large schools of adults across the weir face and in the navigation chamber. However, the total catch consisted of 49 carp, three golden perch, seven bony herring, and one Murray cod. In addition 40 carp were caught from the lock chamber, from among approximately 10,000 carp, that were attracted by leaving the lock gates ajar for about four hours.

The lack of carp and native species in the Williams' cage seems related to unexpected fish behavior. The fish were observed to move downstream as the cage was lowered into place and then approached the cage slowly. Although large schools remained in proximity, very few entered the cage. This avoidance behavior was consistent across 10 days of trials, even after manipulation of local flows (high and low) and during day and night.

It is likely that the avoidance behaviour is associated with the fact that the fish can see the cage when it is located below a weir. This is an issue that requires further research into the construction and operation characteristics of the cage as well as behavioral characteristics of carp and

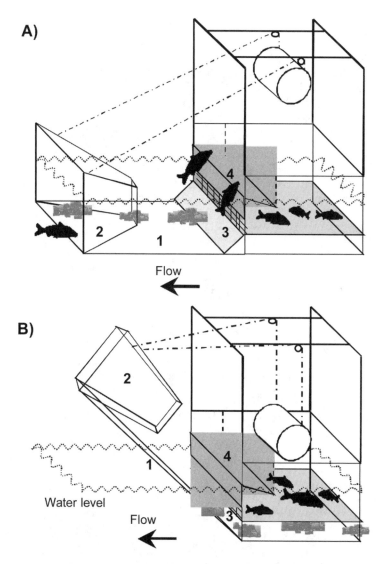

FIG. 11–1. Williams Carp Cage (Stuart et al. 2006a). A, the cage in the operating position. In this position, jumping carp (black fish symbols) are caught and separated from nonjumping Australian native fish (grey fish symbols). B, the cage in the raised position, whereby carp are trapped in the second compartment and native fish can swim through an exit gate below the second compartment. The important components are identified by number as 1) false lifting floor, 2) cone trap, 3) native fish exit gate, and 4) nonreturn slide to prevent carp returning to the native fish exit gate.

FIG. 11–2. Details of the native fish exit gate in the A) closed (operating) position and B) open (raised) position (Stuart et al. 2006a).

TABLE 11-1 **Aggregated results of Williams Carp Cage testing in a fishway (Stuart et al. 2006a).**

Fish Species	Number Entering Fishway	Number Trapped in Jumping Cage	Percentage in Jumping Cage (%)	Size Range (mm)
Carp (*Cyprinus carpio*)	569	517	90.86	244–710
Silver perch (*Bidyanus bidyanus*)	16,014	0	0	92–505
Golden perch (*Macquaria ambigua*)	2,917	2	0.07	94–560
Bony herring (*Nematalosa erebi*)	579	3	0.52	145–302
Murray cod (*Maccullochella peelii peelii*)	77	0	0	123–930
Trout cod (*Macullochella macquariensis*)	1	0	0	301
Freshwater catfish (*Tandanus tandanus*)	1	0	0	250

native fish. Certainly, at the present stage of development, the effectiveness of the Williams cage is limited to fishway applications.

Elimination of Carp from Environmental Flow Releases

This case study illustrates the use of an engineering structure to prevent the passage of carp from an infested lake into a downstream river.

In early 2001, carp were discovered in the Glenelg River in southwest Victoria. Prior to this, the Glenelg River had been considered one of the last carp-free large river systems in southeastern Australia. Subsequent surveys showed that a large carp population exists in the Rocklands Reservoir in the headwaters of the Glenelg River, but the fish has not become well established in the river downstream.

The Glenelg River system supports a diverse fish assemblage with high conservation significance. Twenty species of native freshwater fish and 26 estuarine species have been recorded there (Jackson and Davies 1983, DNRE 2000a, Appendix 7 cited in Arthur Rylah Institute for Environmental Research 2002). Eight species have special conservation significance and are variously listed as protected and/or endangered (e.g., the Victorian Flora and Fauna Guarantee Act in 1988, the Australian and New Zealand Environment and Conservation Council's *List of Threatened Vertebrate Fauna* in 2000, the Commonwealth's Environment Protection and Biodiversity Conservation Act [EPBC Act] in 1999).

An environmental flow may be defined as the provision of water to a river to maintain downstream ecosystems and their benefits where the river is subject to competing water uses and flow regulation (Dyson et al. 2008). The need to release water from Rocklands Reservoir for environmental flows means effective fish screening is crucial to reducing the risk of carp populations becoming established downstream. Since 2001, all environmental flow releases have passed through 2mm fixed mesh screens. These have proven to require high maintenance and to be unable to pass flows greater than about 65 ML/day, which is insufficient to meet the requirements of the river. The challenge in this study was to design and construct self-cleansing screens capable of passing up to 800 ML/day.

Effective carp screening at Rocklands Reservoir requires a number of features:

- Screens should operate within the Rocklands Outlet Channel (as close to the dam wall as possible).
- They must pass flows of 600 ML/day and be easily upgraded to 800 ML/day in the event that infrastructure changes at the dam wall allow improvements to environmental flows.
- They must be able to operate under very low upstream water level conditions.
- All life stages of carp must be screened (including eggs and larvae).
- Operational requirements must be minimal.

The approach utilized continuous deflective separation (CDS) technology, a form of indirect rather than direct screening whereby debris is deflected away from the screen rather than building up against it. The technology has been proven in stormwater applications but was untested in fish screening. For this reason, a major model study was undertaken to assess the technology. Following this study, a prototype system was developed and installed. In this section, the model study is discussed, followed by an evaluation of the subsequent prototype.

Model Study

A single CDS unit of approximately one-fifth the size of those planned for Rocklands Reservoir was used throughout the trials.

The study was done in two stages. Stage 1 involved development of the test arrangement, commissioning, and testing with simulated carp eggs. Stage 2 involved testing of the structure's ability to trap fertilized carp eggs

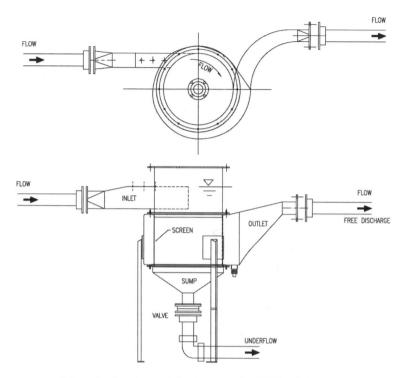

FIG. 11–3. Schematic of continuous deflective separation (CDS) unit. Note that the "Free Discharge" carries bypassed material and the "Underflow" carries captured material. Both pass through filters to ensure the capture of all solid material.

and small larvae, together with organic debris. Tests were conducted over a range of flow conditions representative of the full-scale situation.

The model CDS unit is shown schematically in Fig. 11–3. The bottom outlet was fitted with a valve so that the unit could be drained at the conclusion of each test. The upper and bottom outlets each drained into a filter that would capture any material coming through the unit. The filters were cone-shaped and made from stainless steel mesh with 0.9mm mesh size and supported within a drum. For all stage 2 tests, fine-mesh bags were inserted into the stainless steel mesh filters to ensure the capture of all solid material.

Owing to initial difficulties in sourcing carp eggs because of the time of year, stage 1 tests were run with egg substitutes—sago (tapioca), couscous, and caviar. These products were readily available and exhibited many of the properties of carp eggs, particularly being similar in size, density, and texture.

The stage 2 tests comprised four distinct phases. The first three phases involved testing with fertilized carp eggs alone, fertilized carp eggs attached to vegetation, and carp larvae. The carp eggs had a specific gravity of 1.03 and an average diameter of 1.5mm. The final phase involved testing with fertilized carp eggs once more, but with the outlet from the CDS unit significantly modified to create tranquil discharge conditions. This was done to ensure that observed destruction of the introduced material was due to the conditions within the CDS unit and not due to the turbulent conditions at the outlet.

The study's major requirement was accurate assessment of the unit's trapping ability across a range of flow rates. This required reasonable accuracy in the measurement of flow rates and very high accuracy in the measurement of weights of bypassed and captured eggs and larvae.

A formal testing technique was followed for each test to ensure consistency across all tests and accuracy of individual tests. Full details are provided in Keller and Associates (2005) and Keller et al. (2008).

A total of 34 runs were undertaken. The results are summarized in Table 11–2. The "% Loss" represents the % difference between the weight of the introduced sample and the sum of the weights of the bypassed and captured samples. In most cases, the loss represents the release of the liquid contents of carp eggs that are destroyed during their passage through the CDS unit screen.

The stage 1 tests showed an apparent loss of material ranging between 3% and 16% by weight. The couscous and sago are composed primarily of starch, which dissolves in water, thereby explaining the loss of material.

The stage 2—phase 1 tests are identified as run 13 to run 19. Substantial losses in the mass balance between the total weight of the bypassed and captured samples and the weight of the introduced sample was evident due to egg breakage and consequent passage of their liquid contents through the mesh screen. The bypassed carp eggs suffered extensive damage in their passage through the CDS unit and into the mesh filter. The breakage of the eggs is, in itself, a good outcome as it destroys the viability of any eggs passing through the CDS unit.

Fertilized carp eggs, being sticky, typically attach to vegetation. For this reason, tests were done with fertilized carp eggs attached to synthetic grass. These are identified as stage 2—phase 2 comprising runs 20 to 24. A feature of these results is the high evident loss rate for tests 20, 23, and 24. This was investigated and found to be due to substantial amounts of

TABLE 11–2 **Summary of results of all tests.**

Run	Flow Rate	Egg Type	Initial Weight (g)	% Captured	% Bypassed	% Loss
Stage 1 tests						
1	10	None - set up test				
2	20	None - set up test				
3	24	None - set up test				
4	10	Sago				
5	20	Sago	236.7	103.1	0.0	−3.1
6	10	Couscous	142.5	86.5	0.0	13.5
7	20	Couscous	125.0	84.0	0.0	16.0
8	24	Sago	190.1	93.6	0.0	6.4
9	24	Couscous	142.4	97.2	0.0	2.8
10	24	Caviar	49.0	93.7	1.0	5.3
11	2	Caviar	30.8	95.5	0.3	4.2
12	2	Couscous	85.3	88.3	0.0	11.7
Stage 2 tests – Phase 1						
13	10	Carp	61.1	77.1	0.2	22.7
14	2	Carp	75.0	86.7	0.7	12.7
15	24	Carp	75.0	72.4	7.1	20.5
16	20	Carp	75.0	92.9	5.9	1.2
17	10	Carp	75.0	90.5	1.1	8.4
18	2	Carp	75.0	97.6	0.7	1.7
19	10	Carp	75.0	90.4	0.5	9.1
Stage 2 tests – Phase 2						
20	10	Carp & grass	45.8	59.4	0.4	40.2
21	2	Carp & grass	40.0	84.8	0.5	14.8
22	10	Carp & grass	40.0	97.3	0.8	2.0
23	20	Carp & grass	40.0	68.5	2.5	29.0
24	24	Carp & grass	40.0	67.3	2.5	30.3
Stage 2 tests – Phase 3						
25	2	Larvae	17.3	67.1	1.2	31.8
26	10	Larvae	11.6	68.1	1.7	30.2
Stage 2 tests – Phase 4						
27	2	Carp (natural)	75.0	99.5	0.0	0.5
28	2	Carp (natural)	75.0	84.9	0.5	14.5
29	10	Carp (natural)	75.0	96.1	2.7	1.2
30	10	Carp (natural)	75.0	94.5	4.7	0.8
31	20	Carp (natural)	75.0	90.0	6.4	3.6
32	20	Carp (natural)	75.0	99.7	3.3	−3.1
33	24	Carp (natural)	75.0	101.9	7.1	−8.9
34	2	Carp (natural)	75.0	100.3	0.0	−0.3

the introduced sample being retained within the CDS unit and associated pipework, even after emptying. In demonstrating the performance of the unit, this is not an issue as the bypassed percentages are very low.

The stage 2—phase 3 tests with hatchlings were severely limited because of the small quantity available. For this reason, only two tests were done (runs 25 and 26). Again the results indicate very small bypassed quantities, but large apparent losses. The most probable reason for this loss is that the bypassed component contained only parts of the hatch-lings while the captured component contained whole hatchlings. It is evi-dent that the bypassed hatchlings were severely damaged by their passage through the test rig, releasing fluid contents which were lost in passage through the filter.

However, as for the carp eggs tested in phase 1, it is unclear if the dam-age occurred during passage through the CDS unit or was caused by the force of impact on the mesh filter. If the former is the reason, it is in-consequential since the hatchlings will no longer be viable upon passing downstream. The rig was modified by replacing the outlet pipe with a low-velocity outlet flume, which removed the impact of the bypassed material on the bypass filter.

The final tests were the stage 2—phase 4 tests, identified as runs 27 to 34. The results indicate one large loss and, surprisingly, some negative losses. The reason for the latter is not clear, although it likely represents additional carp egg material that was captured inside the CDS unit from previous tests. Despite the tranquil conditions within the outlet flume, the bypassed sample suffered significant damage in passing through the test rig, suggesting that the damage has occurred within the CDS unit itself at the contra-shear surface of the internal screen.

Fig. 11–4 shows photographs of bypassed eggs (a) and captured eggs (b) obtained during run 15. The difference in appearance of the two sam-ples is obvious.

Bypassed samples were examined under a microscope to determine the degree to which bypassed eggs might be viable. Fig. 11–5 shows a micro-scope photo of the bypassed sample for run 29, revealing some intact but damaged eggs and the remains of broken eggs. A visual estimate of the proportion of relatively intact eggs within the total bypassed sample was approximately 10%. For run 29, the bypassed sample represented 2.7% of the introduced sample. Thus, the percentage of the introduced sample of carp eggs that is not captured by the CDS unit and that survives reason-ably intact was less than 10% of 2.7%, or 0.27%.

Based on the promising results of the model study, a full-scale structure was commissioned and built. An assessment of its performance is pre-sented in the following section.

Prototype Structure

The prototype structure comprised four CDS units in parallel. Fig. 11–6 shows a photograph of the completed structure in operation.

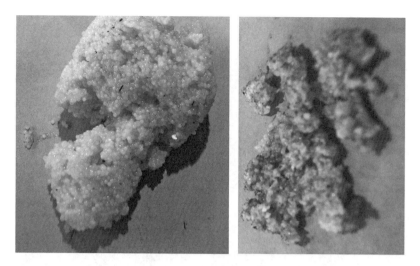

FIG. 11–4. Run 15: captured (A) and bypassed (B) carp eggs.

FIG. 11–5. Microscopic picture of part of the bypassed sample from run 29.

FIG. 11–6. Prototype carp extraction system in operation.

Water from Rocklands Reservoir is directed through the concrete channel from the bottom right of Fig. 11–6, passing through the CDS units before discharging into the Glenelg River at the top left of the figure.

In field tests, eggs were deposited upstream of the CDS units and grab-samples of eggs that had passed through the units were collected downstream. The samples were then examined by microscope to determine estimated proportions of viable and nonviable (broken) eggs.

Five samples were tested. The first four contained carp eggs that had passed through the CDS units. The fifth sample had not passed through the units and was used to provide a visual basis for viable carp eggs. Three different examinations were carried out as follows:

1. The samples were first examined through a microscope to assess the damage to individual eggs due to passage through the CDS unit.
2. Microscopic photographs of each sample were then made. Because of the relatively high-power magnification used, these photographs were made as multiple images by scanning across individual eggs. The images were then digitally stitched together to create a complete photograph of individual eggs.
3. Finally, each sample was individually analyzed to determine counts of viable and nonviable eggs.

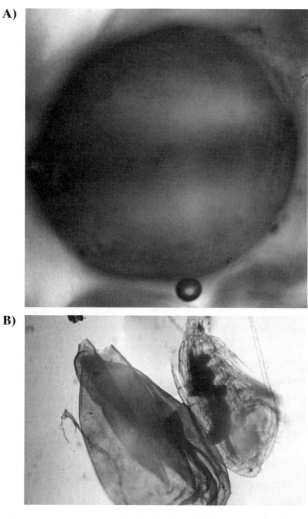

FIG. 11–7. Typical comparison of viable carp egg (sample 5 [a]) with nonviable egg (sample 2 CDS unit 3 50ML/d [b]).

Visual examination of the samples with a microscope revealed substantial damage to the samples that had passed through the CDS units. This was especially clear by comparison with the control sample that had not passed through the units. A typical comparison is presented in Fig. 11–7.

A detailed examination of each sample was carried out to quantify the number of viable and nonviable eggs. The integrated results are presented in Table 11–3.

TABLE 11–3 **Integrated results of quantitative analysis of field samples.**

Sample	Percentage Broken
1	74
2	70
3	58
4	69
Average	68

The results in Table 11–3 indicate that severe damage occurred to carp eggs through passage through the CDS units. However, the results are considered conservatively low for two reasons, as follows:

1. The samples introduced upstream of the CDS units represented about 1kg of carp eggs. Because the samples extracted from downstream of the CDS units were grab samples, there is no indication of the percentage of the sample introduced upstream that passed through the CDS unit. The results from the model tests indicated that very small percentages—always less than 10%—of introduced samples passed through the CDS unit, and, of those that passed through, probably only up to 10% were still viable.
2. Eggs laid naturally are attached as sticky strands on floating vegetation. Again, reference to the model results indicates that the percentage of an introduced sample that passes through the CDS units is further reduced when the eggs are attached to vegetation.

It is clear, then, that the quantitative results are not fully indicative of the total nonviability of carp eggs that enter the CDS units from upstream. The unknown factor in refining the analysis is the percentage of eggs that pass through the units without capture. The model tests indicate that this percentage is very small.

The prototype structure has been operating for five years so far with no evidence of the establishment of carp communities in the Glenelg River.

Summary

Structural means have been shown to be highly effective in controlling the movement of carp along waterways and between water bodies. This study considered two types of structure—the first based on differentiation of fish behavior between indigenous and exotic species, and the second based on nondiscriminatory capture of all biomass.

With the first structure type, there is a clear behavioral difference between indigenous fish that do not jump and carp that do jump. The development of the Williams Carp Cage is based on this difference. Field trials within a fishway indicated that the Williams cage successfully separated 91% of adult carp into a confinement area. In contrast, of 19,641 native fish more than 99.9% passed through the Williams cage and exited the fishway. Testing of the cage in a riverine application was not so successful and this result appeared to be due to fish behavior. The carp were observed to move downstream as the cage was lowered into place and then approached the cage slowly. Although large schools remained in proximity, very few entered the cage. This avoidance behavior was consistent across 10 days of trials.

The trap avoidance behavior displayed by carp was unexpected and, at the present stage of development, limits the effectiveness of the Williams cage to fishway applications. However, usage of the cage may be broadened to assist in the control of other pest species. As noted by Stuart et al. (2006b), several jumping carp species threaten to invade the Mississippi, Missouri, Ohio, and Illinois River basins and the Great Lakes.

The use of the second type of structure is illustrated in a case study whose goal was to avoid spread of carp from an infested reservoir into the Glenelg River while permitting passage of large volumes of water. The structure utilized a stormwater management technology known as continuous deflective separation (CDS) that damaged eggs and hatchlings during passage through a prototype CDS unit and mesh filter. The unit's performance was highly successful, with no more than 0.3% of viable carp eggs and fingerlings surviving passage through the structure. This success has been supported by field studies of grab samples and the fact that after five years of operation with the prototype extraction system, established carp communities have still not been found in the Glenelg River.

Policy Implications

Research into invasive species control and eradication is proceeding on multiple fronts, including commercial harvesting and poisoning, biological control methods, and engineering solutions. In regard to common carp, commercial harvesting and poisoning have been shown to reduce carp numbers temporarily and, occasionally, to eradicate them from isolated areas. Biological control measures include the so-called daughterless technology in which the sex-determination gene is modified, resulting in the production of male offspring only. This technology is still under development and tests to date have shown that the gene is only partially inherited. Furthermore, because of carp's long life cycle, the approach would need to operate over 40–50 years, and be used in conjunction with other short-term control techniques, to have a lasting effect.

Dams, weirs, and other man-made structures are ubiquitous in rivers. These have traditionally been engineered narrowly, with designers concentrating on their economic and social benefits and not considering how they affect such environmental issues as the movement of biomass within the rivers. There is significant potential for targeted man-made aquatic structures to enhance or limit ecological connectivity while maintaining hydraulic connectivity. This can be used to management advantage to control the spread of invasive species.

Engineered structures, two examples of which are presented in this chapter, show great promise in trapping carp. The first structure relies on different behavioral characteristics between carp and native species. Successful application of this technique to other invasive species will require clear differences between the swimming characteristics of the invasive species and the native species.

The application of the stormwater management technology known as continuous deflective separation (CDS) has been tested in the laboratory and in the field and successfully captures all life stages of all biomass. Thus, it is nondiscriminatory and is especially suitable to applications where invasive fish species totally dominate native fish species.

Literature Cited

Arthur Rylah Institute for Environmental Research (ARI). 2002. Control of Common Carp in the Glenelg River Catchment: A Review of Options. Report to Glenelg Hopkins Catchment Management Authority. Heidelberg: ARI.

Australian and New Zealand Environment and Conservation Council (ANZECC). 2000. *ANZECC List of Threatened Vertebrate Fauna*. Canberra: ANZECC.

Balon, E. K. 1995. Origin and domestication of the wild carp, *Cyprinus carpio*: from Roman gourmets to the swimming flowers. *Aquaculture* 129:3–48.

Commonwealth of Australia. 1999. Environment Protection and Biodiversity Con-

servation Act (EPBC Act). Accessed November 22, 2013. http://www.austlii
.edu.au/au/legis/cth/consol_act/epabca1999588/.

Dyson, M., G. Bergkamp, and J. Scanlon, eds. 2008. *Flow: The Essentials of Environmental Flows.* 2nd ed. Gland, Switzerland: International Union for Conservation of Nature.

Grewe, P. 1997. Potential of molecular approaches for the environmentally benign management of carp. In *Controlling Carp: Exploring the Options for Australia*, edited by J. Roberts and R. Tilzey, 119–127. Griffith, NSW: CSIRO Land and Water.

Harris, J. H., and P. C. Gehrke, eds. 1997. *Fish and Rivers in Stress: The NSW Rivers Survey.* Canberra, Australia: NSW Fisheries and the CRC for Freshwater Ecology.

Keller, R. J., and Associates. 2005. Rocklands Reservoir Carp Screening Physical Model Study. Consulting Report for Glenelg-Hopkins Catchment Management Authority.

Keller, R. J., M. Jane, and M. Bain. 2008. Screening Common Carp *Cyprinus Carpio* from Environmental Flow Releases to the Glenelg River, Southwest Victoria. Supplementary Report to the ASFB Darwin Workshop Proceedings 2005.

Koehn, J., A. Brumley, and P. C. Gehrke. 2000. *Managing the Impacts of Carp.* Canberra: Bureau of Resource Sciences.

State of Victoria. 1988. Victorian Flora and Fauna Guarantee Act (1988), Version No. 035, Victorian Parliament.

Stuart, I., J. McKenzie, A. Williams, and T. Holt. 2006a. The "Williams" Cage: A Key Tool for Carp Management in Murray-Darling Basin Fishways. Final Report on Project R3018SPD for Murray-Darling Basin Commission. Heidelberg: ARI.

Stuart, I., A. Williams, J. McKenzie, and T. Holt. 2006b. Managing a migratory pest species: a selective trap for common carp. *North American Journal of Fisheries Management* 26:888–893.

Thresher, R., and N. Bax. 2003. The science of producing daughterless technology; possibilities for population control using daughterless technology; maximising the impact of carp control. In *Proceedings of the National Carp Control Workshop, March 2003, Canberra*, edited by K. L. Lapidge, 19–24. Canberra: Cooperative Research Centre for Pest Animal Control.

CHAPTER TWELVE

Does Enemy Release Contribute to the Success of Invasive Species? A Review of the Enemy Release Hypothesis

Kirsten M. Prior and Jessica J. Hellmann

Introduction

A central question in invasion biology is this: what causes a species to be particularly successful in its introduced range? Once established, many introduced species experience increased fitness or demographic rates in their introduced range compared with their native range. This increased success causes some species to become "invasive," if they negatively affect species, ecosystems, or society (e.g., D'Antonio and Vitousek 1992; Lodge et al. 1994; Prior and Hellmann 2010). The phenomenon of a species becoming invasive as a result of increased success after establishment is referred to as "invasion success" (Colautti et al. 2004; Torchin and Mitchell 2004). The introduction of the European green crab, *Carcinus maenas*, to North America provides an example of invasion success where individuals exhibited greater biomass in North America compared with individuals from the native range in Europe (Torchin et al. 2001). Understanding factors that cause invasion success is critical for the management of invasive species.

Eradicating or removing populations of established invasive species is a significant challenge for resource managers. Removing populations or controlling populations of invasive species to some acceptable level often requires ongoing management efforts that can be logistically difficult and expensive, and that have environmental impacts (Louda et al. 1997; Byers et al. 2002). The goals of control programs are often to reduce invasive species to levels that minimize their impacts, whether economic (e.g., in agricultural systems) or environmental (e.g., in natural systems) (Clewley et al. 2012). Applying the most effective control strategy is beneficial for managers to reduce the costs and nontarget impacts of control measures. To accomplish this, basic ecological information about what factors control populations of invasive species is needed, yet this information is often lacking (Byers et al. 2002).

Invasion involves a series of stages, each containing ecological filters that species must pass through before progressing to the next stage (Richardson et al. 2000). Invasive species must be picked up, be transported to, and colonize a new region, where they may successfully establish and spread to new locations (Kolar and Lodge 2001). Factors such as propagule pressure (the number and frequency of introduction events) and the biological characteristics of the introduced species influence the probability that species will make it through the early stages of the invasion process, such as colonization. Biotic interactions, abiotic conditions, and the characteristics of the invader influence the probability that a species will make it through the later stages, such as successfully establishing populations and becoming invasive or exhibiting invasion success (Kolar and Lodge 2001; Catford et al. 2008).

Once a species is established, success will depend on niche opportunities of their recipient environment. Opportunities that promote success, for example, provide some favorable combination of resources, enemies, competitors, mutualists, and environmental conditions. Niche opportunities vary between environments, such that restricted niche opportunities can provide resistance to invasion (i.e., "biotic resistance"), while high niche opportunities promote invasion success (Shea and Chesson 2002; Levine et al. 2004; Mitchell et al. 2006).

Escape from enemies (e.g., predators, parasites, and pathogens) is one example of a high niche opportunity that can promote invasion success. The enemy release hypothesis (ERH) is a popular explanation of invasion success (Elton 1958; Maron and Vilà 2001; Keane and Crawley 2002). The ERH posits that some invasive species lose their enemies and benefit from a release from enemy control. Loss of native enemies can occur anywhere along the invasion route. Also, enemies from the introduced range could switch to attacking introduced species, but may be slow to do so or be less effective enemies (Elton 1958; Keane and Crawley 2002; Colautti et al. 2004). The ERH is based on the following conditions: introduced species lose enemies during the transport or invasion processes, and enemies significantly negatively affect species demographics or fitness in their native range. If enemy loss occurs and enemies are important, then release from enemy control will enable increased demographic success (Maron and Vilà 2001; Colautti et al. 2004).

The ERH provides the underpinnings for biological control. Classical biological control is predicated on the assumption that enemies limit populations of species in their native range and that reintroducing specialized, natural enemies should limit populations of invasive species in their introduced areas (Keane and Crawley 2002). Biological control is advocated as a good alternative to more conventional control methods, such as herbicides, given that it can be relatively inexpensive and has the potential for self-perpetuating, long-term control (McFadyen 1998). While there are some good examples of the successful control of invasive species, some biological control agents fail to control target pests (McEvoy and Coombs 1999; Myers and Bazely 2003). Also, biological control comes with the risk of agents having negative effects on nontarget, native species (e.g., Louda et al. 1997). Evaluating the potential for enemies to control and specialize on target species is an important goal of biological control programs.

Despite the importance of the ERH, it is often accepted uncritically from observations of enemy loss or when biological control agents are successful (Keane and Crawley 2002). For example, many studies have found that invasive species are attacked by fewer enemies or enemy species in their introduced range and infer that these species are benefitting from a release from enemy control (Colautti et al. 2004; Mitchell and Power 2003; Torchin et al. 2003). However, if enemies do not also negatively affect species demographics or fitness in their native range (an additional condition of the ERH), then a loss of enemies likely has little to do with a species' increased success (Maron and Vilà 2001; Colautti et al. 2004; MacDonald and Kotanen 2010; Williams et al. 2010; Prior and Hellmann 2013). A more correct and complete way to examine the ERH is to measure, by manipulating enemy pressure, the effect of enemies on species fitness or demographics between the native and introduced range. One can conclude that enemy release facilitates a species' increased success if enemies have a larger effect on species fitness or demographics in their native range compared with their introduced range (Colautti et al. 2004; Hierro et al. 2005; Callaway et al. 2004; DeWalt et al. 2004; Williams et al. 2010; Prior and Hellmann 2013).

Because invasion success is a biogeographical phenomenon where populations of species in the introduced range are on average more successful than those in the native range, a common approach to studying the role of enemies in invasion is to conduct "biogeographical comparisons" of enemy prevalence, richness, or effects in both regions (sensu Colautti et al. 2004). We surveyed the literature for studies that examine the ERH using this biogeographical approach to review the current evidence for the ERH. Colautti et al. (2004) present a similar review of ERH studies and we add on seven years of published studies to their review. Other literature reviews and quantitative analyses of the ERH have been conducted, but these focus on single taxonomic groups such as plants (e.g., Liu and Stiling 2006) and marine invertebrates (Torchin 2002).

Here, we provide a detailed consideration of the importance of the ERH as a potential cause of invasion success. Specifically, we report the results of a new literature review of studies that take a biogeographical approach to examine the role of enemies in facilitating the success of invasive species in multiple trophic levels and ecosystems. We use this review to highlight gaps in the available evidence for the ERH. We draw upon existing literature to discuss the likely extent of the ERH as an explanation for the success of invasive species. Finally, we discuss the implications of the ERH for the management of invasive species. Understanding the fundamental elements of this hypothesis is an essential step in guiding the management of invasive species, and we point out specific ways that our understanding falls short of the enthusiasm of this hypothesis as a widespread explanation for invasion.

Review of ERH Biogeographical Comparison Studies

We reviewed studies that addressed the ERH by means of a "biogeographical comparison" approach. These are of two types: first, those that are "observational" in nature that compare enemy richness or prevalence (i.e., abundance or attack rates) between species' native and introduced ranges; second, those that are "experimental" that compare the effect of enemies on prey performance between species' native and introduced ranges.

Another common approach to examine the ERH is to conduct "community comparison" studies that measure the prevalence or effects of enemies on native, introduced, and/or invasive species in the introduced range (sensu Colautti et al. 2004). We did not include these studies in our analysis because they ask a slightly different question regarding invasion success: do invasive species undergo competitive release mediated by enemies? As such, these studies do not assess invasion success as it is defined here: release from enemies in the native range, or success relative to native populations. Community comparison ERH studies have been reviewed elsewhere (Liu and Stiling 2006).

Searches were conducted using ISI Web of Science for the term "enemy release hypothesis." Studies were also used that were reviewed by and that cited Colautti et al. (2004) and Torchin and Mitchell (2004). These are two reviews that have assessed the ERH for all organisms. The term "biotic resistance hypothesis" (BRH) (enemies provide resistance to invasion) was also searched so as to not bias studies to those that found evidence in support of the ERH. Studies that were reviewed by and that cited reviews by Levine et al. (2004) and Parker et al. (2006) (two reviews of the BRH) were also included. We also used references that were cited in and cited studies in underrepresented groups (e.g., all groups other than terrestrial plants). Searches ended in September 2011.

We included studies that examined the ERH by conducting observational surveys of enemy prevalence (i.e., overall attack rates or abundance) or enemy species richness (i.e., number of enemy species) between species' native and introduced ranges. These surveys were in the form of field surveys, or literature or database surveys. Some studies examined multiple introduced species. Studies were also included in our review if they examined the ERH by conducting enemy exclusion experiments in species' native and introduced ranges. These studies measured the effect of enemies on prey performance. Response variables included species growth rates, biomass, survival, population size, and population growth. These studies were often field studies that excluded enemy communities in both regions. Some plant pathogen studies were conducted in greenhouses where plants were grown in soil from both regions that contained intact enemy communities.

Studies were assigned to an ecosystem type (terrestrial, marine, freshwater) and prey type (plant, invertebrate, vertebrate). If one study evaluated the ERH for species in multiple ecosystem or prey types, support for the ERH was evaluated separately for these groups within a study. This occurred in two studies; Torchin 2002 conducted surveys on marine invertebrates and one marine vertebrate, and Torchin et al. 2003 conducted surveys on terrestrial vertebrates, marine invertebrates, and freshwater vertebrates. Enemy type (pathogens, parasites, invertebrates, and vertebrates), type of study (survey, experiment, review), distance introduced (intercontinental vs. intracontinental introduction) were also recorded for each study (Table 12–A1 in appendix).

A study supported the ERH if it found a reduction in enemy prevalence, enemy richness, or effects on prey performance in the introduced range compared with the native range (+; Table 12–A1). A study did not support the ERH if there were no differences between regions (n.s.; Table 12–A1). Finally, a study found support in the opposite direction of the ERH if any of these responses were higher in the introduced range (–; Table 12–A1). The numbers of studies finding positive support from observational and experimental studies of the ERH were tallied for the whole data set and for each ecosystem/prey type group. If studies found no support, negative support, or equivocal support (e.g., found increased richness but decreased prevalence), they were not counted as evidence in support of the ERH.

Our search yielded 55 studies that evaluated the ERH (Table 12–A1). Of these studies, 43 were observational, 17 were experimental, and 5 conducted both observational surveys and experiments. Of the observational studies, 2 conducted reviews, 31 were field surveys, and 10 used a mixture of reviews and field surveys. Of the observational studies 37 out of 47 (79%) found support for the ERH. Of the experimental studies 10 out of 17 (59%) supported the ERH. The majority of studies evaluated species that have been introduced between continents (intercontinental introductions) (43 studies) compared with intra-continental introductions (9 studies).

The majority of studies were conducted on introduced terrestrial species, with a strong bias towards terrestrial plants (Fig. 12–1). Fewer studies evaluated the ERH for marine organisms. We found no biogeographical studies for freshwater plants or invertebrates. The majority of studies that evaluated the ERH by taking an observational approach did so in terrestrial plant systems (Fig. 12–2a). All but 3 of the studies that have evaluated the ERH by conducting biogeographical experiments have done so in terrestrial plant systems (Fig. 12–2b). Two studies (by Vermeij et al. [2009] and Forslund et al. [2010]) experimentally tested the ERH for marine algae, and one study by Prior and Hellmann (2013) did so for a terrestrial insect.

In most ecosystem and prey groups there was evidence from observational studies in support of the ERH. Observations of enemy loss were supported in over 50% of cases in all groups except for marine

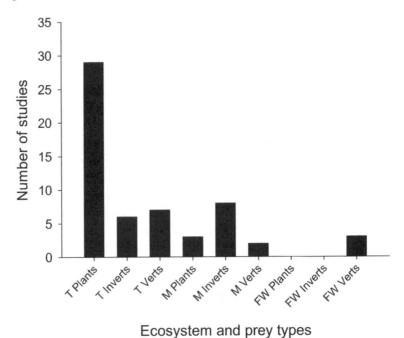

FIG. 12–1. Number of studies that examined the ERH by conducing "biogeographical comparisons" of enemy prevalence, enemy species richness, or the effect of enemies on prey performance in species' native and introduced ranges in different ecosystem types and prey groups. FW, freshwater; M, marine; T, terrestrial.

invertebrates (Fig. 12–2a). Support for the ERH from experimental studies, on the other hand, was more equivocal. The ERH was supported in 8 out of 14 cases for terrestrial plants, not supported in the one terrestrial insect study, and supported in both of the marine plant cases (Fig. 12–2b; Table 12–A1).

Does Enemy Loss Cause Invasion Success?

A critical issue with the current evidence for the ERH is that the majority of evidence comes from observational studies of enemy prevalence or richness. These types of studies address only one aspect of the ERH—enemy loss. The majority of these studies (79%) observed enemy loss, suggesting that it may be a common phenomenon for many introduced species. It is incorrect to assume that observations of enemy loss, however, provide sufficient evidence that release from enemy control is causing invasion success. There are multiple reasons for this. First, observations of enemy attack rates or species richness alone do not represent the overall effect of the enemy community on prey performance. Second, this approach does not allow for the assessment of the role of enemies relative to other factors that may be important in driving invasion success. Given that most ERH studies are observations of enemy loss, and enemy loss may not always translate into release from enemy control, the importance of this highly cited hypothesis is likely currently overstated.

One reason that observations of enemy loss are an insufficient assessment of the ERH is that a lower enemy abundance or a loss of enemy species will not always translate into an enemy community that is less effective at controlling prey populations. For example, some studies found no change in or higher enemy species richness and/or prevalence in species' introduced ranges compared with their native ranges (Memmott et al. 2000; Poulin and Mouillot 2003; Cripps et al. 2006; Prior and Hellmann 2013). Higher abundance or richness of enemies in introduced regions can occur if generalist enemies from alternative prey switch to novel prey that are not well adapted for defenses against these enemies (Cripps et al. 2006; Parker et al. 2006; Ward-Fear et al. 2009). However, more enemy species or higher enemy prevalence does not necessarily translate into higher enemy effects as enemy species that are lost could be more effective enemies than those that are gained. Cripps et al. (2006), for example, recorded overall higher abundance of phytophagous insects on the invasive weed,

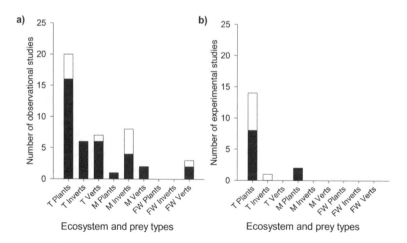

FIG. 12–2. Number of studies that found positive support of the ERH from a) observational surveys, and b) experimental studies in different ecosystem types and prey groups. Black bars represent the number of studies that found positive support for the ERH out of all of the studies that examined the ERH. FW, freshwater; M, marine; T, terrestrial.

Lepidium draba, in its introduced range in North America. Species richness, diversity, and evenness, however, were higher in this plant's native range. Species with specialized feeding modes, such as root-feeders and gall-formers, were absent in the introduced range but were replaced with a higher prevalence of generalist sap suckers. The specialist feeders likely have a higher impact on overall plant fitness compared with the generalists. Thus, the overall effectiveness of the enemy community could be higher in the native range even though overall damage levels were lower. Measuring the overall effect of the enemy community on prey performance between the two regions is, therefore, essential to properly assess the ERH.

Another reason for observations of enemy loss being insufficient evidence to support the ERH is that these observations do not take into account the relative effect of enemies compared with other potentially important factors such as resources, competitors, mutualists, or the environment (Shea and Chesson 2002; Mitchell et al. 2006). Enemy attack rates may be lower in species' introduced ranges, but if enemies do not limit populations in the native range, and some other factor is limiting, then enemy loss will not be an important driver of demographic success (Maron and Vilà 2001; Prior and Hellmann 2013). MacDonald and Kotanen (2010), for example, examined the effects of pathogens, herbivores, and disturbance on the performance of invasive ragwort, *Ambrosia artemisiifolia,* in its native range. They found that enemies had no to little effect on plant performance compared with disturbance. Thus, although ragwort has lost many enemy species in its introduced range (Genton et al. 2005), escape from enemies is unlikely to provide ragwort with a significant demographic advantage in its introduced range. Similarly, Williams et al. (2010) examined the relative effect of specialist enemies and disturbance on the population growth of houndstonge, *Cynoglossum officinale,* in its native and introduced range. While they found the effect of enemies was reduced, they found that increased disturbance has a much larger effect on vital life history stages and in turn on population growth in the introduced range. The authors conclude that enemies contribute relatively little to this species invasion success compared with differences in disturbance between the two regions.

Is the Potential for Enemy Release Equal for Different Types of
Invasive Species?

Another issue with the current evidence for the ERH is that the majority of studies have examined this hypothesis for terrestrial plants. Furthermore, virtually all studies that have experimentally tested the ERH using a biogeographical approach have done so in terrestrial plant systems (except see Vermeij et al. 2009; Forslund et al. 2010; Prior and Hellmann 2013). Thus, what we currently understand about the ERH from empirical studies is highly biased toward this group of invaders. Given the paucity of studies in other systems, we can say little about the general prevalence of this mechanism in facilitating the success of all types of invasive species.

It should not be expected that the potential for enemy release is similar for all types of invasive species for multiple reasons. First, the extent of enemy loss is likely to be different for invasive species depending on the types of enemies they harbor. Second, the importance of enemies in controlling species in their native range should not be expected to be similar for all types of invasive species.

Enemy loss is likely to be different for different types of invasive species given that they harbor different types of enemies. For example, specialist invertebrates that feed on one or a few related species are common enemies of terrestrial plants and invertebrates (Strong et al. 1984). Specialists are more likely to be lost than generalist enemies as generalists from the introduced range are more likely to switch from alternative prey (Cornell and Hawkins 1993; Keane and Crawley 2002; Parker et al. 2006). Given that specialists share a long coevolutionary history with their prey, it is less likely that they will have the ability to switch from alternative prey in the native range (Cornell and Hawkins 1993; Keane and Crawley 2002). However, there is some evidence that specialist enemies have switched from closely related native congeners to introduced species (Cornell and Hawkins 1993; Creed and Sheldon 1995; Jobin et al. 1996). Furthermore, specialists often have strong effects on prey fitness and demographics (Strong et al. 1984) and their loss could, therefore, significantly contribute to a species' increased success. Many studies that we reviewed specifically documented the loss of specialist enemies from invasive terrestrial plant species (Table 12–A1; Memmott et al. 2000; Fenner and Lee 2001; Van der Putten et al. 2005; Bruun 2006; Cripps et al. 2006; Hansen et al. 2006).

There is also substantial evidence to support the loss of disease organisms such as parasites and pathogens (Table 12–A1). Many studies have found that introduced plants commonly lose plant pathogen enemies. One comprehensive review examined 473 plants that were introduced to North America and found that they lost over 90% of their native fungal and viral pathogens (Mitchell and Power 2003). In our review, we found that 10 out of 12 studies found evidence for pathogen loss for invasive terrestrial plants (Table 12–A1; e.g., Beckstead and Parker 2003; Knevel et al. 2004;

Roy et al. 2011). Additionally, studies that experimentally tested the ERH for plant pathogens largely found evidence in support of the ERH (e.g. Reinhart et al. 2003; Callaway et al. 2004; Reinhart and Callaway 2004; Van Grunsven et al. 2009; except see Beckstead and Parker 2003; Lewis et al. 2006; Volin et al. 2010). Torchin et al. (2003) found that for 26 host species of mollusks, crustaceans, fishes, birds, mammals, amphibians, and reptiles, half of their parasite species were missing from their introduced range and overall parasitism rates were lower. We found that 14 out of 19 studies provided evidence for parasite loss (Table 12–A1). Parasites are known to have large impacts on host populations (Hudson et al. 1998). Thus, given that their likelihood of loss is high, they are likely important agents of enemy release.

Invasive species that are dominated by generalist enemy communities, on the other hand, are not as likely to lose enemies (Keane and Crawley 2002; Parker et al. 2006). Marine and freshwater plants, for example, are largely preyed on by generalist herbivores and may not benefit from enemy release (Hay 1991; Shurin et al. 2006). However, given the lack of biogeographical comparison studies in many generalist-dominated systems (Fig. 12–1; Table 12–A1) it is premature to conclude that enemy loss does not occur in these instances. In fact, of the three marine plant studies all found evidence of enemy loss. Vermeij et al. (2009) found that although generalist vertebrate herbivores preferred introduced alga over native algae in Hawaii, overall herbivore attack rates were still lower in Hawaii (introduced range) compared with the Caribbean (native range). Forslund et al. (2010) and Wikstrom et al. (2006) found lower herbivory on the algae, *Fucus evanescens*, in the introduced range because populations of this algae were more resistant to herbivores. In another generalist-dominated community, Shwartz et al. (2009) found that the rose-ringed parakeet, *Psittacula krameri*, undergoes lower enemy pressure in its introduced range in the United Kingdom because the predators in this region are too large to prey on eggs in nest cavities. In general, there are many circumstances in which generalist enemies can be lost—for example, if they are introduced to species depauperate communities such as disturbed habitats or islands or to areas where enemy communities are vastly different and unlikely to be able to adapt to attacking introduced prey (Vermeij et al. 2009; Shwartz et al. 2009). Enemy loss should not be ruled out for invasive species that harbor generalist enemies, and we need more biogeographical studies of invasive species in generalist-dominated systems to understand how common and in what circumstances enemy loss occurs.

Another condition of the ERH, that enemies have negative effects on species in their native range, should also not be expected to be similar for all types of invasive species. Community ecology theories, for example, suggest that there is heterogeneity in top-down effects for species in different trophic levels and ecosystems (Hairston et al. 1960; Fretwell 1977; Sih et al. 1985; Menge and Sutherland 1987; Shurin et al. 2002; Shurin et al. 2006). Further, a meta-analysis of enemy effects on prey performance in species' native ranges revealed systematic differences in enemy effects among ecosystems and prey group, suggesting that enemy release is a more likely explanation of invasion success for some invasive species than for others on the basis of this important condition of the ERH (Prior 2011).

Potential for Enemy Release for Different Prey Groups and Ecosystems

Given the paucity of ERH studies in certain prey groups and ecosystems, it is premature to make general conclusions about the prevalence of this mechanism of invasion success for all types of invasive species. Prior (2011) provides a qualitative framework to help make predictions about which types of invasive species have the most potential to benefit from enemy release based on 1) the extent of enemy loss they experience, and 2) the strength of native top-down control (see Prior 2011 for an extensive discussion of this framework).

An invasive species has a high potential of experiencing enemy release if enemy loss is high and enemy effects are strong in the native range. Here, we define enemy loss as the difference between the effectiveness of the native enemy community and the introduced enemy community. A species has a moderate potential of experiencing enemy release if enemy loss is high, but enemy effects in the native range are weak; or if enemy loss is low, when native enemy effects are high. Finally, an invasive species will have a low potential to experience enemy release if enemy loss is low and native enemy effects are weak, or if enemy loss does not occur, and/or if enemies do not affect prey performance in the native range (Prior 2011; Fig. 12–3).

We can qualitatively predict the potential for invasive species from particular prey groups and ecosystem types to experience enemy release (Fig. 12–3). For example, Prior (2011) found that terrestrial insects experienced strong native enemy effects. If their enemies are specialists that are likely to be lost, there is a high potential that they will experience enemy

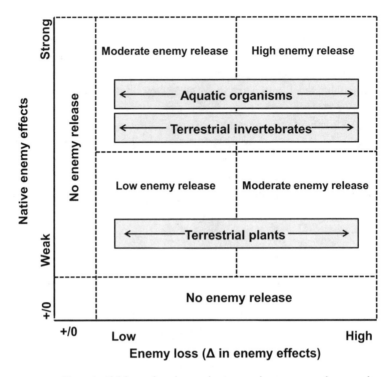

FIG. 12–3. The potential for an invasive species to experience enemy release can be predicted from 1) the extent of enemy loss (native enemy effects—introduced enemy effects) and 2) the strength of enemy effects in the native range. Aquatic organisms and terrestrial insects undergo strong enemy effects in the native range and their potential for enemy release will be moderate to high depending on the extent of enemy loss. Terrestrial plants are under weak enemy effects in their native range and their potential to experience enemy release is low to moderate depending on their extent of enemy loss. (Figure adapted from Prior 2011.)

release. If terrestrial insects harbor mostly generalist enemies in their native range, then there is lower potential for species in this group to experience enemy release. Prior (2011) also found that terrestrial plants experienced weak enemy effects. If they harbor specialist enemies, there is a moderate potential that they will experience enemy release, but a lower potential if they harbor generalist enemies. Finally, Prior (2011) found that many aquatic species were under strong native enemy effects. If aquatic invasive species are introduced into enemy depauperate communities or into a community of enemies that are not well adapted to attacking them, then there is a high potential for them to experience enemy release.

Introduction Distance

The majority of studies that have examined the ERH have done so for species that have been transported over long distances (i.e., intercontinental introductions). Species that are introduced or expand over shorter distances (i.e., intracontinental introductions) may receive less benefit from enemy escape opportunities because these species are more likely to move into locations where recipient communities contain similar or closely related species. As well, interacting species in these instances are more likely to catch up to their introduced counterparts (Mitchell et al. 2006; Mueller and Hellmann 2008; except see Engelkes et al. 2008; Phillips et al. 2010). Species transported over long distances, on the other hand, will be introduced into communities with which they share little coevolutionary history. In unrelated communities fewer enemies will be able to shift from alternative prey and this could lead to a higher potential for release (Mitchell et al. 2006; Strauss et al. 2006). A recent study compared enemy escape opportunities for intracontinental and intercontinental plants species. Contrary to their expectations, both types of introduced species equally benefitted from enemy release (Engelkes et al. 2008). Further, we found that all of the studies of intracontinental introductions found evidence for enemy loss (Table 12–A1).

Case Study: Intracontinental Introduction of a Terrestrial Insect, *Neuroterus saltatorius*

We present a case study of a rigorous examination of the ERH in a unique context: for a species in a higher trophic level that has undergone a short-distance, poleward introduction. Results from this study will be used to discuss some of the issues with the current evidence for the ERH.

Neuroterus saltatorius, the jumping gall wasp, is an oak-gall wasp (Hymenoptera: Cynipidae) whose native distribution occurs from northwest Texas to Washington state (Duncan 1997). Oak-gall wasps form galls on oak trees in which their developing larvae gain protection from abiotic conditions and enemies. This species was introduced to Vancouver Island, British Columbia, where it is reported to be outbreaking on its host plant Garry oak, *Quercus garryana,* by the Canadian Forest Service (Duncan 1997). This introduction is an example of a short-distance, poleward introduction where this species was introduced into a similar habitat (oak savannah ecosystem), with similar abiotic conditions (Prior and Hellmann 2013), where it interacts with many of the same species as it does in its native range including its host plant *Q. garryana*. The main enemies of *N. saltatorius* are a suite of parasitoid wasps (Smith 1995).

In its introduced range, *N. saltatorius* is considered invasive as it occurs at much higher densities in oak patches than it does in its native range in Washington State (Prior and Hellmann 2013). Like many gall-formers, this species is patchily distributed within an oak patch (Egan and Ott 2007) and its within-patch distribution is patchier with fewer trees affected by this species in its native range. When *N. saltatorius* occurs at high densities, hundreds of galls can be found on a single leaf. These high densities can cause foliar necrosis of large continuous portions of the leaf (Duncan 1997). The fitness consequences of *N. saltatorius* to *Q. garryana* are not known; however, it has been implicated as a potential factor for the decreased regeneration of *Q. garryana* on the Island (MacDougall et al. 2010). At high densities, this species also has negative fitness consequences for a co-occurring oak-specialist butterfly, the Propertius duskywing, *Erynnis propertius* (Prior and Hellmann 2010).

The prevalence of parasitoids (parasitoid attack rates) is 30% lower in Island populations of *N. saltatorius* compared with Washington populations, suggesting that enemy loss may be occurring for this invasive species (Prior and Hellmann 2013). The number of parasitoid species attacking *N. saltatorius* in the two regions is similar. Seven parasitoid species and one lethal inquiline attack this species in its native range and 8 parasitoids in its introduced range. The composition of the parasitoid species community is different, however, with differences in the relative abundance of species between regions. Taken together, there is observational evidence of increased demographic success and enemy loss in this system.

The ERH predicts that enemies will have a larger effect on prey demographics in their native range compared with their introduced range. That is, enemy release occurs if: $\Delta S_{Native} > \Delta S_{Introduced}$; where ΔS is the effect of the parasitoid community on gall wasp survivorship measured as the difference in survivorship in the absence of parasitoids (i.e., experimental exclosures) and in the presence of parasitoids (i.e., controls) (Fig. 12–4a). In this system, enemy effects were actually higher in the introduced range compared with the native range ($\Delta S_{Native} < \Delta S_{Introduced}$) (Prior and Hellmann 2013; Fig. 12–4b). These higher enemy effects in the introduced range occurred because background mortality in the absence of enemies was higher in the native range and lower in the introduced range (Fig. 12–4b). This result suggests that while a loss of parasitoid pressure likely

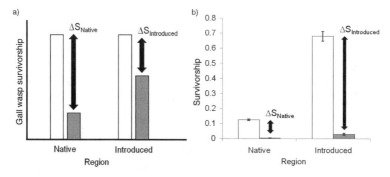

FIG. 12–4. The ERH predicts that enemy effects on prey performance will be higher in the native range and lower in the introduced range. a) In the gall wasp system the ERH would be supported if: $\Delta S_{Native} > \Delta S_{Introduced}$, where ΔS reflects the effect of enemies on gall wasp survivorship measured as the difference in survivorship between experimental exclosures (white bars) and controls (gray bars). b) The ERH is rejected in this system as, $\Delta S_{Native} < \Delta S_{Introduced}$. This result occurs because there is lower background mortality in exclosures in the introduced range compared to exclosures in the native range. The bars in (b) represent the mean enemy effects of multiple sites in the native (n=3) and the introduced (n=5) range. Error bars represent ± S.E. (Figure adapted from Prior and Hellmann 2013.)

plays a part in this species' increased success, other factors are also limiting this species in its native range and facilitating its success in its introduced range. For example, gall-formers have intimate relationships with their host plants (Stone et al. 2002) and host plant populations could be more resistant to or less suitable in the native range where these species share a long coevolutionary history. Oaks in the introduced range may be less resistant or more suitable, facilitating the outbreaks of *N. saltatorius*.

This case highlights that observations of enemy loss are not a sufficient proxy for enemy release and that experimental studies that measure the effect of enemies in both regions are needed to properly assess the ERH. Although enemy loss was observed, release from enemy control is not the driving mechanism of invasion success because some other factor, such as host plant suitability, is important in limiting this species' demographic success in its native range (also see Williams et al. 2010). Future studies of invasion success should aim to manipulate multiple potentially important factors in both regions to uncover their relative importance in driving invasion success.

This study provides an example of a short-distance expansion. There was overlap in the enemy community, as expected for a species moving over short distances; however, there were also lower overall attack rates.

This could be because parasitoids are locally adapted to alternative hosts in the introduced range (e.g., Menéndez et al. 2008). Even though enemy loss did not translate into release, *N. saltatorius* is undergoing increased demographic success. This suggests that species moving over short distances, such as those undergoing climate-driven range expansions, have the potential to become "invasive," but not always as in this case due to enemy release.

Finally, this study is a test of the ERH for a species in a higher trophic level than plants. The ERH has primarily been applied to and tested in invasive plants; however, it is also a central hypothesis for the biological control of outbreaking invertebrates (Roy et al. 2011). Phytophagous insects are some of the most numerous invasive species worldwide (Kenis and Branco 2010), and enemy release, especially from parasitoid enemies, is often invoked as an explanation for their success (Cornell and Hawkins 1993; Menéndez et al. 2008). Prior (2011) predicts that terrestrial insects will have a moderate to high potential of experiencing enemy release based on the extent of enemy loss (Fig. 12–3). In the case of *N. saltatorius*, enemies did negatively affect this species in its native range, but enemy effects were higher in the native range. This could be because enemy loss was relatively low (i.e., there was high overlap in the enemy community) and because other factors also caused significant changes in survivorship between the two regions. This finding is consistent with the prediction that terrestrial insects that harbor generalist enemies have a moderate potential of experiencing enemy release. Further tests of the ERH are needed for invasive species in different trophic levels and ecosystems to back up claims of its importance as an explanation for invasion success.

Conclusions

Our review revealed that there are many studies that have addressed the ERH, yet there are currently gaps in the types of studies in which this hypothesis has been evaluated. In summary, most studies are observational studies of enemy loss, are biased toward one group of organisms, and focus on long-distance introductions. Enemy release is claimed to be an important driver of invasion success. However, given the few studies that have properly examined this hypothesis we cannot currently confirm that enemy release is important for many types of invasive species. More studies taking a biogeographical experimental approach to properly evaluate the ERH are needed, especially for invasive species such as terrestrial insects and aquatic organisms that have a high potential of benefitting from enemy release.

Policy Implications

Enemy release provides the underpinnings for biological control (Keane and Crawley 2002). If enemy release is important, then reestablishing a link between an invasive species and its native enemies should provide an effective means of control. The aim of classical biological control is to introduce coevolved specialist enemies from a species' native range to reduce populations of invasive species to a level that reduces their ecological or economic effects (Hulme 2006). Despite the importance of the ERH to this management practice, surprisingly few studies have examined the role of enemies in controlling invasive species in their native range.

While many biological control agents are successful, some fail to control target species (Myers and Bazely 2003). There are many reasons for these failures. One is that if enemy release is not driving invasion success, then reintroducing enemies may not be a particularly effective strategy for control (Maron and Vilà 2001). For example, the cinnabar moth, *Tyria jacobaeae*, is a biological control agent that has been introduced to control ragweed, *Senecio jacobaea*, in North America (McEvoy et al. 1993). The cinnabar moth is a seed herbivore, reducing seed set of ragweed; however, ragweed is not limited by seed set, but rather by seed sites (Crawley and Gillmann 1989; McEvoy et al. 1993). Thus, even though ragweed has lost many enemies, including the cinnabar moth, enemy loss has little to do with the invasion success of this species and reintroducing enemies may not be an effective management strategy.

Conducting rigorous studies to uncover what factors control invasive species is a time-consuming and expensive endeavor. Waiting for such studies to be conducted is often not a feasible option for managers assigned with the task of controlling invasive species. Nonetheless, it is incorrect to assume that enemy release ubiquitously causes the success of invasive species. A better understanding of the importance of this mechanism and other mechanisms of invasion success would greatly improve management of established species. Halting control efforts until such studies are conducted is not the answer; managers and scientists who conduct basic research in ecology and evolution could benefit from working together on these issues. Taking steps to understand the importance of enemies and other factors in driving success is in the best interest of managers to reduce costs, wasted efforts, and potential nontarget impacts.

Acknowledgments

Ideas for this work were conceived while KMP was in the Department of Biological Sciences at the University of Notre Dame. This chapter was written while KMP was a Postdoctoral Fellow in the Department of Ecology and Evolutionary Biology at the University of Toronto. This work was partially supported by the Office of Science (Biological and Environmental Research), US Department of Energy (grant DE-FG02–05ER to JJH), and by the University of Toronto and the Ontario Ministry of Economic Development and Innovation (to KMP).

TABLE 12–A1 **List of studies that have examined the enemy release hypothesis (ERH) using a biogeographical comparison approach**

Invasive Species	Enemy Type (s)[a]	Observational Evidence[b]		Experimental Evidence[b]	Distance[c]	Response (E/P)[d]	Type of Study[e]	Study
		Prevalence	Richness					
Terrestrial Plants								
1 tree	P/I	+/+			Inter	damage (P)	S	Adams et al. 2009
1 forb	P			n.s.	Inter	biomass (P)	E	Andonian et al. 2011
1 grass	P		+	n.s.	Inter	richness (E)/germination (P)	S/E	Beckstead & Parker 2003
1 forb	P/I		+/n.s.		Inter	richness (E)	R	Brunn 2006
1 forb	P			+	Inter	biomass (P)	E	Callaway et al. 2004
1 forb	I	+			Inter	damage (P)	S	Cripps et al. 2010
1 forb	I	-	+		Inter	abundance (E)/richness (E)	S	Cripps et al. 2006
13 forbs	P/I			*+/−	Inter (OI)	survival (P)	E	DeWalt et al. 2004
1 forb	I	+			Inter (CI)	damage (P)	S	Fenner et al. 2001
1 forb	P/I		+		Inter	damage (P)	S	Genton et al. 2005
1 forb	I	n.s./+			Intra	richness (E)	S	Hansen et al. 2006
1 grass	P	+	+	**+/−	Inter	richness (E)/growth (P)	S/E	Knevel et al. 2004
1 forb	I	+		n.s.	Inter	damage (P)/growth (P)	E	Lewis et al. 2006
1 shrub	I	+			Inter	abundance (E)	S	Memmott et al. 2000
473 plants	P		+		Various	richness (E)	R	Mitchell & Power 2003
1 tree	I	+			Inter (OI)	damage (P)	S	Norghauer et al. 2011
2 trees	P			+	Inter	biomass (P)	E	Reinhart et al. 2004
1 tree	P			+	Inter	biomass (P)	E	Reinhart et al. 2003
1 tree	P	+	+	+	Inter	prevalence (E)/richness (E)/mortality (P)	S/E	Reinhart et al. 2010
1 grass	P/I/V	+/n.s./n.s.	+	+	Inter	damage (P)/population growth (P)	S/E	Roy et al. 2011
1 grass	I		+		Inter	richness (E)	S	van der Putten et al. 2005
1 forb	P			+	Inter	biomass (P)	E	Van Grunsven et al. 2009
1 forb	P			+	Intra	biomass (P)	E	Van Grunsven et al. 2010

(continued)

TABLE 12–A1 (continued)

Invasive Species	Enemy Type(s)[a]	Observational Evidence[b] Prevalence	Observational Evidence[b] Richness	Experimental Evidence[b]	Distance[c]	Response (E/P)[d]	Type of Study[e]	Study
Terrestrial Plants								
138 plants	P		+		Inter	richness (E)	S/R	Van Kleunen et al. 2009
1 forb	I	+			Inter	damage (P)	S	Vilá et al. 2005
1 fern	P			-	Inter	growth rate (P)	E	Volin et al. 2010
1 forb	I			+	Inter	survival (P)	E	Williams et al. 2010
1 forb	P/I	+			Inter	damage (P)	S	Wolfe 2002
1 tree	P	+			Intra	damage (P)	S	Zocca et al. 2008
Terrestrial Insects								
87 insects	I	+	+		Various	parasitism rate (E)/richness (E)	S/R	Cornell & Hawkins 1993
1 moth	I	+			Intra	parasitism rate (E)	S	Grabenweger et al. 2010
1 butterfly	I	+	+		Intra	parasitism rate (E)/richness (E)	S	Menéndez et al. 2008
1 gall wasp	I	+	n.s.	-	Intra (CI)	parasitism rate (E)/richness (E)/survival (P)	S/E	Prior & Hellmann 2013
1 gall wasp	I		+		Intra	richness (E)	S	Schönrogge et al. 1995
1 ant	P	+	+		Inter	prevalence (E)/richness (E)	S/R	Yang et al. 2010
Terrestrial Vertebrates								
1 reptile	Par		+		Inter	richness (E)	S/R	Burke et al. 2007
1 bird	Par	n.s.	n.s.		Inter (OI, CI)	prevalence (E)/richness (E)	S	Ishtiaq et al. 2006
1 rodent	Par		+		Inter (OI)	richness (E)	S	Lopez-Darias et al. 2008
30 birds	Par		+		Inter (CI)	richness (E)	S/R	MacLeod et al. 2010
1 amphibian	Par	+			Intra	prevalence (E)	S	Phillips et al. 201
1 bird	Par	+			Inter	prevalence (E)	S	Shwartz et al. 2009
4 vertebrates	Par	+	+		Various	prevalence (E)/richness (E)	S/R	Torchin et al. 2003

	Enemy type[a]	[b]	[b]	Distance introduced[c]	Response measured[d]	Type of study[e]	Reference
Marine Plants							
2 algae	I/V		+	Inter (OI)	growth rate (P)	E	Vermeij et al. 2009
1 algae	I		+	Inter	biomass (P)	E	Forslund et al. 2010
1 algae	I	+		Inter	abundance (E)/richness (E)	S	Wikstrom et al. 2006
Marine Invertebrates							
1 snail	Par		+	Inter	richness (E)	S/R	Blakeslee & Byers 2008
2 crabs	Par	+	+	Inter	richness (E)	S/R	Blakeslee et al. 2009
1 mussel	Par	n.s.		Inter	prevalence (E)	S	Calvo-Ugarteburu & McQuaid 1998
1 amphipod	Par	n.s.	n.s.	Inter	prevalence (E)/richness (E)	S	Slothouber et al. 2010
1 crab	Par/I/V	+/n.s.		Inter	prevalence (E)/damage	S	Torchin et al. 2001
9 invertebrates	Par	+	+	Various	prevalence (E)/richness (E)	S/R	Torchin 2002
10 invertebrates	Par	+	+	Various	prevalence (E)/richness (E)	S/R	Torchin et al. 2003
1 amphipod	Par	n.s.	n.s.	Intra	prevalence (E)/richness (E)	S	Wattier et al. 2007
Marine Vertebrates							
1 marine fish	Par	+	+	Various	prevalence (E)/richness (E)	S/R	Torchin 2002
1 marine fish	Par	+	+	Inter (OI)	prevalence (E)/richness (E)	S	Vignon et al. 2009
Freshwater Vertebrates							
2 fw fish	Par	+	+	Inter	prevalence (E)/richness (E)	S	Kvach & Stepien 2008
2 fw fish	Par	n.s.		Inter	richness (E)	S	Poulin & Mouillot 2003
6 fw fish	Par	+		Various	prevalence (E)/richness (E)	S/R	Torchin et al. 2003

[a] Enemy types (P = pathogen, I = invertebrate, V = vertebrate, Par = Parasite).

[b] Evidence for the ERH (+ = lower prevalence, richness, or effects in the introduced range. n.s. = no difference. - = higher prevalence, richness, or effects in the introduced range).

[c] Distance introduced (Inter = intercontinental, Intra = intracontinental, OI = oceanic island also considered Inter, CI = continental island also considered Intra).

[d] E = response measured on enemy community, P = response measured on prey.

[e] Type of study (S = observational field survey, R = literature or database review, E = enemy exclusion experiment).

* DeWalt et al. 2004 found evidence for enemy release in open habitats, but not forested habitats.

** Knevel et al. 2004 found evidence for enemy release in some sites, but not others.

Literature Cited

Adams, J. M., W. Fang, R. M. Callaway, D. Cipollini, and E. Newell. 2009. A cross-continental test of the enemy release hypothesis: leaf herbivory on *Acer platanoides* (L.) is three times lower in North America than in its native Europe. *Biological Invasions* 11:1005–1016.

Andonian, K., J. L. Hierro, L. Khetsuriani, P. Becerra, G. Janoyan, D. Villarreal, and L. Cavieres. 2011. Range-expanding populations of a globally introduced weed experience negative plant-soil feedbacks. *PLOS ONE* 6:e20117.

Beckstead, J., and I. M. Parker. 2003. Invasiveness of *Ammophilia arenaria*: release from soil-borne pathogens? *Ecology* 84:2824–2831.

Blakeslee, A. M. H., and J. E. Byers. 2008. Using parasites to inform ecological history: comparisons among three congeneric marine snails. *Ecology* 89:1068–1078.

Blakeslee, A. M. H., C. L. Keogh, J. E. Byers, A. M. Kuris, K. D. Lafferty, and M. E. Torchin. 2009. Differential escape from parasites by two competing introduced crabs. *Marine Ecology Progress Series* 393:83–96.

Bruun, H. H. 2006. Prospects for biocontrol of invasive *Rosa rugosa*. *BioControl* 51:141–181.

Burke, R. L., S. R. Goldberg, C. R. Bursey, S. L. Perkins, and P. T. Andreadis. 2007. Depauperate parasite faunas in introduced populations of *Podarcis* (Squamata: Lacertidae) lizards in North America. *Journal of Herpetology* 41:755–757.

Byers, J. E., S. Reichard, J. M. Randall, I. M. Parker, C. S. Smith, W. M. Lonsdale, I. A. E. Atkinson, et al. 2002. Directing research to reduce the impacts of non-indigenous species. *Conservation Biology* 16:630–640.

Callaway, R. M., G. C. Thelen, A. Rodriguez, and W. E. Holben. 2004. Soil biota and exotic plant invasion. *Nature* 427:731–733.

Calvo-Ugarteburu, G., and C. D. McQuaid. 1998. Parasitism and introduced species: epidemiology of trematodes in the intertidal mussels *Perna perna* and *Mytilus galloprovincialis*. *Journal of Experimental Marine Biology and Ecology* 220:47–65.

Catford, J. A., R. Jansson, and C. Nilsson. 2008. Reducing redundancy in invasion ecology by integrating hypotheses into a single theoretical framework. *Diversity and Distributions* 15:22–40.

Clewley, G. D., R. Eschen, R. H. Shaw, and D. J. Wright. 2012. The effectiveness of classical biological control of invasive plants. *Journal of Applied Ecology* 49:1287–1295.

Colautti, R. I., A. Ricciardi, I. A. Grigorovich, and H. J. MacIsaac. 2004. Is invasion success explained by the enemy release hypothesis? *Ecology Letters* 7:721–733.

Cornell, H. V., and B. A. Hawkins. 1993. Accumulation of native parasitoid species on introduced herbivores: a comparison of hosts as natives and hosts as invaders. *American Naturalist* 141:847–865.

Crawley, M. J., and M. P. Gillmann. 1989. Population dynamics of cinnabar moth and ragwort in grassland. *Journal of Animal Ecology* 58:1035–1050.

Creed, R. P., and S. P. Sheldon. 1995. Weevils and watermilfoil: did a North American herbivore cause the decline of an exotic plant? *Ecological Applications* 5:1113–1121.

Cripps, M. G., G. R. Edwards, G. W. Bourdot, D. J. Saville, H. L. Hinz, and S. V. Fowler. 2010. Enemy release does not increase performance of *Cirsium arvense* in New Zealand. *Plant Ecology* 209:123–134.

Cripps, M. G., M. Schwarzländer, J. L. McKenney, H. L. Hinz, and W. J. Price. 2006. Biogeographical comparison of arthropod herbivore communities associated with *Lepidium draba* in its native, expanded and introduced ranges. *Journal of Biogeography* 33:2107–2119.

D'Antonio, C. M., and P. M. Vitousek. 1992. Biological invasions by exotic grasses, the grass/fire cycle, and global change. *Annual Review of Ecology and Systematics* 23:63–87.

DeWalt, S. J., J. S. Denslow, and K. Ickes. 2004. Natural-enemy release facilitates habitat expansion of the invasive tropical shrub *Clidemia hirta*. *Ecology* 85: 471–483.

Duncan, R. W. 1997. Jumping Gall Wasp. Forest Pest Leaflet 80, Pacific Forestry Centre, Natural Resources Canada, Canadian Forest Service, Victoria, British Columbia, Canada.

Egan, S. P., and J. R. Ott. 2007. Host plant quality and local adaptation determine the distribution of a gall-forming herbivore. *Ecology* 88:2868–2879.

Elton, C. S. 1958. *The Ecology of Invasions of Animals and Plants*. London: Methuen.

Engelkes, T., E. Morriën, K. J. F. Verhoeven, T. M. Bezemer, A. Biere, J. A. Harvey, L. M. McIntyre, W. L. M. Tamis, and W. H. van der Putten. 2008. Successful range-expanding plants experience less above-ground and below-ground enemy impact. *Nature* 456:946–948.

Fenner, M., and W. G. Lee. 2001. Lack of pre-dispersal seed predators in introduced Asteraceae in New Zealand. *New Zealand Journal of Ecology* 25:95–99.

Forslund, H., S. A. Wikström, and H. Pavia. 2010. Higher resistance to herbivory in introduced compared to native populations of a seaweed. *Oecologia* 164: 833–840.

Fretwell, S. D. 1977. The regulation of plant communities by food chains exploiting them. *Perspectives in Biology and Medicine* 20:169–185.

Genton, B. J., P. M. Kotanen, P. O. Cheptou, C. Adolphe, and J. A. Shykoff. 2005. Enemy release but no evolutionary loss of defence in a plant invasion: an intercontinental reciprocal transplant experiment. *Oecologia* 146:404–414.

Grabenweger, G., P. Kehrli, I. Zweimuller, S. Augustin, N. Avtzis, S. Bacher, J. Freise, et al. 2010. Temporal and spatial variations in the parasitoid complex of the horse chestnut leafminer during its invasion in Europe. *Biological Invasions* 12:2797–2813.

Hairston, N. G., F. E. Smith, and L. B. Slobodkin. 1960. Community structure, population control, and competition. *American Naturalist* 44:421–425.

Hansen, S. O., J. Hattendorf, R. Wittenberg, S. Y. Reznik, C. Nielsen, H. P. Raven, and W. Nentwig. 2006. Phytophagous insects of giant hogweed *Heracleum mantegazzianum* (Apiaceae) in invaded areas of Europe and in its native area of the Caucasus. *European Journal of Entomology* 103:387–395.

Hay, M. E. 1991. Marine-terrestrial contrasts in the ecology of plant chemical defenses against herbivores. *Trends in Ecology and Evolution* 6:632–365.

Hierro, J. L., J. L. Maron, and R. M. Callaway. 2005. A biogeographical approach to plant invasions: the importance of studying exotics in their introduced and native range. *Journal of Ecology* 93:5–15.

Hudson, P. J., A. P. Dobson, and D. Newborn. 1998. Prevention of population cycles by parasite removal. *Science* 282:2256–2258.

Hulme, P. E. 2006. Beyond control: wider implications for the management of biological invasions. *Journal of Applied Ecology* 43:835–847.

Ishtiaq, F., J. S. Beadell, A. J. Baker, A. R. Rahmani, Y. V. Jhala, and R. C. Fleischer, R. 2006. Prevalence and evolutionary relationships of haematozoan parasites in native versus introduced populations of common myna *Acridotheres tristis. Proceedings of the Royal Society B* 273:587–594.

Jobin, A., U. Schaffner, and W. Nentwig. 1996. The structure of the phytophagous insect fauna on the introduced weed *Solidago altissima* in Switzerland. *Entomologia Experimentalis et Applicata* 79:33–42.

Keane, R. M., and M. J. Crawley. 2002. Exotic plant invasions and the enemy release hypothesis. *Trends in Ecology and Evolution* 17:164–170.

Kenis, M., and M. Branco. 2010. Impact of alien terrestrial arthropods in Europe. In *BioRisk*. Vol. 4, *Alien Terrestrial Arthropods of Europe*, edited by A. Roques and D. Lees, 51–7. Bulgaria: Pensoft.

Knevel, I. C., T. Lans, F. B. J. Menting, U. M. Hertling, and W. H. van der Putten. 2004. Release from native root herbivores and biotic resistance by soil pathogens in a new habitat both affect the alien *Ammophila arenaria* in South Africa. *Oecologia* 141:502–510.

Kolar, C. S., and D. M. Lodge. 2001. Progress in invasion biology: predicting invaders. *Trends in Ecology and Evolution* 16:199–204.

Levine, J. M., P. B. Adler, and S. G. Yelenik. 2004. A meta-analysis of biotic resistance to exotic plant invasions. *Ecology Letters* 7:975–989.

Lewis, K. C., F. A. Bazzaz, Q. Liao, and C. M. Orians. 2006. Geographic patterns of herbivory and resource allocation to defense, growth, and reproduction in an invasive biennial, *Alliaria petiolata. Oecologia* 148:384–395.

Liu, H., and P. Stiling. 2006. Testing the enemy release hypothesis: a review and meta-analysis. *Biological Invasions* 8:1535–1545.

Lodge, D. M., M. W. Kershner, J. E. Aloi, and A. P. Covich. 1994. Effects of an omnivorous crayfish (*Orconectes rusticus*) on a freshwater littoral food web. *Ecology* 75:1265–1281.

Lopez-Darias, M., A. Ribas, and C. Feliu. 2008. Helminth parasites in native and invasive mammal populations: comparative study on the Barbary ground squirrel *Atlantoxerus getulus* L. (Rodentia, Sciuridae) in Morocco and the Canary Islands. *Acta Parasitologica* 53:296–301.

Louda, S. M., D. Kendall, J. Connor, and D. Simberloff. 1997. Ecological effects of an insect introduced for the biological control of weeds. *Science* 277: 1088–1090.

MacDonald, A. A., and P. M. Kotanen. 2010. The effects of disturbance and enemy exclusion on performance of an invasive species, common ragweed, in its native range. *Oecologia* 162:977–986.

MacDougall, A.S., A. Duwyn, and N. T. Jones. 2010. Consumer-based limitations drive oak recruitment failure. *Ecology* 91:2092–2099.

MacLeod, C. J., A. M. Paterson, D. M. Tompkins, R. P. Duncan. 2010. Parasites lost: do invaders miss the boat or drown on arrival? *Ecology Letters* 13:516–527.

Maron, J. L., and M. Vilà. 2001. When do herbivores affect plant invasion? Evidence for the natural enemies and biotic resistance hypotheses. *Oikos* 95:361–373.

McEvoy, P. B. and E. M. Coombs. 1999. Biological control of plant invaders: regional patterns, field experiments, and structured population models. *Ecological Applications* 9: 387–401.

McEvoy, P. B., N. T. Rudd, C. S. Cox, and M. Huso. 1993. Disturbance, competition, and herbivory effects on ragwort *Senecio jacobaea* populations. *Ecological Monographs* 63:55–75.

McFadyen, R. E. C. 1998. Biological control of weeds. *Annual Review of Entomology* 43:369–393.

Memmott, J., S. V. Fowler, Q. Paynter, A. W. Sheppard, and P. Syrett. 2000. The invertebrate fauna on broom, *Cytisus scoparius*, in two native and two exotic habitats. *Acta Oecologia* 21:213–222.

Menéndez, R., A. González-Megías, O. T. Lewis, M. R. Shaw, and C. D. Thomas. 2008. Escape from natural enemies during climate-driven range expansion: a case study. *Ecological Entomology* 33:413–421.

Menge, B. A., and J. P. Sutherland. 1987. Community regulation: variation in disturbance, competition, and predation in relation to environmental stress and recruitment. *American Naturalist* 130:730–757.

Mitchell, C. E., A. A. Agrawal, J. D. Bever, G. S. Gilbert, R. A. Hufbauer, J. N. Klironomos, J. L. Maron, et al. 2006. Biotic interactions and plant invasions. *Ecology Letters* 9:726–740.

Mitchell, C.E., and A. G. Power. 2003. Release of invasive plants from fungal and viral pathogens. *Nature* 421:625–627.

Mueller, J. A., and J. J. Hellmann. 2008. An assessment of invasion risk from assisted migration. *Conservation Biology* 22:562–567.

Myers, J. H., and D. Bazely. 2003. *Ecology and Control of Introduced Plants*. Cambridge: Cambridge University Press.

Norghauer, J. M., A. R. Martin, E. E. Mycroft, A. James, and S. C. Thomas. 2011.

Island invasion by a threatened tree species: evidence for natural enemy release of mahogany (*Swietenia macrophylla*) on Dominica, Lesser Antilles. *PLOS ONE* 6:e18790.

Parker, J. D., D. E. Burkepile, and M. E. Hay. 2006. Opposing effects of native and exotic herbivores on plant invasions. *Science* 311:1459–1461.

Phillips, B. L., C. Kelehear, L. Pizzatto, G. P. Brown, D. Barton, and R. Shine. 2010. Parasites and pathogens lag behind their host during periods of host range-advance. *Ecology* 91:872–881.

Poulin, R., and D. Mouillot. 2003. Host introductions and the geography of parasite taxonomic diversity. *Journal of Biogeography* 30:837–845.

Prior, K. M. 2011. Novel community interactions following species' range expansions. PhD diss., University of Notre Dame.

Prior, K. M., and J. J. Hellmann. 2010. Impact of an invasive oak-gall wasp on a native butterfly: a test of plant-mediated competition. *Ecology* 91:3284–3293.

Prior, K. M., and J. J. Hellmann. 2013. Does enemy loss cause release? A biogeographical comparison of parasitoid effects on an introduced insect. *Ecology* 94:1015–1024.

Reinhart, K. O., and R. M. Callaway. 2004. Soil biota facilitate exotic *Acer* invasions in Europe and North America. *Ecological Applications* 14:1737–1745.

Reinhart, K. O., A. Packer, W. H. van der Putten, and K. Clay. 2003. Plant-soil biota interactions and spatial distribution of black cherry in its native and invasive ranges. *Ecology Letters* 6:1046–1050.

Reinhart, K. O., T. Tytgat, W. H. van der Putten, and K. Clay. 2010. Virulence of soil-borne pathogens and invasion by *Prunus serotina*. *New Phytologist* 186:484–495.

Richardson, D. M., P. Pyšek, M. Rejmánek, M. G. Barbour, F. D. Panetta, and C. J. West. 2000. Naturalization and invasion of alien plants: concepts and definitions. *Diversity and Distributions* 6:93–107.

Roy, H. E., L. J. Lawson Handley, K. Schönrogge, R. L. Poland, and B. V. Purse. 2011. Can the enemy release hypothesis explain the success of invasive alien predators and parasitoids? *BioControl* 56:451–468.

Schönrogge, K., G. N. Stone, and M. J. Crawley. 1995. Spatial and temporal variation in guild structure: parasitoids and inquilines of *Andricus quercuscalicis* (Hymenoptera: Cynipidae) in its native and alien ranges. *Oikos* 723:51–60.

Shea, K., and P. Chesson. 2002. Community ecology theory as a framework for biological invasions. *Trends in Ecology and Evolution* 17:170–176.

Shurin, J. B., E. T. Borer, E. W. Seabloom, K. Anderson, C. A. Blanchett, B. Broitman, S. D. Cooper, and B. S. Halpern. 2002. A cross-ecosystem comparison of the strength of trophic cascades. *Ecology Letters* 5:785–791.

Shurin, J. B., D. S. Gruner, and H. Hillebrand. 2006. All wet or dried up? Real differences between aquatic and terrestrial food webs. *Proceedings of the Royal Society B* 273:1–9.

Shwartz, A., D. Strubbe, C. J. Butler, E. Matthysen, and S. Kark. 2009. The effect of enemy-release and climate conditions on invasive birds: a regional test using the rose-ringed parakeet (*Psittacula krameri*) as a case study. *Diversity and Distributions* 15:310–318.

Sih, A., P. Crowley, M. McPeek, J. Petranka, and K. Strohmeier. 1985. Predation, competition, and prey communities: a review of field experiments. *Annual Review of Ecology, Evolution, and Systematics* 16:269–311.

Slothouber Galbreath, J. G. M., J. E. Smith, J. J. Becnel, R. K. Butlin, and A. M. Dunn. 2010. Reduction in post-invasion genetic diversity in *Crangonyx pseudogracilis* (Amphipoda: Crustacea): a genetic bottleneck or the work of hitch-hiking vertically transmitted microparasites? *Biological Invasions* 12:191–209.

Smith, J. L. 1995. Life history, survivorship, and parasitoid complex of the jumping gall wasp, *Neuroterus saltatorius* (Edwards) on Garry oak, *Quercus garryana* Douglas. Master's thesis, University of Victoria.

Stone, G. N., K. Schönrogge, R. J. Atkinson, D. Bellido, J. Pujade-Villar. 2002. The population biology of oak gall wasps (Hymenoptera: Cynipidae). *Annual Review of Entomology* 47:633–668.

Strauss, S. Y., C. O. Webb, and N. Salamin. 2006. Exotic taxa less related to native species are more invasive. *Proceedings of the National Academy of Sciences USA* 103:5841–5845.

Strong, D. R., J. H. Lawton, and R. Southwood. 1984. Insect on plants: community patterns and mechanisms. Cambridge, MA: Harvard University Press.

Torchin, M. E. 2002. Parasites and marine invasions. *Parasitology* 124:S137–S151.

Torchin, M. E., K. D. Lafferty, A. P. Dobson, V. J. McKenzie, and A. M. Kuris. 2003. Introduced species and their missing parasites. *Nature* 421:628–630.

Torchin, M. E., K. D. Lafferty, and A. M. Kuris. 2001. Release from parasites as natural enemies: increased performance of a globally introduced marine crab. *Biological Invasions* 3:333–345.

Torchin, M. E., and C. E. Mitchell. 2004. Parasites, pathogens, and invasions by plants and animals. *Frontiers in Ecology and the Environment* 2:183–190.

Van der Putten, W. H., G. W. Yeates, H. Duyts, C. Schreck Reis, and G. Karssen. 2005. Invasive plants and their escape from root herbivory: a worldwide comparison of the root-feeding nematode communities of the dune grass *Ammophila arenaria* in natural and introduced ranges. *Biological Invasions* 7:733–746.

Van Grunsven, R. H. A., F. Bos, B. S. Ripley, C. M. Suehs, and E. M. Veenendaal. 2009. Release from soil pathogens plays an important role in the success of invasive *Carpobrotus* in the Mediterranean. *South African Journal of Botany* 75:172–175.

Van Grunsven, R. H. A., W. H. Van der Putten, T. M. Bezemer, F. Berendse, and E. M. Veenendaal. 2010. Plant-soil interactions in the expansion and native range of a poleward shifting plant species. *Global Change Biology* 16:380–385.

Van Kleunen, M., and M. Fischer. 2009. Release from foliar and floral fungal pathogen species does not explain the geographic spread of naturalized North American plants in Europe. *Journal of Ecology* 97:385-392.

Vermeij, M. J. A., T. B. Smith, M. L. Dailer, and C. M. Smith. 2009. Release from native herbivores facilitates the persistence of invasive marine algae: a biogeographical comparison of the relative contribution of nutrients and herbivory to invasion success. *Biological Invasions* 11:1463–1474.

Vilà, M., J. L. Maron, and L. Marco. 2005. Evidence for the enemy release hypothesis in *Hypericum perforatum*. *Oecologia* 142:474–479.

Volin, J. C., E. L. Kruger, V. C. Volin, M. F. Tobin, and K. Kitajima. 2010. Does release from natural belowground enemies help explain the invasiveness of *Lygodium microphyllum*? A cross-continental comparison. *Plant Ecology* 208:223–234.

Ward-Fear, G., G. P. Brown, M. Greenlees, and R. Shine. 2009. Maladaptive traits in invasive species: in Australia, cane toads are more vulnerable to predatory ants than are native frogs. *Functional Ecology* 23:559–568.

Wikstrom, S. A., M. B. Steinarsdottir, L. Kautsky, and H. Pavia. 2006. Increased chemical resistance explains low herbivore colonization of introduced seaweed. *Oecologia* 148:593–601.

Williams, J. L., H. Auge, and J. L. Maron. 2010. Testing hypotheses for exotic plant success: parallel experiments in the native and introduced ranges. *Ecology* 91:1355–1366.

Wolfe, L. M. 2002. Why alien invaders succeed: support for the escape-from-enemy hypothesis. *American Naturalist* 160:705–711.

Yang, C., Y. Yu, S. M. Valles, D. H. Oi, Y. Chen, D. Shoemaker, W. Wu, and C. Shih. 2010. Loss of microbial (pathogen) infections associated with recent invasions of the red imported fire ant *Solenopsis invicta*. *Biological Invasions* 12:3307–3318.

Zocca, A., C. Zanini, A. Aimi, G. Frigimelica, N. La Porta, and A. Battisti. 2008. Spread of plant pathogens and insect vectors at the northern range margin of cypress in Italy. *Acta Oecologia* 33:307–313.

Where To from Here? Policy Prospects at International, National, and Regional Levels

The first three sections of this book covered ways that science and human perceptions drive responses to invasive species threats, how invasive species introduction can be understood and managed, and how established invaders can be controlled. While these chapters offer some hope for future invasion management, they also make clear the lack of effective policy for reducing harm caused by invasive species. Research results bear this out—invasive species continue to become established around the globe, and studies show that the number of established non-native species in many ecosystems is increasing faster now than ever before. If the impacts from invasive species are to be reduced, policy needs to be more comprehensive and cohesive. Section 4, the final multichapter section of the book, features four chapters that are directed specifically at policy issues and prospects. Each chapter concludes with a section titled "Looking Forward" in which the authors make recommendations for future policy.

The first of these (chapter 13), by Stanley (Stas) Burgiel, provides a broad framework for understanding at a global scale how different agencies, jurisdictions, and levels of government currently address invasive species issues. Next, Clare Shine (chapter 14) describes invasive species policy in the European Union, and efforts there to implement new approaches. The EU came together largely as a trading bloc to promote easy movement of goods and people. Breaking down trade barriers has benefits and costs, with one of the latter being the removal of restrictions that could be used to prevent the (potentially costly) movement of invasive species. The work described in this chapter is relevant to trading blocs across the globe.

Marc Miller (chapter 15) provides an overview of the development and current state of US law as it relates to invasive species. He shows that although invasive species are known as damaging across many sectors (e.g., agriculture, fisheries), there exists no effective and broad legal framework for addressing them. Instead, many agencies operate their own programs specific to their own interests. Finally, Joel Brammeier and Thomas Cmar (chapter 16) describe invasions in the North American Great Lakes that have been facilitated through ballast water and connecting waterways. They suggest specific policies and actions that would help protect the Great Lakes from future invasions.

The final, concluding chapter of the book (chapter 17), written by the editors, follows this section and looks broadly at ways forward for invasive species policy.

From Global to Local: Integrating Policy Frameworks for the Prevention and Management of Invasive Species

Stanley W. Burgiel

The views expressed in this article are those of the author and do not necessarily represent the views of, and should not be attributed to, the National Invasive Species Council or any other US government agency.

Terms

- Pathway: the means by which nonnative species are introduced into a new region.
- Sector: a commercial or noncommercial field sharing common characteristics, particularly from a perspective of policy and law (e.g., environment, trade, transport, human health, agriculture and livestock).
- Scale: the relative geographic dimension at which invasive species policy and management efforts are applied (e.g., international, regional, national, local)

Introduction

Invasive species are clearly recognized as a global problem that ultimately has its direct impacts at the local level (McNeely 2001; Mack et al. 2000; Bright 1999). Conversely, efforts to manage invasive species at the site level may be pointless without widening one's frame of reference to include adjacent areas and sources of new introductions. Working from these assumptions, this chapter will examine three intersecting lenses that can be used to analyze policy frameworks addressing invasive species: pathways for introduction, sector, and scale.

Pathways, the means by which invasive species are introduced into a new region, are the key chokepoints for stopping their movement. Pathways can be associated with traded goods, their packaging, and vehicles as well as the movement of individuals and their belongings. Given that pathways for introduction intersect with such a wide range of issues, policy responses have similarly been fragmented across a range of legal sectors. In general, commercial practices and their associated laws and policies evolved within particular fields such as agriculture, transport, or human health. This siloed approach superseded concerns about invasive species, which can easily intersect a number of sectors. Policy responses have often been grafted atop a broad range of commercial and noncommercial sectors with their own respective laws, policies, and institutions. While this analysis will not specifically define or assess particular pathways, it will outline the relation between various international institutions and agreements and the different movement vectors for invasive species.

Sector refers to the set of issues covered by international agreements and institutions, such as environment, trade, transport, human health, and agriculture and livestock. Significant attention will be paid to the role of trade in the movement of invasive species associated with commodities, travelers, and modes of transport. The trade agreements embodied in the World Trade Organization (WTO), such as the General Agreement on Tariffs and Trade and the Agreement on the Application of Sanitary and Phytosanitary Measures (SPS Agreement), are arguably the most important set of rules governing how to minimize the risks associated with the introduction and spread of invasive species. This section will also look at the role of international standard-setting bodies in the area of plant and animal health, as well as the role of environmental agreements in providing a more complete picture for the consideration of the ecological impacts of invasive species.

Finally, the concept of geographic scale will be used to show how policy issues move up and down the chain from the local site to the international level. Scale is particularly important when looking at the formation of policy frameworks and the role of institutional mechanisms that set the parameters for what countries can do to manage the risks of invasive species.

Effective prevention efforts require coordination across pathways for the intentional and unintentional introduction of invasive species, sectors of policy and law (trade, environmental, health), and geographic scale. Examples in this chapter will highlight these three approaches to looking at invasive species and how they contribute to the formulation of policies, particularly at the international level. The chapter will close with some reflections and recommendations on the relative benefits and disadvantages of policy making at these scales.

Pathways

Pathways can generally be defined as the means by which a potentially invasive species is moved and/or introduced into a new habitat. Such introductions can be intentional in order to meet a perceived economic or social objective, such as horticultural plantings, biofuel stocks, or aquaculture species. Alternatively, introductions can be unintentional where an invasive species hitchhikes on a commodity (e.g., agricultural pests), within associated packaging (e.g., wood-boring beetles in solid wood packaging material), in shipping conveyances (e.g., sea and air shipping containers), and on vehicles transporting goods (e.g., ship hulls and ballast water).

Analyses have depicted the suite of known pathways by different criteria. For example, the pathways group of the US Aquatic Nuisance Species Task Force and National Invasive Species Council (NISC) divided pathways into transport, living industry, and miscellaneous categories (NISC 2005). Hulme et al. look at the management of intentional and unintentional introductions using a framework that splits the issue across the commodity in question, the vector for movement, and dispersal mechanisms (Hulme et al. 2008). However they are categorized, many pathways are covered within the context of international law. International efforts to reduce the risk of agricultural pests and livestock diseases have a long history and are broadly embedded within the WTO's SPS Agreement and more directly within the work of the International Plant Protection Convention (IPPC) and World Animal Health Organization (OIE). Similarly, the International Maritime Organization has developed the International Convention for the Control and the Management of Ships' Ballast Water and Sediments to reduce risks associated with the movement of aquatic invasive species by ships.

Taking a different approach, the Convention on Biological Diversity has specifically identified gaps in the international legal framework that require the development of guidance for countries. While these may be areas lacking international rules or codes of conduct, some of them fall within the substantive scope of existing agreements and institutions. The following table lists some of the pathway gaps as well as relevant institutions where applicable.

Gaps and Inconsistencies in the International Legal Framework (Shimura et al. 2010; CBD 2006)

- Animals (not plant pests), including pets, live bait, live food fish (OIE, IPPC, WTO/SPS
- Agreement, United Nations Food and Agriculture Organization—FAO)
- Marine biofouling (International Maritime Organization—IMO)
- Civil air transport (International Civil Aviation Organization—ICAO)
- Aquaculture/mariculture (FAO)
- Conveyances (IPPC sea and air containers)
- Interbasin water transfers and canals
- Emergency relief
- Development assistance
- Military activities

While some of these pathway gaps are widely recognized (e.g., aquaculture and movement of live animals), others have not traditionally been associated with introductions. For example, emergency relief and disaster assistance can lead to the unintentional introduction and spread of invasive species, such as the introduction of the larger grain borer (*Prostephanus truncatus*) and parthenium weed (*Parthenium hysterophorus*) in Africa with food aid and the spread of cogongrass by emergency equipment after Hurricane Katrina (Hoggard 2006; Murphy and Cheesman 2006). More generally, development assistance has also been implicated in both the intentional introduction of invasive species through aquaculture (tilapia, perch), soil stabilization efforts (*Prosopis juliflora*), and biomass development (water hyacinth—*Eicchornia crassipes*) (Obiri 2011; Gutierrez and Reaser 2005). In response, some institutions like the Inter-American Development Bank and the African Development Bank have started considering invasive species within their safeguard policies for lending and project development. Finally, the movement of military vehicles from one area of operation to another has provided classic examples of introductions. Movement of military cargo and equipment has been associated with both the introduction of the brown tree snake (*Boiga irregularis*) onto Guam where it has decimated native vertebrate fauna as well as the introduction of the Formosan termite (*Coptotermes formosanus*) into Louisiana and other coastal towns at the end of World War II (LSUAgCenter 2012; Rodda et al. 1992; Savidge 1987).

This catalog of pathway gaps has raised a key question about the appropriate role of the Convention on Biological Diversity (CBD), an environmental agreement focused on biodiversity conservation, in facilitating efforts in other sectors of law or in areas arguably left uncovered by existing international agreements. While there is no easy answer and countries differ in their political views of the CBD's role, this does highlight the challenges of the patchwork approach to addressing pathways at the international level. This complexity of regulating pathways will be further highlighted in the following section, which explores the various areas of international law and policy relevant to invasive species.

Sectors

As mentioned above, a wide range of fields overlays the pathways by which invasive species move. At the international level these fields can loosely be categorized as trade, transport, animal and plant health, environment, and human health. Again there are overlaps, as trade links to both the transportation mechanism for moving goods as well as to health and sanitary concerns around commercial animal and plant products. However, this division fits nicely with the relative weighting of these agreements as elements of international law to be implemented at the national level. Agreements that are highly enforceable (e.g., subject to dispute settlement with carrots and sticks for compliance) are frequently described as "hard" law, whereas agreements with fewer mechanisms to enforce national implementation are seen as "soft" law (Abbott and Snidal 2000).

Table 13–1 highlights different sectors and some of the international agreements that fall within them. The list is organized along a gradient from hard through soft.

Trade is arguably the most important "sector" in the transfer of invasive species between countries. Trade by its nature involves the movement of goods and people around the world and is a major driver of economic growth. Globalization, a term which is viewed as synonymous with the international expansion of trade in the world economy, also connotes the reduction of economic, political, cultural, and physical barriers to the exchange of goods and ideas. A body of conventional wisdom has evolved linking invasive species, trade and globalization with this basic tenet:

Growth in international trade = increased invasive species

Many experts have identified a correlation between increasing international trade, industrial growth, and increasing introductions of invasive

TABLE 13-1 **International legal and institutional framework by sector.**

Sector	Framework	Type of Framework
Trade	• World Trade Organization (WTO) • General Agreement on Tariffs and Trade (GATT) • SPS Agreement	Enforceable/Binding
Animal & plant health	• International Plant Protection Convention (IPPC) • World Animal Health Organization (OIE) • UN Food and Agriculture Organization (FAO)	
Transport	• International Maritime Organization (IMO) • International Civil Aviation Organization (ICAO)	
Human health	• World Health Organization (WHO)	
Environment	• Convention on Biological Diversity (CBD) • Ramsar Convention on Wetlands of International Importance	Nonenforceable/Voluntary

species and their impacts (Aukema et al. 2011; Aukema et al. 2010; Essl et al. 2011; Holmes et al. 2009). More goods and commodities are being introduced into international trade, which includes variety (e.g., tropical fruits, exotic pets) as well as volume. As more countries enter the global market there are correspondingly more importers as well as exporters, which increases the range of ecosystems in these countries that can be source and destination for new invasive species. Finally, faster transit times due to increased air freight, faster ships, shorter and more efficient shipping routes, and expedited customs and quarantine processes may improve the chances of invasive species surviving transit (Burgiel et al. 2006; Ruiz and Carlton 2003; Jenkins 1996).

Prevention efforts, whether through inability or failure, have not kept pace with this increased risk of invasion. Many nations, particularly developing countries, lack the capacity to effectively address such risks through laws and regulations, responsible institutions, and scientific knowledge about existing native and nonnative species. Many developed countries also lack comprehensive systems addressing the suite of environmental, economic, and social aspects of biological invasions.

From an environmental perspective, there is arguably a lack of (or lesser) concern about the ecological impacts of invasive species as compared with impacts on agriculture, livestock, or public health. This is evident in the range of tools and political views that put economic concerns ahead of environmental ones (although arguments about the economic value of ecosystem services are gaining traction) (World Bank 2011; Vilà et al. 2009; TEEB 2008; Costanza et al. 1997). Finally, the trade-environment discourse of the mid-1990s and early 2000s voiced a concern about the WTO's "chill factor," where fear of invoking a WTO dispute was allegedly sufficient to prevent countries from setting more stringent national requirements to protect the environment from invasive species and other drivers of ecological destruction. Passage of time may have assuaged such fears, particularly given that the monolithic façade of the WTO has shown itself vulnerable to internal divisions and difficulties in finalizing the Doha round of trade negotiations.

Countering this conventional wisdom is also a growing body of "unconventional wisdom," suggesting that there are some bright spots amid this larger picture of a global increase in biological invasions. First, the risk of introductions may decline with increased trade, particularly in high-value sectors that could suffer major financial losses because of invasive species. For example, livestock companies and governments are very attentive to animal diseases like bovine spongiform encephalopathy (BSE), foot and mouth disease, and avian influenza, which have majorly disrupted national economies, local livelihoods, and food security. Similarly, fruit flies, pests of citrus and stone fruits, and a host of other crop pests concern agriculturists, and drive the usage of pesticides and integrated pest management techniques.

An obvious correlation here is improvement in the efforts and understanding of national governments, particularly in areas of national economic significance. Participation in regional bodies addressing plant and animal health is growing and many countries are developing national frameworks or strategies on invasive species. However, despite such progress, self-regulation and engagement directly with the private sector may often be more successful in the short and medium term than pursuit of comprehensive international standards or national regulatory systems.

Another piece of unconventional wisdom is that rigorous national prevention measures that require more stringent procedures to ensure "clean" commodities may adversely affect trade, particularly for countries lacking the resources to meet that higher bar. This tension has become a fundamental issue within the trade arena. On the one hand, the WTO recognizes the sovereign right of countries to protect themselves to whatever degree they see fit and justifiable. On the other hand, a major goal of the WTO's Doha Round of negotiations is to facilitate market access for developing countries, which sometimes clashes with more stringent SPS requirements. This can partially be addressed by focusing on measures that are least trade restrictive or least costly, harmonizing such SPS requirements and building capacity in those developing countries. However, the fundamental tension between protection of natural resources and increased market access still remains.

With this as background, the WTO's SPS Agreement is likely the most important international tool for regulating the risks associated with invasive species, even though the agreement itself doesn't use the term once. The SPS Agreement sets a framework for how countries can protect plant, animal, and human life or health (which includes the environment) in a manner that has the least impact possible on trade with other countries. It thereby establishes two important ways for nations to meet this objective. First, countries can establish their own SPS regulations, which require adherence to a number of requirements related to scientific basis, transparency, least-trade restrictiveness, and other aspects related to the conduct of risk analyses. Such measures can also include provisions for emergencies, when there is insufficient time for due process without incurring serious risk to the threatened resource (WTO 2010).

Alternatively, countries can implement national regulations that conform to existing international standards for plant and animal health as well as food safety, developed by the IPPC, OIE, or Codex Alimentarius (the three international standard setting bodies recognized by the WTO). In such cases, countries are not required to develop risk analyses or scientific justifications, as the standards in question have already been vetted and approved by the international community. Adherence to such international standards also has the benefit of harmonizing requirements across a number of countries, thereby simplifying the situation for companies and national agencies that would otherwise have to deal with multiple systems with different requirements. Two downsides to this approach are that such international standards do not provide comprehensive coverage of the risks or pathways associated with invasive species. For example, the OIE covers diseases that affect animals, but does not address animals that might prove invasive themselves. Additionally, many experts regard such international standards as a lower bar (or least common denominator) of protection that may not sufficiently protect a country against the full range of associated invasive species risks. That said, in a world of capacity gaps and resource needs, "plug and play" standards can be a dramatic improvement over the alternative of a country's doing nothing.

The key piece that makes these two options so effective is that their implementation is subject to dispute settlement under the WTO. Several cases under the WTO have looked at how risks are assessed, types of risk mitigation measures, and overall national processes for providing SPS approval for the import of goods, such as genetically modified organisms. The potential for such scrutiny of national processes, along with the threat of trade sanctions in cases of violation, is a significant motivator for countries to stick to the rules. The scope of the SPS agreement on all aspects of plant and animal health, along with the WTO's attention to compliance through a credible dispute settlement system, is what puts the WTO atop the list in terms of hard law relevant to invasive species risks.

Shifting away from the WTO, the next international line of defense against invasive species are those agreements addressing animal and plant health—specifically the IPPC and OIE. Generally, these institutions have focused on traditional agriculture and livestock issues. However, they have gradually moved into broader environmental concerns. For example, the IPPC expanded its guidance on pest risk assessments to include environmental impacts, and its pathway-focused efforts on solid wood packaging materials and sea containers address impacts on the full range of plant species. The OIE maintains lists of priority animal viruses and diseases for national notification, and it recently added two diseases impacting amphibians—*Batrachochytrium dendrobatidis* (chytrid fungus) and ranavirus (Kahn and Pelgrim 2010; WTO 2006).

As discussed above, these institutions are responsible for developing international standards that can then be implemented by member countries without further requirements to meet their obligations under the WTO SPS Agreement. Thus, they have significant value and weight from the perspective of international law in the areas that they have developed guidance. However, that stature is dependent on their recognition by the WTO. Additionally, they do not provide comprehensive coverage of the risks to plant and animal health, and for both of those reasons could be considered "medium" in strength as international law.

Transport-related agreements present a slightly different picture as they are oriented less with a particular commodity or product, and more with modes of transportation used for the transit of commercial goods and people. The International Maritime Organization's International Convention for the Control and Management of Ships' Ballast Water and Sediments sets out the requirements for ballast water management on different ship types/tonnages, as well as specific standards for both ballast water exchange and microbial concentrations in treated ballast water (IMO 2004). Many regard the standard, particularly on microbial, as setting a low bar (hence interest by US agencies to enact stricter requirements), and the agreement does include provisions for inspecting vessels and addressing violations. Consequences could include fines, processing delays, or refusal for vehicles to enter a particular port or waters. This provides a useful incentive for promoting compliance among ship owners and operators.

International governance over other transport-related pathways is less rigorous. IMO's Marine Environment Protection Committee has been developing guidance on biofouling of ship hulls, sea chests, and other parts. Similarly, the International Civil Aviation Organization has addressed linkages to aviation and invasive species on a few occasions. However, work under these two organizations (and other complementary processes) has not progressed beyond preliminary, nonbinding guidance for the biofouling and civil aviation pathways (ICAO 2004; ICAO 2001).

While the World Health Organization (WHO) is not an international treaty in itself (nor does it comprise a set of agreements such as the WTO), it is responsible for developing norms and other guidance for member states that address a range of human health issues. There are definitional debates about whether outbreaks of infectious diseases affecting humans can also be considered invasive species. (For example, the CBD's definition of invasive alien species refers only to impacts on biodiversity, and to date the WHO Secretariat has chosen not to participate in the international liaison group convened by the CBD.) Such issues aside, it is clear that the WHO along with the international public health community has developed a powerful early detection/notification and rapid response network that can mobilize national action and cooperation, expertise, and funds in the event of an outbreak of an infectious disease. The impetus for compliance comes not from enforcement measures but a self-interest in the health of one's populace that is directly tied to the broader connection between emerging infectious diseases and human health throughout a globalized world.

At the bottom of the spectrum are those soft law agreements that lack operative enforcement provisions or that provide nonbinding guidance to their member countries. The CBD is the prime example; however, its weakness in enforceability also serves as a strength in terms of the breadth of guidance that it can provide. Discussions under the WTO and within the relevant standard-setting agreements are subject to a high degree of scrutiny precisely because countries will be held accountable for not adhering to any agreed-upon rules. The more voluntary nature of the CBD has arguably enabled it to venture into a number of areas where these other agreements dare not tread. For example, it conducted the aforementioned review of gaps and inconsistencies in the international framework and has

developed a series of guiding principles for countries and other stakeholders to consider as they develop their national invasive species systems and
strategies. It has also started looking at some of these gaps by convening
groups of international experts to discuss the necessary guidance on these
unregulated pathways, such as the movement of animals through the pet,
aquarium, and live food trades.

The array of different instruments addressed above presents a number
of difficulties for coordinating work around invasive species, most specifically how to address and respect the particular mandate of each institution and the rules of its governing body. The CBD has started to serve
as a facilitator by convening the secretariats of the major relevant international agreements and institutions in a group called the international
liaison group on invasive alien species. This effort includes a focus on how
to improve coverage of existing gaps within the current mandate of participating institutions, as well as looking at how to share resources and
amplify synergies, particularly in the area of capacity building. Perhaps
more importantly, it serves as a mechanism for international secretariats
to communicate directly on a more regular basis. Even with such efforts
to coordinate, countries are still the final arbiters of interpreting and integrating guidance coming out of these international bodies into their national systems. This dynamic shifts the attention directly to the differing
policy functions and activities taking place at different geographic scales.

Scale

The previous section looking at the different sectors of law and policy
focused specifically at the international level and provides a segue into
a broader discussion about scale. It goes without saying that there is an
interplay between national policies and international legal regimes as they
work to inform and define each other. This iterative process can also link
in the regional and site levels, forming a chain that extends up from the
local resource or habitat threatened by an invasive species all the way to
the international level and then back down again. This section on scale
will look at each of these levels as depicted in Fig. 13–1 by briefly revisiting
the international level and then proceeding with a closer look at the policy
linkages associated with the regional, national, and local scales.

The fundamental role of international policy and law is to establish
the norms by which countries and their citizens behave around issues
of common interest. The range of international agreements described

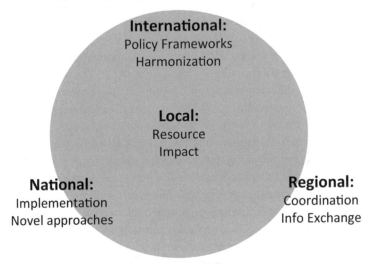

FIG. 13–1. Integrating approaches to prevention.

previously establishes a framework by which countries can act to reduce the risks posed to their economies, agricultural systems, and ecosystems. The focus on standards and harmonization of rules across countries sets a common playing field that countries can adopt and with which business actors and other nongovernmental entities can interact with some degree of predictability. It is certainly not a comprehensive or uniform system as some countries will pursue more rigorous or protective national systems and others may lack the capacity or political will to maintain even the basic floor of international standards. Despite such discrepancies the rules for how countries should act are clearly delineated. Additionally, international law and policy evolve as countries or groups thereof (including regional formulations such as the European Union—EU) propose and promote processes for developing guidance in new areas. Solid wood packaging and ballast water are two examples from the recent past, and the future may see standards and/or guidance around hull fouling and sea containers.

Shifting down in scale, the regional level is particularly important in the interplay between governments. The concept of region used herein refers to a geographic scale above the nation-state, such as continents or oceanic basins. Absent major physiogeological boundaries, geographic proximity allows for the spread of invasive species from one country to another. Within regions, major hubs for transport or trade can facilitate this spread, thereby reinforcing the need for countries to work together to identify, share information on, and coordinate around common priorities and invasive species risks. The spread of the cactus moth (*Cactoblastis cactorum*), which feeds on opuntia, a genus including ~200 species of prickly pear cacti, provides a key example. From its introduction into the southeastern Caribbean as a biocontrol in the early 1960s, the cactus moth has island hopped throughout the Caribbean basin, around the Gulf States of the United States, and potentially into Mexico (the center of genetic origin of opuntia), where it could potentially decimate populations of these important cultural, economic, and ecological species (Arroyo et al. 2007; Zimmerman and Sandi y Cuen 2005). Given such dynamics around the spread of invasive species, a country has a strategic interest in the biosecurity of its neighbors, which may be the strongest (or weakest) line of defense against the threat of new biological invasions. In fact, Mexico has provided financial support to the US government to look at methods to control the spread of the cactus moth in an effort to prevent it from reaching Mexican borders.

From a policy perspective, the regional level hosts a range of institutions that maintain formal and informal links to national governments and work around invasive species. The EU and other regionally based intergovernmental institutions provide a broad forum for the associated countries to identify and address common problems related to invasive species. The EU has engaged in an extensive process to develop a regional strategy on invasive species that could include core responsibilities and obligations for its member states (building on a looser arrangement developed under the Council of Europe and the Bern Convention on the Conservation of European Wildlife and Natural Habitats that sets out broad guidance while lacking active enforcement mechanisms) (Council of the European Union 2009). Invasive species have similarly been a focus of discussions under the South Pacific Regional Environment Programme (SPREP), the Secretariat for the Pacific Community (SPC), and the African Union's New Partnership for Africa's Development (NEPAD). While many would argue that NEPAD has not met its potential, in the Pacific SPREP and SPC have played a critical role in building national capacity, developing strategic priorities, and supporting collaborative initiatives such as the Pacific Ant Prevention Programme.

In a similar vein, free trade agreements have also provided a highly appropriate basis on which to look at invasive species issues. The Commission on Environmental Cooperation formed alongside the North American Free Trade Agreement has consistently addressed invasive species issues, including linkages to trade, tools for assessing the risk of aquatic species, and means to collate and share information and expertise across the continent (CEC 2008). Other US free trade agreements have included invasive species within their environmental capacity building efforts, such as the Central American Free Trade Agreement and potentially the TransPacific Partnership, a free trade agreement involving nine countries around the Pacific Rim that is currently under negotiation. The Australia–New Zealand Closer Economic Relations Trade Agreement is an example where the two countries are working to harmonize and streamline quarantine and inspection procedures.

The plant and animal health sectors also have regional institutions, some of which have proven quite effective in advancing national priorities up to the international level. For example, within the framework of the IPPC there are a number of regional plant protection organizations covering Asia and the Pacific, Europe and the Mediterranean, North America, Africa, the Caribbean, Central America, and the Southern Cone and Andean regions of South America. Given resources and the capacity of national members, some of these are more active than others (e.g., the North American Plant Protection Organization [NAPPO] and the European and Mediterranean Plant Protection Organization [EPPO]). By example, when the United States started intercepting pests on wood packaging materials arriving from China in the late 1990s, it established its own national measures to address the risk. The United States then worked collaboratively with Canada and Mexico to formulate such measures into a regional standard for the continent through NAPPO in 2001. This ultimately led to the development of an international standard on solid wood packaging material under the IPPC (APHIS 2003; IPPC 2002).

Leaving aside linkages to trade, the development of regional standards, and support for cooperative activities, perhaps the most basic role that can be played at the regional level is the sharing of information. Under the Inter-American Biodiversity Information Network a series of national nodes and databases across the Americas has been developed to gather and disseminate information on invasive species along with tools to assess their introduction and spread. Under the FAO, a series of regional networks focused on forests has also been established to share information. These initiatives include the Asia-Pacific Forest Information Network, the

Forest Invasive Species Network for Africa, the Near East Network on Forest Health and Invasive Species, and the Southern Cone Network on Invasive Alien Species and Forest Ecosystems.

Regional coordination and institutions do have their difficulties and obstacles, the least of which is adequate resources and capacity particularly in developing regions of the world. Additional political, cultural, and language differences within regions pose obstacles. For example, across the Caribbean there is no single intergovernmental entity that covers all the countries given the multitude of languages and historical affiliations, the status of overseas territories, and political issues around the relationship between the United States and Cuba. That said, regional-level entities and efforts, whether intergovernmental or nongovernmental, are an important means of sharing information and an incubator of policy priorities and guidance.

Many, following a state-centric paradigm, would argue that the national level is the most important given the sovereign role of countries to develop and implement their policy priorities. Certainly, without national action regional and international efforts would be severely hampered if not meaningless, while individually states could still protect their borders without such supranational institutions. However, the gains of regional and international cooperation outweigh the high resource costs of countries going it alone, as evidenced by the existing web of policy ties across geographic scales.

Nation-states or countries are the point of policy definition and implementation, which comes through a process of risk identification and prioritization. Implementation of regulations and activities focused on prevention can be targeted at potential risks before they reach the border, at the border through quarantine and inspection programs, and finally after the border through early detection/rapid response measures as well as eradication and control efforts. Also, despite the trend toward harmonization/homogenization described in the discussion on international policy, the national level shows a broad diversity of approaches and expertise.

New Zealand is renowned for its comprehensive biosecurity policies, and Australian work on risk analysis procedures has been replicated broadly. South Africa's Working for Water program is lauded for its combination of employment and livelihood programs with invasive species control efforts. Japan and Israel have both taken novel approaches toward the control of animal imports, particularly those intended for the pet trade. Finally, the United States has a wealth of experience across the board from policy development on invasive plants and animals, application of biocontrols and early detection methodologies, and basic research on invasive species. These examples highlight but a small portion of the breadth of responses by national governments in addressing the risk of invasive species. A broader list of national responsibilities would also include coordination across agencies and with local authorities and other stakeholders, strategic planning, development of legal systems, infrastructure for customs and quarantine, baseline research, funding, public education, and management activities.

Finally, the finest point in scale is the local or site level. The local level is on the one hand the point of impact for an introduced invasive species, and on the other hand the original source of that particular species somewhere else in the world. Without an adverse impact on a particular resource or habitat an invasive species would be regarded simply as a nonnative species of little concern. Control, management, and restoration activities by their nature occur at the site level, and have been a longstanding core of invasive species management efforts. However, the conventional wisdom that prevention is cheaper and more effective than responding to every incursion on the map supports a broader, outward-looking approach to minimize the potential for those introductions. We can thereby increase our effectiveness at preventing new introductions by progressively working back up the geographic scale through national, regional, and international policy efforts. That said, looking back to the local level, it's the experience with adverse impacts at the local level that informs and motivates the public, industry, and government leaders to take actions in their respective spheres.

Looking Forward

The main thrust of this analysis has been to show how invasive species policy and law is divided and intersects across sectors and pathways at one level, and then by geographic scale from another perspective. Advancing policy efforts and ensuring their implementation and efficacy can be a delicate dance that plays off of the relative political cachet of different governmental and intergovernmental efforts. Motivating government agencies and the secretariats of these international institutions to work together and outside of their traditional silos has been the biggest obstacle to date, but these discussions are now happening. Legal jurisdictions, governance, and available resources are likely to become the new stumbling blocks, but progressive initiatives in the form of regional collaboration and integrated national programs can serve as bright spots to help guide the way. This ongoing dialogue and our ability to mitigate the risk of invasive species from a policy perspective thereby require considering a number of key steps and lessons.

At the international level:
1. Plug the holes in the international framework through the development of guidance, keeping in mind the relative benefits of both soft and hard law approaches.
2. Increase coordination across international agreements in different sectors to build synergies at the policy level and enhance capacity building efforts. This can also help break through the policy silos across ministries at the national level.

At the regional level:
3. Increase information exchange and collaboration to minimize the spread of invasive species at the regional level. Identify and strengthen the weakest links in the chain of defense.
4. Identify and elevate policy priorities and management responses to the international level, where a global response is advantageous and appropriate.

At the national level:
5. Develop national invasive species strategies to identify the full suite of policy and management needs and priorities, which can be used to guide the further development of tools, infrastructure, and other measures.
6. Implement existing international standards as a basic floor and take advantage of rights under SPS Agreement to take stricter measures that meet national priorities.

At the site level:
7. Highlight the value of local resources and habitats and the consequences of invasive species that have adversely impacted or threaten those resources to educate and motivate public opinion and political will. Emphasize that prevention at a broader scale is the best defense at a smaller scale.

Literature Cited

Abbott, K. W., and D. Snidal. 2000. Hard and soft law in international governance. *International Organization* 54:421–457.

Animal and Plant Health Inspection Service (APHIS). 2003. *Importation of Solid*

Wood Packaging Material: Final Environmental Impact Assessment. Washington, DC: USDA.

Arroyo, H. S., J. C. Tovar, J. O. Cordoba, and C. A. Aguilera. 2007. Impacto económico y social en caso de introducción y establecimiento de la palomilla del nopal (*Cactoblastis cactorum* berg) en México. Vienna: International Atomic Energy Agency.

Aukema, J. E., B. Leung, K. Kovacs, C. Chivers, K. Britton, J. Englin, S. J. Frankel, et al. 2011. Economic impact of non-native forest insects in the continental United States. *PLOS ONE* 6(9):e24587. doi:10.1371/journal.pone.0024587.

Aukema, J. E., D. G. McCullough, B. Von Holle, A. M. Liebhold, K. Britton, and S. J. Frankel. 2010. Historical accumulation of nonindigenous forest pests in the continental United States. *BioScience* 60(11):886–897.

Bright, C. 1999. Invasive species: pathogens of globalization. *Foreign Policy* (116):50–64.

Burgiel, S., G. Foote, M. Orellana, and A. Perrault. 2006. Invasive alien species and trade: integrating prevention measures and international trade rules. Washington, DC: CIEL, Defenders of Wildlife.

Commission on Environmental Cooperation (CEC). 2008. The North American Mosaic: An Overview of Key Environmental Issues—Invasive Species. Montreal: CEC.

Convention on Biological Diversity (CBD). 2006. Decision VIII/27: Alien species that threaten ecosystems, habitats or species (Article 8) (h): Further consideration of gaps and inconsistencies in the international regulatory framework. Secretariat of the Convention on Biological Diversity, Montreal.

Costanza, R., R. d'Arge, R. de Groot, S. Farberk, M. Grasso, B. Hannon, K. Limburg, et al. 1997. The value of the world's ecosystem services and natural capital. *Nature* 387:253–260.

Council of the European Union. 2009. A Mid-Term Assessment of Implementing the EU Biodiversity Action Plan and Toward an EU Strategy on Invasive Alien Species: Council Conclusions (11412/09, ENV465). Brussels, Belgium: Council of the European Union.

Essl, F., S. Dullinger, W. Rabitsch, P. E. Hulme, K. Hülber, V. Jarošík, I. Kleinbauer, et al. 2011. Socioeconomic legacy yields an invasion debt. *Proceedings of the National Academy of Sciences USA* 108:203–207.

Gutierrez, A. T., and J. K. Reaser. 2005. Linkages between development assistance and invasive alien species in freshwater systems in Southeast Asia. Washington, DC: USAID Asia and Near East Bureau.

Hoggard, R. 2006. Observations concerning the spread of non-native plants in the wake of hurricane events. *Wildland Weeds* (Winter):10.

Holmes, T. P., J. E. Aukema, B. Von Holle, A. M. Liebhold, and E. Sills. 2009. Economic impacts of invasive species in forests: past, present, and future. In *The Year in Ecology and Conservation Biology*, edited by R. S. Ostfeld and W. H. Schlesinger. *Annals of the New York Academy of Sciences* 1162:18–38.

Hulme, P. E., S. Bacher, M. Kenis, S. Klotz, I. Kühn, D. Minchin, W. Nentwig, et al. 2008. Grasping at the routes of biological invasions: a framework for integrating pathways into policy. *Journal of Applied Ecology* 45:403–414.

International Civil Aviation Organization (ICAO). 2001. Resolution 33–18: Preventing the introduction of invasive alien species. Montreal: ICAO.

International Civil Aviation Organization (ICAO). 2004. Resolution 35–19: Preventing the introduction of invasive alien species. Montreal: ICAO.

International Maritime Organization (IMO). 2004. International convention for the control and management of ship's ballast water and sediments. London: IMO.

International Plant Protection Convention (IPPC). 2002. Guidelines for regulating wood packaging material in international trade. Rome: FAO.

Jenkins, P. 1996. Free trade and exotic species introductions. *Conservation Biology* 10:1.

Kahn, S., and W. Pelgrim. 2010. The role of the World Trade Organization and the "three sisters" (the World Organisation for Animal Health, the International Plant Protection Convention and the Codex Alimentarius Commission) in the control of invasive alien species and the preservation of biodiversity. *Revue Scientifique et Technique, Office International des Epizooties* 29(2):411–417.

LSUAgCenter. 2012. Formosan subterranean termites. Accessed April 9, 2012. http://text.lsuagcenter.com/en/environment/insects/Termites/termites/Formosan-Subterranean-Termites.htm.

Mack, R. N., D. Simberloff, W. M. Lonsdale, H. Evans, M. Clout, and F. Bazzaz. 2000. Biotic invasions: causes, epidemiology, global consequences and control. *Issues in Ecology* 5:1–20.

McNeely, J. A. 2001. The Great Reshuffling: Human Dimensions of Invasive Alien Species. Gland, Switzerland: IUCN.

Murphy, S. T., and O. D. Cheesman. 2006. *The Aid Trade: International Assistance as Pathways for the Introduction of Invasive Alien Species*. Biodiversity Series Paper No. 69. Washington, DC: World Bank.

National Invasive Species Council (NISC) Pathways Work Team. 2005. Focus Group Conference Report and Pathways Ranking Guide, June 21–22, 2005.

Obiri, J. F. 2011. Invasive plant species and their disaster-effects in dry tropical forests and rangelands of Kenya and Tanzania. *Journal of Disaster Risk Studies* 3(2):417–428.

Rodda, G. H., T. H. Fritts, and P. J. Conry. 1992. Origin and population growth of the brown tree snake, *Boiga irregularis*, on Guam. *Pacific Science* 46:46–57.

Ruiz, G. M., and J. T. Carlton. 2003. Invasion vectors: a conceptual framework for management. In *Invasive Species: Vectors and Management Strategies*, edited by G. M. Ruiz and J. T. Carlton, 459–504. Washington, DC: Island.

Savidge, J. A. 1987. Extinction of an island forest avifauna by an introduced snake. *Ecology* 68:660–668.

Shimura, J., D. Coates, and J. K. Mulongoy. 2010. The role of international organizations in controlling invasive species and preserving biodiversity. *Revue Scientifique et Technique, Office International des Epizooties* 29(2):405–410.

TEEB (The Economics of Ecosystems and Biodiversity). 2008. The Economics of Ecosystems and Biodiversity: An Interim Report. Cambridge: European Communities.

Vilà, M., C. Basnou, P. Pysek, M. Josefsson, P. Genovesi, S. Gollasch, W. Nentwig, et al. 2009. How well do we understand the impacts of alien species on ecosystem services? A pan-European, cross-taxa assessment. *Frontiers in Ecology and Environment* 8:135–144.

World Bank. 2011. *The Changing Wealth of Nations: Measuring Sustainable Development in the New Millennium.* Washington, DC: World Bank.

World Trade Organization (WTO). 2006. OIE international standards related to invasive alien species: Communication from the World Organization for Animal Health (OIE) (G/SPS/GEN/732). Geneva: WTO.

World Trade Organization (WTO). 2010. Sanitary and phytosanitary measures. WTO Agreements Series. Geneva: WTO.

Zimmerman, H. G., and M. P. Sandi y Cuen. 2005. The status of *Cactoblastis cactorum* (Lepidoptera: Pyralidae) in the Caribbean and the likelihood of its spread to Mexico (TC MEX/5/029 project). Vienna: International Atomic Energy Agency.

CHAPTER FOURTEEN

Developing Invasive Species Policy for a Major Free Trade Bloc: Challenges and Progress in the European Union

Clare Shine

Abbreviations

CBD: Convention on Biological Diversity
DAISIE: Delivering Alien Invasive Species Inventories for Europe
EPPO: European and Mediterranean Plant Protection Organization
EU: European Union
Guiding Principles: CBD *Guiding Principles for the prevention, introduction and mitigation of impacts of alien species that threaten ecosystems, habitats or species* (annexed to Decision VI/23 adopted at CBD COP6 in 2002 (The Hague, Netherlands)).
IEEP: Institute for European Environmental Policy
IPPC: International Plant Protection Convention
IS: invasive species
OIE: World Organisation for Animal Health
PRA: Pest risk analysis
WTO: World Trade Organization

1. Invasive Species as an EU Concern

This chapter focuses on the European Union (EU) at a pivotal moment of change. For ten years, decision makers within the world's biggest free trade bloc have been grappling with the challenge of how to address invasive species as a major driver of global environmental change. Several Member States have adopted their own domestic measures, while others have held back pending an EU-level decision. EU policy efforts have now gathered momentum, culminating in a formal commitment in 2011 to develop dedicated legislation applicable to the now 28 Member States (following Croatia's accession on July 1, 2013). The following sections explain the EU context and the specific challenges and features that will shape any future EU legislation. However, it remains to be seen whether the European financial crisis will dilute political will to reach agreement on a robust IS framework.

1.1 Overview of the EU as a Free Trade Bloc

The European Union (EU) is a regional economic integration organization with 28 Member States and 506.8 million inhabitants.[1] The combined territory of these countries covers 1,707,787 square miles and comprises the world's second-longest coastline after Canada.

EU Member States collectively operate a single internal market with a single external border. The EU currently accounts for 19 percent of world imports and exports.[2] Competence for trade policy is centralized with the European Trade Commissioner, both for international representation in the World Trade Organization (WTO) and with individual trading partners. Competence for environmental policy is shared between the EU and Member States: national environmental policy must comply with requirements laid down in specific EU directives or regulations.

As in other parts of the world, European lifestyles lead to species' introductions beyond their natural range, intentionally or by accident. Globalization—opening new trade routes, trading with new partners, diversifying the range of commercial products, expanding tourism—increases opportunities for organisms to be moved between continents and into, within, and from the EU.

In the last decade, excluding the impacts of the 2008–2010 financial crises, international trade grew at an average 12 percent / year[3] and annual GDP growth for EU-27 was 1.3–3.9 percent.[4] WTO International Trade Statistics show that merchandise trade within regions still dominates world trade, particularly in Europe where in-region trade accounted for 72 percent of European trade in 2009 (cf. 52 percent of Asia's exports remained within Asia and 48 percent of North America's exports remained in North America).[5]

EU transport pathways are also expanding. Taking shipping as an example, more than 90% of world trade is carried by sea: by 2018, the world fleet could increase by nearly 25% with volumes nearly doubling compared with 2008. Prior to the current economic crisis, EU maritime transport was predicted to grow from 3.8 billion tonnes in 2006 to 5.3 billion tonnes in 2018. Forty percent of intra-European freight is already carried by short-sea shipping and over 400 million sea passengers pass through European ports each year.[6]

1.2 Invasive Species Impacts and Trends in Europe

Many species intentionally introduced in the past underpin primary production systems central to Europe's economy. They provide employment opportunities and some are highly appreciated in society (e.g., ornamental plants, pet animals, exotic birds, game, fish for angling and aquaculture). Many introduced organisms do not spread. Among the ones that do, most remain in human-influenced habitat.

For these reasons, European concern focuses on the *subset* of nonnative species whose introduction and/or spread may threaten biological diversity and/or socioeconomic interests, including human livelihoods and public health. Consistent with usage throughout this book, these organisms are referenced in this chapter as "invasive species" (IS).

The EU-funded DAISIE research program[7] shows that IS occur in all taxonomic groups and all ecosystem types and affect the EU's continental landmass, islands, and seas. IS can trigger wholesale ecosystem changes, disrupting key services (Vilà et al. 2010). Invasive plants can be an important factor in weakening the stability of agricultural and forestry habitats, making them more vulnerable to pest outbreaks.

In some cases IS may have positive impacts, sometimes giving rise to conflicts of interest.[8] However, these are generally outweighed by documented negative impacts to the following:

- the economy and local livelihoods (e.g., lost yield, food insecurity, reduced water availability, land degradation, erosion);
- public health and well-being (e.g., allergies, skin problems, transmission of human diseases such as Chikungunya and salmonellosis, introduction of potentially dangerous animals such as poisonous snakes);
- biodiversity (e.g., extinction or displacement of species at the species and the genetic level through competition, transmission of diseases or hybridization; and/or alteration and threats to habitats and ecosystems);

- ecosystem services (e.g., water quality and retention, destabilization of river banks, erosion, changed nutrient cycles leading to changed food chains and/or disruption of plant-pollinator interactions) (Kettunen et al. 2009).

The DAISIE research indicates a strong correlation between economic growth and introduction rates. In Europe as a whole, the number of discoveries of IS grew by 76% over the period 1970–2007 (McGeoch et al. 2010). The *rate* of new introductions is considered to be still increasing for all taxonomic groups except mammals. The *cumulative number* of nonnative species is increasing for all groups including mammals, with one new mammal introduced per year (DAISIE data, presented in Hulme et al. 2008).

In monetary terms, lost output due to IS, health impacts, and expenditure to repair IS damage has already cost EU stakeholders at least 12 billion EUR / year over the past 20 years, of which costs identified for key economic sectors have been estimated at over 6 billion EUR / year (Kettunen et al. 2009).

These figures are significant underestimates for three reasons. First, information on ecological and economic impacts is available for only about 10 percent of the nearly 11,000 alien species present in Europe (Vilà et al. 2010). Second, data is scarce for key sectors like fisheries and forests and almost nonexistent for tourism. And third, because assigning a monetary value on possibly irreparable damage to environmental public goods is notoriously difficult, such damage is often simply left out of the calculation.

Biological invasions are predominantly human-induced processes. In addition to globalization, factors that contribute to the expansion of IS include global environmental change and chemical and physical disturbance to species and ecosystems (McNeely et al. 2001). Opportunities for introduced species to establish and spread in the EU are thus likely to increase in areas affected by environmental degradation caused by pollution, habitat loss, land use change, and extreme weather events.

Climate change impacts (e.g., warming temperatures, changes in CO_2 concentrations) are difficult to predict with certainty. However, European research suggests that they are likely to alter species' distributions and provide unaided pathways for dispersal by making it easier for the following to occur:

- species translocated to regions near their natural range to establish populations;
- nonnative species that are currently benign to become invasive for the first time;
- already-invasive species to increase their range further—or to become less of a threat (Capdevila-Argüelles and Zilletti 2008; Walther et al. 2009).

The interaction of IS and climate change has direct implications for the economy and society (Burgiel and Muir 2010). The European Commission recognizes that climate change could increase the spread of serious infectious vector-borne diseases, including zoonoses (diseases transmitted from animals to humans); threaten animal well-being; and impact plant health by favoring new or migrant harmful organisms that could affect trade in animals, plants, and their products (EC 2009).

2. Mandate for an EU Strategy and Early Warning System

2.1 Rationale for a Coordinated EU Policy Response

Invasive species are internationally recognized as one of the five pressures directly driving ecosystem change and biodiversity loss (MA 2005). EU and Member State policies have long been in place to address the other four: habitat change, overexploitation, pollution, and climate change. However, EU Member States have struggled to agree on the most effective way to take regionally coordinated action to prevent and manage biological invasions.

The EU has many entry points and extensive porous borders. Once a new organism enters from a third country, it may be freely moved within the internal market unless specific restrictions apply. As trade, border control, and customs are EU competencies, measures to identify and address IS risks associated with trade and transport pathways need to be jointly developed.

Territory within the EU encompasses very diverse biogeographic and climatic zones. Managing existing problem species, and preventing intra-EU introductions/spread of potentially invasive organisms, needs to be handled at the appropriate scale across political and administrative boundaries. Coordination between neighboring units on common threats is needed to avoid wasted efforts or investments on one side of a border.

Looking outwards, the EU is a major commercial exporter as well as a significant provider of external development assistance to non-EU third countries. Its policies need to address IS risks associated with export trade pathways, which can provide a source of potentially invasive species to other regions of the world, and ensure that EU-supported development programs do not contribute to the introduction or spread of IS.

Economic arguments also support action at EU level. "Prevention is better than cure" is the mantra of IS policy-making, given the financial and technical constraints of taking effective control action once an IS has become widespread.[9] EU-commissioned research into the costs of IS policy options (Shine et al. 2009) has demonstrated that inaction or delayed action leads to more serious impacts, costs more to the EU economy and societies, and damages the function and resilience of Europe's ecosystems.

As parties to the Convention on Biological Diversity (CBD), the EU and its Member States are committed to take prevention or management actions for species that threaten ecosystems, habitats, or species.[10] Of the 174 European species listed as critically endangered by the IUCN Red List, 65 are in danger because of introduced species. Well-known examples of threatened native species include European mink *Mustela lutreola* and Ruddy duck *Oxyura leucocephala*.

Much of Europe's globally significant biodiversity is found in isolated islands. Vulnerable to introduced marine pests, these islands have suffered disproportionately from invasions that threaten endemic and endangered species, particularly seabirds, and rare habitat types. The EU recognizes that effective action in these overseas territories is vital to its credibility regarding biodiversity conservation.[11] However, the legislative picture is complex. The EU's Outermost Regions[12] are subject to EU law and form part of the internal market, despite their biogeographic isolation, whereas the Overseas Countries and Territories have variable legal status and are free to adopt locally driven IS prevention measures[13] (RSPB 2007, Soubeyran 2008).

Although some Member States have developed lively awareness-raising campaigns, IS risks and impacts still have low visibility in the EU as a whole. In a 2010 survey commissioned by the European Commission on *Attitudes of Europeans towards the issue of biodiversity*, only 3 percent of respondents selected IS as the most important threat to biodiversity (the top three threats identified across the EU as a whole were air and water pollution [27%], manmade disasters, such as oil spills or industrial accidents [26%]; and intensive farming, deforestation and overfishing [19%]).[14]

2.2 Commitment of EU Institutions to Take Action

Europe's IS policy landscape changed radically soon after the adoption of the CBD Guiding Principles, when the forty member countries of the Council of Europe formally endorsed the *European Strategy for Invasive Alien Species* (Genovesi and Shine 2004). This nonbinding instrument was developed after three years of stakeholder consultations and provided a catalyst for motivated countries to strengthen their domestic IS policies and educational activities. Biannual follow-up meetings of the Council of Europe's IS Expert Working Group served to evaluate progress, share lessons learnt, agree on technical guidance on key risks, and develop voluntary codes of conduct for specific pathways (e.g., horticulture, the pet trade).

The European Commission did not participate directly in the Council of Europe process. However, in 2006, EU institutions formally agreed to develop a strategy "to substantially reduce the impacts of invasive alien species and alien genotypes and to establish an early warning system,"[15] and called for this strategy to be aligned with the CBD Guiding Principles[16] and to take account of the Council of Europe's Strategy.

This EU commitment triggered various technical studies and an online consultation.[17] In 2008 the Commission published a Communication *Towards an EU Strategy on Invasive Species* (EC 2008) which outlined four policy options. These ranged from (A) business as usual and (B) maximizing use of existing approaches and voluntary measures to (C) targeted amendment of existing legislation and/or (D) a comprehensive, dedicated EU legal framework.

The EU's Committee of the Regions and the European Economic and Social Council supported the development of dedicated legislation and emphasized the need for urgent and immediate action.[18]

In 2009, the Environment Council of the Council of the European Union—comprising the Environment Ministers of the EU's then 27 Member States—formally supported the development of a comprehensive framework to cover prevention and information exchange; early detection, warning, and rapid response, including prevention of spread and eradication; monitoring, control, and long-term containment; and restoration of damaged biodiversity where feasible. The Council called for a biogeographic approach and the specific circumstances of islands and ultraperipheral regions to be taken into account and for EU and national policies on trade, agriculture, forestry, aquaculture, transport, and tourism to integrate IS prevention.[19]

In 2010, the Environment Council identified the lack of efficiently targeted instruments to tackle IS as one reason why the EU had missed its target to halt biodiversity loss by 2010.[20]

In May 2011, the European Commission published its Biodiversity Strategy to 2020 (EC 2011) for reversing biodiversity loss[21] and accelerating the EU's transition toward a resource efficient and green economy. This provides a delivery framework for the global commitments in the CBD Strategic Plan and Aishi Targets[22] as well as the EU's own policy objectives. For invasive species, the 2020 Strategy (§3.4) notes that

- IS threats are likely to increase unless robust action is taken at all levels to control the introduction and establishment of these species and address those already introduced;
- although the challenges posed by IS are common to many Member States, there is no dedicated, comprehensive EU policy to address them with the exception of legislation concerning the use of alien and locally absent species in aquaculture;
- this gap should be filled with a dedicated EU legislative instrument that could tackle outstanding challenges relating inter alia to IS pathways, early detection and response, and containment and management of IS (see Part 4 below).

3. Existing EU Policy Baseline and Main Challenges

3.1 International and Regional Context for EU Action

When formulating action to address IS threats, the EU has to consider the patchwork of multilateral agreements, standards, and guidance applicable in different sectors within its competence. This fragmented policy landscape—and its relationship to the international trade regime—has been extensively studied through the Global Invasive Species Programme, the CBD, and other bodies (see, e.g., Shine 2007; Burgiel et al. 2006; Shine et al. 2005) and is further discussed in chapter 13 (Stas Burgiel).

The term *invasive species* is sufficiently broad to encompass alien pests and diseases of plants and animals. In the field of animal and plant health, EU frameworks seek convergence with standards developed by the World Organization for Animal Health (OIE) and the International Plant Protection Convention (IPPC). These are both recognized as standard-setting bodies by the WTO Agreement on the Application of Sanitary and Phytosanitary Measures. OIE and IPPC have traditionally focused on preventing and managing threats to farmed livestock and agricultural crops but the need to address risks to wild species and the uncultivated environment is progressively recognized.

Pioneering action has been taken by the relevant IPPC regional body, the European and Mediterranean Plant Protection Organization (EPPO), which has 50 member countries. Working closely with the Council of Europe's IS Expert Group, EPPO has progressively mainstreamed biodiversity considerations in its program on invasive alien plants.[23] It set up a dedicated Panel on Invasive Alien Species to identify pests that could present a risk to natural plant communities and propose management options. The EPPO decision support scheme, "Pest Risk Analysis for quarantine pests," now covers biodiversity risks and was recently used for pathway-scale risk analysis for aquatic plants.[24] The Panel is now developing prioritization criteria to support a common approach to identifying the highest-risk invasive plant species in Europe.

The EU itself is not an EPPO member but several Member States play a proactive role on the Panel (e.g., Germany, Sweden, United Kingdom) and have argued for synergies between national plant health and IS frameworks. The European Commission is funding large-scale research to streamline and improve pest risk analysis techniques.[25]

In the transport sector, the only binding instrument for an invasion vector is the International Maritime Organization (IMO) International Convention for the Control and Management of Ships' Ballast Water and Sediments 2004. Ratification is a national competency but the EU provides support through the European Maritime Safety Agency. This agency monitors international, regional, and subregional developments to help Member States and the Commission identify any need for further EU-level action to promote effective ballast water management on board ships in European waters and ensure a coherent approach within different European regions.[26] Ballast water risks are also addressed through the four regional seas conventions around Europe.[27]

3.2 Scope of Existing EU Policies and Funding to Tackle IS Threats

The EU policy baseline reflects the complexity at the international level. IS provisions are scattered across a range of instruments (for details, see Shine et al. 2010), which adds to the challenge of coordinating a cross-sectoral policy response.

The EU Animal Health Strategy and Action Plan (EC 2007) aims to establish a modernized single regulatory framework with improved incentives for responsibility- and cost-sharing and better border and on farm biosecurity. It covers the health of all EU animals (food, farming, sport, pets, entertainment, zoos), including wild animals and animals in research that entails a risk of their transmitting zoonoses. Environmental risks associated with the import, release, or escape of alien animals are not addressed but the Strategy provides a basis to regulate the import and intra-EU movement of animals that are vectors of diseases that could affect native biodiversity.[28]

The EU Plant Health Regime[29] is an open system that allows movement of all plants and plant products unless import bans or intra-EU controls are imposed to prevent introduction/spread of "harmful organisms." These are defined as "pests of plants or of plant products, which belong to the animal or plant kingdoms, or which are viruses, mycoplasmas or other pathogens." The main focus is agriculture and forestry although several plant pests with biodiversity impacts are listed.[30] The Regime has not been used to assess risks associated with intentional introductions of, for example, fast-growing species for afforestation or biofuel cultivation. It was recently evaluated for its ability to take account of emerging threats linked to globalization and climate change and better align its concepts and scope with IPPC, EPPO, and other instruments.[31] The Evaluation recommended expanding the regime's scope to cover harmful organisms with environmental impacts and possibly the natural spread of harmful organisms not linked to trade-related movement, for example, including as a consequence of climate change (FCEC 2010).

Both the animal and plant health regimes establish EU-wide rules for border inspections and common documentation and tracking systems to accompany intra-EU movements of cleared commodities.[32] Mandatory reporting, early warning, and rapid response systems are enabled through the Animal Disease Notification System and EUROPHYT. Intra-EU movement controls (i.e., derogating from free movement of goods within the Single Market) may be imposed in limited situations to prevent or minimize disease or pest incursions. EU cofinancing is available to reduce the cost of disease/pest spread and minimize barriers to intra-EU trade. An EU Veterinary Emergency Team was established in 2007 and may intervene when a Member State or third country requests assistance. There is no equivalent for plant health.

The Wildlife Trade Regulation[33] is used to regulate EU trade in protected species of wild fauna and flora. Interestingly, it provides a legal basis to suspend the import of *"live specimens of species for which it has been established that their introduction into the natural environment of the Community presents an ecological threat to wild species of fauna and flora indigenous to the Community."* Intra-EU movement and holding (e.g., for captive breeding and rearing) of such species may be regulated, but this has not been done for the four invasive animal species currently banned for import.[34] However, the Regulation does not cover rapid response or control and has no stand-alone provisions to address trade in wildlife species that are native in some parts of the EU and alien and potentially invasive in others. Official EU information materials do address welfare and escape risks associated with live specimens of exotic animals and the potential ecological threats posed to EU biodiversity.

The Aquaculture Regulation[35] is a pioneering EU instrument to address the ecological risks of introductions by a specific sector. It provides a legally binding framework for Member States to assess and minimize possible impacts of alien and locally absent species used in aquaculture. In so doing, it contributes to aquaculture's sustainable development. Unlike the animal, plant, and wildlife trade regimes, the Regulation establishes a decentralized and closed (white list) system, except for Annex IV species that have been used in aquaculture for a long time in certain parts of the EU.[36] Member States have primary responsibility for risk assessments, permitting and follow-up measures, based on mandatory criteria.[37] The precautionary principle is embedded through a risk-based distinction between open and closed facilities, provisions for pilot release, contingency planning, monitoring, and rapid response if an introduced species or nontarget organism becomes invasive. Where a neighboring country may be affected, prior consultation is required and the European Commission has override powers to confirm, amend, or cancel a permit, following scientific and stakeholder consultations.

The Habitats Directive[38] and Birds Directive[39] underpin EU biodiversity policy through the Natura 2000 network of protected sites and a strict system of species and habitat protection. Both directives contain an explicit prevention obligation regarding introductions of nonnative species into the wild (though key terms are not defined). In practice, however, they have not been consistently applied across Member States and have proved ineffective in preventing the continued introduction and spread of IS on European territory. The Directives do not address the keeping of IS in containment or captivity, nor explicitly provide for surveillance / rapid response, although many control and management programs have been supported by the LIFE and LIFE+ financial instruments.

The Water Framework Directive and Marine Strategy Framework Directive[40] establish common frameworks for Member States to assess, monitor and improve the ecological status of their inland, coastal, and marine waters, cooperating at the river basin or regional sea level. However, only the Marine Directive explicitly includes IS impacts in the list of pressures affecting the quality of waters to be monitored. Several Member States do address IS impacts when implementing the Water Framework Directive but there is wide variation in the scope of national monitoring and the way IS data is used in ecological status classification (ECOSTAT 2009).

Relevant horizontal (cross-cutting) measures include the following:

- the Environmental Liability Directive,[41] which establishes a framework based on the "polluter pays" principle for environmental damage or for an imminent threat of such damage caused by pollution of a diffuse character, where it is possible to establish a causal link between the damage and the activities of individual operators;
- EU support for risk assessment through the European Food Safety Authority. This was created to mobilize and coordinate scientific resources to provide high-quality and independent scientific advice and review risk assessments carried out by other actors. It includes specific panels on animal health and plant health and works on risk assessments for genetically modified organisms that may be relevant to alien genotypes[42];
- EU financial instruments that support IS prevention/management measures. Through the EU Financial Instrument for the Environment (LIFE) and Framework Research programs, the EU has contributed to financing almost 300 projects addressing IS for a total budget exceeding 132 million EUR over the last 15 years. However, this is not suited to rapid response as the selection procedure takes about 12 months (Scalera 2010);
- The European Agricultural Fund for Rural Development provides opportunities to support IS control as part of Good Agricultural and Environmental Condition (GAEC) measures within cross-compliance. Generally, however, IS considerations are poorly integrated in EU programs funded with the major budget lines (Scalera 2008).
- EU policies to address landscape connectivity and climate change adaptation (EC 2009) and promote renewable energy[43] (which could include cultivation and afforestation of fast-growing species or genotypes for biofuel/biogas production within the context of climate change mitigation).

3.3 Challenges for the Future EU Strategy

LEGAL UNCERTAINTY IN THE CONTEXT OF THE SINGLE MARKET.   Only one EU instrument (the Aquaculture Regulation) currently sets out binding criteria to guide Member State measures to regulate movement / holding of potential IS with ecological impacts. In the absence of clear standards for other pathways, Member States wanting to adopt domestic prevention measures must respect the principle of free movement of goods within the Single Market. The Treaty on the Functioning of the European Union prohibits quantitative restrictions on imports and measures having equivalent effect unless necessary for the protection of health of humans, animals, and plants.[44] Any such measures must be proportional and not constitute a means of arbitrary discrimination or a disguised restriction on trade between Member States.

To date, very few European Court of Justice (ECJ) rulings have interpreted these provisions with regard to IS. This legal uncertainty has deterred some Member States from adopting national measures potentially affecting intra-EU trade, even for known risks (e.g., Sweden). However, other countries—including Belgium, France, Ireland, Spain, and the United Kingdom—have significantly expanded domestic measures to regulate trade-related pathways, each using their own procedures and criteria for risk assessment and screening. In parallel, some subnational jurisdictions have taken the lead where national measures are slow to be adopted: for example, the Spanish Autonomous Community of Valencia banned trade in water hyacinth and other known invasive plants used in horticulture, even when these were legally sold in neighboring local jurisdictions.

GAPS IN LEGISLATIVE SCOPE.   Taxonomic coverage is weak for nonnative animals and for nonnative plants that do not qualify as diseases or pests. Existing legislative instruments do not address IS risks associated with captive-bred specimens and coverage is not explicit at the level of subspecies and genotypes.

INCONSISTENT TERMINOLOGY AND LACK OF DATA.   The lack of agreed definitions or criteria affects the development of indicators and the comparability of existing databases (Genovesi et al. 2013). Lack of data on both invasive and native species means that species' invasiveness tends to be underestimated and detection of impacts is more difficult (McGeoch et al. 2010).

LACK OF A UNIFIED INFORMATION AND EARLY WARNING SYSTEM SUPPORTED BY RISK ASSESSMENT.   For IS affecting biodiversity, there is no equivalent of the EU animal/plant health information, early warning, and rapid response systems. Impacts on biodiversity and ecosystem functions are not explicitly addressed in existing EU frameworks except for aquaculture.

Contingency planning and rapid response for environmental threats is essentially a matter for national discretion. There are no targeted policies to protect vulnerable ecosystems.

MONITORING AND MANAGEMENT GAP.   There are no EU instruments or criteria to monitor the status and spread of IS within the EU nor a comprehensive inventory of IS monitoring schemes. The EU also lacks a prioritized threat list and a common approach to managing IS already established in one part of EU territory to prevent spread to other parts of the EU. However, there are some promising examples of more localized transboundary or biogeographic coordination, aligned with the ecosystem approach. For example, island-wide IS coordination is now in place between the Republic of Ireland and Northern Ireland, using shared risk-based listing for trade controls and common IS management measures for transboundary rivers.

4. Progress toward Dedicated EU Legislation on Invasive Species

Target 5 of the EU Biodiversity Strategy to 2020 (EC 2011) provides that *"By 2020, Invasive Alien Species and their pathways are identified and prioritised, priority species are controlled or eradicated, and pathways are managed to prevent the introduction and establishment of new IAS."*
    The Strategy proposes a two-prong approach to meet this target:

- Action 15: Strengthen the EU Plant and Animal Health Regimes. The Commission will integrate additional biodiversity concerns into the Plant and Animal Health regimes by 2012.
- Action 16: Establish a dedicated instrument on Invasive Alien Species. The Commission will fill policy gaps in combating IS by developing a dedicated legislative instrument by 2012.

Two EC Directorates-General (Health & Consumers (DG SANCO) and Environment (DG ENV)) have led legislative development in consultation with other key services. Proposals for detailed measures are then followed by regulatory impact assessment. Revision of the plant and animal health regimes were launched first, followed by the IS instrument to close legislative gaps.
    The Environment Directorate-General constituted three working groups on Prevention, Early Warning and Rapid Response, and Control

and Management, with members drawn from governments, business and civil society. Their starting point was the technical study prepared by the Institute for European Environmental Policy (Shine et al. 2010), which integrated the findings of the European Environment Agency's feasibility study for a Europe-wide IS information and early warning system (Genovesi et al. 2010).

The IEEP study made concrete proposals for components requiring a legislative basis, taking account of existing EU measures and financial instruments. It envisaged a two-tier approach, combining a streamlined framework of principles and measures to guide Member State prevention and management efforts with EU-level measures for the most serious threats. It also analyzed possible costs of administrative systems needed to implement such components and how such costs would be distributed between the EU and Member States under different scenarios.

Four strategic goals were suggested to guide legislative development:

- regional coordination on risk assessment, building on available protocols, best practices, and capacity for application at species, pathway, and/or biogeographic level. This should support consideration of biodiversity and ecosystem impacts and, where possible, socioeconomic impacts linked to cost-benefit analysis;
- a structured framework for pathway management focused on prevention and rapid response at appropriate biogeographic scales and clearly allocating roles and responsibilities at each stage. This should be closely linked to the proposed EU/Europe-wide information and early warning system to make it possible to exclude or respond promptly to incursions before species become problematic;
- a regionally coherent approach for managing established IS and ecological restoration, taking account of climate change as a future driver of IS spread. Recommended measures include coordinated action plans, linked to ecosystem-based approaches under the existing EU nature, freshwater, and marine directives; targeted eradication actions for isolated islands; and maintaining or restoring resilient ecosystems to improve adaptation capacity to climate change and continued supply of ecosystem services;
- building EU awareness, responsibility, and incentives, based on a partnership approach. A coherent framework should be aligned with the general legislative principles of EU environmental policy (precaution, prevention, rectifying pollution at source and "polluter pays").[45] Measures could include campaigns to increase issue visibility at political, industry, and consumer levels; voluntary codes of conduct and other initiatives to support technical innovation and species substitution; market-based instruments, including development or extension of certification schemes to address key pathways; improved cost recovery

and liability mechanisms; and leverage of EU funding instruments to support IS mainstreaming across all key sectors.

With regard to terminology, the IEEP study emphasized the need to reach a common operational understanding. "Nonnative/alien" concepts need to be capable of application at the appropriate biogeographic scale (i.e., not limited by political or administrative boundaries). Any definition used in future legislation will need to cater for species that are native in part of the EU/a single country and alien (and potentially invasive) in other parts of the EU/the country concerned. Consideration should also be given to how to treat the expansion of species ranges within EU territory, including but not limited to altered distribution linked to climate change.

Broad consensus exists about the need for prioritization criteria to target IS interventions at the highest risks and allocate limited resources based on feasibility of outcomes. A mechanism is needed to categorize "IS of EU concern,"[46] for example, IS that affect EU-protected species / habitats or have significant environmental, economic, and/or health impacts that adversely affect other EU policy objectives.

This requires the development of a system of species lists, based on risk assessment, which could be annexed to the future regulatory instrument. Species listed in the highest risk category could be subject to mandatory prevention, monitoring, and/or management measures, either at EU-wide level or in defined biogeographic areas. This would provide a legal basis to enforce response measures where needed and avoid the "weakest link" problem, whether the required action was local (e.g., isolated islands), transboundary (e.g., shared river basins), or subregional (e.g., European marine waters).

To be efficient and cost-effective, such a system needs a flexible and rapid decision-making procedure to ensure that regulatory listing is responsive to new threats or changes in target species / ecosystems. Preliminary proposals for a biogeographic approach to listing (NOBANIS 2010) will require further technical analysis to assess its feasibility for regulatory purposes.

5. Looking Forward

In February 2013, the EU finally proposed concrete legislative measures in the form of a draft *Regulation Of The European Parliament And Of*

The Council on the Prevention And Management Of The Introduction And Spread Of Invasive Alien Species.[47] After years of preparatory work and discussions, Member States will have to decide whether to give political backing to binding rules to tackle expanding pathways and curb the exponential cost of policy inaction.

The proposed EU Regulation would establish a legal framework for three types of interventions: prevention, early warning and rapid response, and management. A list of invasive alien species of Union concern will be drawn up with Member States using risk assessments and scientific evidence. Implementation will require cooperation between EU directorates to reach a common understanding of key terms and concepts as well as coordination between services at EU external borders. Agreed goals and criteria can provide a clearer basis to manage policy trade-offs. The impacts of undetected or uncontrolled introductions and spread affect EU biodiversity, ecosystem functions, livelihoods, and business. However, levels of public awareness and decision-maker concern about IS issues vary widely between Member States. Some island countries and countries with islands have been amongst those pushing hard for progress, but the financial crisis will inevitably increase pressure to avoid additional regulatory burdens and cost to industry.

A key step would be to give Member States greater legal certainty about the domestic measures they may adopt to regulate species holding/movement without infringing the rules of the Single Market. Individual countries need to improve cross-sectoral coordination to tackle known and emerging pathways more efficiently and also cooperate strategically across borders.

Making best use of scarce resources will require more systematic mainstreaming of IS at all levels. Opportunities to integrate IS prevention and management into sectoral activities supported by the main EU financial instruments should be maximized.

Proportionate, targeted, and scientifically justified action—consistent with relevant WTO agreements—needs to be facilitated through a regionally coordinated risk assessment platform and the development of a European information and early warning system. Costing carried out for a European feasibility study (Genovesi et al. 2010) suggests that the overall cost of an EU system would provide economies of scale and be cheaper for Member States than setting up and running discrete national systems, particularly if such a system were harmonized with existing plant/animal health information systems.

Improved horizon scanning, data supply, and technical capacity are critical to support rapid eradication and help prioritize management programs at subregional or EU level. These should be tailored to the appropriate biogeographic scale and provide high environmental protection for the most ecologically vulnerable areas, including isolated islands.

Delivery mechanisms for the future EU framework should encourage innovation above and beyond minimum standards laid down by the new legislation. The EU can support research and development on techniques to reduce risks associated with key pathways and could also leverage economies of scale by supporting shared protocols for regional application in the place of unilateral measures.

The EU should work toward a coherent framework of incentives and deterrents to promote responsible practices and distribute the costs and benefits of IS interventions more equitably. Some Member States already have promising examples of voluntary codes and good practices that could be scaled up for wider use at little cost. The potential to apply existing EU legislation on the polluter pays principle, environmental liability, and environmental crime to IS-related damage should be clarified with regard to high-risk pathways and/or negligent or illegal conduct.

All envisaged measures will require much greater political will, stakeholder engagement, and public understanding. The EU can centrally harmonize awareness-building and information campaigns at least for certain target audiences, notably at entry points into the EU.

Notes

1. http://epp.eurostat.ec.europa.eu/portal/page/portal/eurostat/home/, accessed July 7, 2013.

2. http://ec.europa.eu/trade/about/ (covers EU-27), accessed July 7, 2013.

3. A conservative estimate based on the WTO 2009 trade statistics, excluding the impacts of the 2008–2010 financial crises. http://www.wto.org/english/res_e/statis _e/its2009_e/its09_world_trade_dev_e.pdf.

4. Eurostat—growth rate of GDP volume, excluding the impacts of the 2008– 2010 financial crises: http://epp.eurostat.ec.europa.eu/tgm/table.do?tab=table&in it=1&plugin=1&language=en&pcode=tsieb020.

5. http://www.wto.org/english/res_e/statis_e/its2010_e/section1_e/its10_high lights1_e.pdf, accessed October 14, 2011.

6. One hundred thousand vessels of 500 dwt and more compared with 77,500 vessels in 2008: total capacity is expected to reach more than 2,100 million dwt in 2018 (up from 1,156 million dwt in 2008). Source: OPTIMAR Study, LR Fairplay Research Ltd & Partners (September 2008).

7. DAISIE (*Delivering Alien Species Inventories for Europe*) (http://www
.europe-aliens.org), supported under the Sixth EU Research Framework
Programme.

8. For example, the introduced Pacific oyster *Crassostrea gigas* can act as an
"ecosystem engineer" to stabilize beaches but also make them less attractive/accessible to tourists

9. This is enshrined in the EU Animal Health Strategy (see §3.2 below) and
the "three-stage hierarchy" endorsed by Parties to the Convention on Biological
Diversity in *Guiding Principles for the prevention, introduction and mitigation of
impacts of alien species that threaten ecosystems, habitats or species* (Annex to CBD
Decision VI/23, 2002: http://www.cbd.int/decision/cop/?id=7197). This provides
that prevention of unwanted introductions is the most cost-effective, efficient, and
least environmentally damaging approach, followed by eradication where feasible
or long-term containment/control.

10. Article 8(h) CBD mandates each Contracting Party, as far as possible and
as appropriate, to prevent the introduction of, control, or eradicate those alien species which threaten ecosystems, habitats or species.

11. Communication on Biodiversity: Halting the Loss of Biodiversity by 2010—
and beyond (COM (2006) 216 Final).

12. These include French Guiana, Guadeloupe, Martinique, Réunion Island,
and St. Martin (France); Azores, Madeira (Portugal); Canary Islands (Spain).

13. These currently include Greenland (Denmark); French Polynesia, French
Southern and Antarctic Lands, Mayotte, New Caledonia, Saint-Pierre and Miquelon, Wallis and Futuna (France); Aruba, Netherlands Antilles (Netherlands); Ascension Island, British Antarctic Territory, British Indian Ocean Territory, British
Virgin Islands, Cayman Islands, Falkland Islands, Montserrat, Pitcairn Islands,
Saint Helena, Tristan da Cunha, South Georgia and the South Sandwich Islands,
Turks and Caïcos Islands (UK).

14. Source: *Flash Eurobarometer 290 (Gallup Organisation 2010)*. Note that
respondents could select only one threat from a list.

15. Communication on Halting the Loss of Biodiversity by 2010 and Beyond
(COM(2006)216) and Action Plan (SEC(2006)621); Council Conclusions of 18 December 2006 and of 3 March 2008; European Parliament Committee on the Environment, Public Health and Food Safety (Report of 28 March 2007); Opinion
of the Committee of the Regions of 6 December 2006; Opinion of the European
Economic and Social Committee of 15 February 2007.

16. *Supra*, note 10.

17. *Supra*, note 1.

18. Opinion of the Committee of the Regions on A New Impetus for Halting
Biodiversity Loss (DEVE-IV-039, 80th Plenary Session, 17–18 June 2009); Opinion
of the European Economic and Social Committee on the Communication—*Towards an EU Strategy on Invasive Species* (NAT/433 Invasive Species dated 11 June
2009).

19. Council Conclusions on a midterm assessment of implementing the EU Biodiversity Action Plan and Towards an EU Strategy on Invasive Alien Species (2953rd Environment Council meeting, June 25, 2009).

20. Council Conclusions on Biodiversity: Post-2010 EU and global vision and targets and international ABS regime (15 March 2010), building on the Communication Options for an EU vision and target for biodiversity beyond 2010 (COM (2010) 4 final, adopted January 19, 2010).

21. Only 17 % of habitats and species and 11 % of key ecosystems protected under EU legislation are currently in a favourable state (http://www.eea.europa .eu /publications/eu-2010-biodiversity-baseline/).

22. Adopted at the tenth meeting of the CBD Conference of the Parties (Nagoya, Japan, October 2010).

23. http://www.eppo.org/INVASIVE_PLANTS/ias_plants.htm.

24. http://www.eppo.org/QUARANTINE/plants/pathway_analysis/aquarium _plants.htm, accessed October 19, 2011.

25. PRATIQUE (Development of more efficient risk analysis techniques for pests and pathogens of phytosanitary concern). EU Seventh Research Framework Programme 2007–2013: https://secure.fera.defra.gov.uk/pratique/index.cfm.

26. http://www.emsa.europa.eu/main/ballast-water/involvement.html.

27. Respectively covering the North-East Atlantic, the Baltic, the Mediterranean, and Black Seas.

28. For example, the EU-wide ban on import of wild birds was adopted to prevent transmission of avian flu (Commission Regulation (EC) No 318/2007 of 23 March 2007 laying down animal health conditions for imports of certain birds into the Community and the quarantine conditions thereof).

29. Based on the Plant Health Directive on protective measures against the introduction into the Community of organisms harmful to plants or plant products and against their spread in the Community (2000/29/EC as amended) and complemented by control regulations for some pests.

30. For example, Pinewood nematode (*Bursaphelenchus xylophilus*), Asian Longhorned Beetle (*Anoplophora glabripennis*), Citrus Longhorned Beetle (*A. chinensis*), the fungus *Ceratocystis fagacearum* that causes North-American oak wilt, or *Phytophthora ramorum* that threatens native shrubs and trees.

31. http://ec.europa.eu/food/plant/strategy/index_en.htm.

32. Respectively the Common Veterinary Entry document and the "plant passport."

33. Council Regulation 338/97/EC, Commission Regulation (EC) No. 865/2006 laying down detailed rules for its implementation: see generally http://ec.europa .eu /environment/cites/legis_wildlife_en.htm.

34. Red eared slider (*Trachemys scripta elegans*); American bullfrog (*Lithobates catesbeianus*); painted turtle *(Chrysemys picta)*; American ruddy duck *(Oxyura jamaicensis)*.

35. Council Regulation concerning use of alien and locally absent species in aquaculture (No.708/2007 of 11 June 2007) (OJ L168/1 of 28.06.2007).

36. The current list includes the ten most farmed species in the EU, some of which are highly invasive (e.g., rainbow trout *Oncorhynchus mykiss;* Pacific or Japanese oyster, *Crassostrea gigas*). Member States that wish to *restrict* the use of such species must justify this by environmental risk assessment. Conditions for adding additional species to Annex IV are laid down in an implementing regulation.

37. The Preamble states that "potential risks, which may in some cases be far reaching, are initially more evident locally. The characteristics of local aquatic environments throughout the Community are very diverse and MS have the appropriate knowledge and expertise to evaluate and manage the risks to the aquatic environments falling within their sovereignty or jurisdiction."

38. Council Directive 92/43/EEC on the conservation of natural habitats and of wild fauna and flora.

39. Directive 2009/147/EC of the European Parliament and of the Council of 30 November 2009 on the conservation of wild birds (codified version of Council Directive 79/409/EEC).

40. EU Water Framework Directive (2000/60/EC) ; EU Marine Strategy Framework Directive (2008/56/EC).

41. Council Directive 2004/35/EC on environmental liability with regard to the prevention and remedying of environmental damage.

42. http://www.efsa.europa.eu/.

43. Directive 2009/28/EC of 23 April 2009 on the promotion of the use of energy from renewable sources.

44. Article 36 Treaty on the Functioning of the EU (ex Article 30, EC Treaty).

45. Article 191, Treaty on the Functioning of the European Union (following the Treaty of Lisbon, this replaces but is largely the same as Article 174, Treaty establishing the European Community, with an added focus on combating climate change).

46. Term used under the Animal Health Strategy (EC 2007).

47. COM/2013/0620 final - 2013/0307 (COD) (text and explanatory notes available at http://ec.europa.eu/environment/nature/invasivealien/).

Literature Cited

Burgiel, S., G. Foote, M. Orellana, and A. Perrault. 2006. *Invasive Alien Species and Trade: Integrating Prevention Measures and International Trade Rules.* Washington, DC: Center for International Environmental Law and Defenders of Wildlife.

Burgiel, S., and A. Muir. A. 2010. *Invasive Species, Climate Change and Ecosystem-Based Adaptation: Addressing Multiple Drivers of Global Change.* Washington, DC: Global Invasive Species Programme (GISP).

Capdevila-Argüelles, L., and B. Zilletti. 2008. A Perspective on Climate Change and Invasive Alien Species. Convention on the Conservation of European Wildlife and Natural Habitats. T-PVS/Inf (2008) 5 rev. Strasbourg, 16 June 2008.

EC. 2007. A new Animal Health Strategy for the European Union (2007–2013) where "Prevention is better than cure." Communication from the Commission (COM 539 (2007) final).

EC. 2008. *Towards an EU Strategy on Invasive Species*. European Commission. Brussels, 3.12.2008 (COM (2008) 789. http://eur-lex.europa.eu/LexUriServ/LexUriServ.do?uri=COM:2008:0789:FIN:EN:PDF).

EC. 2009. White Paper "Adapting to climate change in Europe—options for EU action" (COM (2009) 147 final) adopted by the European Commission on 1 April 2009.

EC. 2011. Communication from the Commission to the European Parliament, the Council, the Economic and Social Committee and the Committee of the Regions. Our life insurance, our natural capital: an EU Biodiversity Strategy to 2020 (Brussels, 3.5.2011 COM(2011) 244 final). http://ec.europa.eu/environment/nature/biodiversity/comm2006/pdf/2020/1_EN_ACT_part1_v7%5b1%5d.pdf.

ECOSTAT. 2009. Ecostat Workshop on alien species and the EC Water Framework Directive. Final Report of a workshop, prepared by Alison Lee, UK Joint Nature Conservation Committee (Ispra, Italy 17–18 June 2009).

FCEC. 2010. Evaluation of the Community Plant Health Regime. Final Report by the Food Chain Evaluation Consortium carried out under DG SANCO Framework Contract for evaluation and evaluation related services—Lot 3: Food Chain (awarded through tender no 2004/S 243-208899). Available with supporting documents and cost questionnaires on http://ec.europa.eu/food/plant/strategy/index_en.htm.

Genovesi P., S. Butchart, M., McGeoch, and D. Roy. 2013. Monitoring trends in biological invasion, its impact and policy responses. In *Biodiversity Monitoring and Conservation: Bridging the Gaps between Global Commitment and Local Action*, edited by B. Collen, N. Pettorelli, J. Baillie, and S. Durant. Oxford: Wiley-Blackwell.

Genovesi P., R. Scalera, S. Brunel, W. Solarz, and D. Roy. 2010. Towards an information and early warning system for invasive alien species (IAS) threatening biodiversity in Europe. European Environment Agency, Technical Report 5/2010. 52 pp.

Genovesi, P., and C. Shine. 2004. European strategy on invasive alien species. *Nature and Environment* 137. Strasbourg: Council of Europe. 67 pp.

Hulme, P., S. Bacher, M. Kenis, S. Klotz, I. Kühn, D. Minchin, W. Nentwig, et al. 2008. Grasping at the routes of biological invasions: a framework for integrating pathways into policy. *Journal of Applied Ecology* 45:403–414.

Kettunen, M., P. Genovesi, S. Gollasch, S. Pagad, U. Starfinger, P. ten Brink, and C. Shine. 2009. Technical support to EU strategy on invasive species (IAS)—Assessment of the impacts of IAS in Europe and the EU (Final draft report for the European Commission). Institute for European Environmental Policy (IEEP), Brussels, Belgium.

MA. 2005. *Ecosystems and Human Well-being: Synthesis*. Millennium Ecosystem Assessment. Washington, DC: Island.

McGeoch, M.A., S. H. M. Butchart, D. Spear, E. Marais, E. J. Kleynhans, A. Symes, J. Chanson, and M. Hoffmann. 2010. Global indicators of biological invasion: species numbers, biodiversity impact and policy responses. *Diversity and Distributions* 16:95–108.

McNeely, J. A., H. A. Mooney, L. E. Neville, P. Schei, and J. K. Waage, eds. 2001. *A Global Strategy on Invasive Alien Species*. Gland, Switzerland: IUCN.

NOBANIS. 2010. Recommendations from the workshop: Developing an early warning system for invasive alien species (IAS) based on the NOBANIS database. Proceedings of a workshop in Waterford, Ireland, 1–2 June 2010, organized by the European Network on Invasive Alien Species and the European Environment Agency. http://www.nobanis.org/files/Workshop%20recommendations _FINAL.pdf.

RSPB. 2007. Costing Biodiversity Priorities in the UK Overseas Territories. Final report 2 April 2007. http://www.rspb.org.uk/Images/Costing_biodiversity_in _UKOTs_tcm9–273309.pdf.

Scalera, R. 2008. EU funding for management and research of invasive alien species in Europe (prepared as support for a pilot project on "Streamlining European 2010 Biodiversity Indicators (SEBI2010)," Contract no. 3603/B2007. EEA.53070).

Scalera R. 2010. How much is Europe spending on invasive alien species? *Biological Invasions* 12(1):173–177.

Shine, C. 2007. Invasive species in an international context: IPPC, CBD, European Strategy on Invasive Alien Species and other legal instruments. *EPPO Bulletin* 37:103–113.

Shine, C., M. Kettunen, P. Genovesi, F. Essl, S. Gollasch, W. Rabitsch, R. Scalera, U. Starfinger, and P. ten Brink. 2010. Assessment to support continued development of the EU Strategy to combat invasive alien species. Final Report for the European Commission. Institute for European Environmental Policy (IEEP), Brussels, Belgium.

Shine, C., M. Kettunen, A. Mapendembe, P. Herkenrath, S. Silvestri, and P. ten Brink. 2009. Technical support to EU strategy on invasive species (IAS)—Analysis of the impacts of policy options/measures to address IAS. Final report for the European Commission. UNEP-WCMC (Cambridge) / Institute for European Environmental Policy (IEEP), (Brussels, Belgium). 101 pp. + Annexes.

Shine, C., N. Williams, and F. Burhenne-Guilmin. 2005. Legal and institutional frameworks for invasive alien species. In *Invasive Alien Species: A New Synthesis*, edited by H. A. Mooney, J. McNeely, L. E. Neville, P. J. Schei, and J. K. Waage, 233–284. Washington, DC: Island.

Soubeyran, Y. 2008. Espèces exotiques envahissantes dans les collectivités françaises d'outre-mer. Etat des lieux et recommandations. Collection Planète Nature. Comité français de l'UICN, Paris, France.

Vilà, M., C. Basnou, P. Pyšek, M. Josefsson, P. Genovesi, S. Gollasch, W. Nentwig, et al. 2010. How well do we understand the impacts of alien species on ecosystem services? A pan-European, cross-taxa assessment. *Frontiers in Ecology and Environment* 8(3):135–144. doi:10.1890/080083.

Walther, G. R., A. Roques, P. E. Hulme, M. T. Sykes, P. Pyšek, I. Kühn, M. Zobel, et al. 2009. Alien species in a warmer world: risks and opportunities. *Trends in Ecology and Evolution* 24:686–693.

CHAPTER FIFTEEN

There Ought to Be a Law! The Peculiar Absence of Broad Federal Harmful Nonindigenous Species Legislation

Marc L. Miller

Introduction

Over the past twenty years, harmful invasive species have moved from a topic of concern for "invasion" biologists who specialize in the movement of species to being a topic on the radar of federal and state governments, businesses, nongovernmental organizations, and scholars sprinkled throughout various disciplines.[1] Yet even as harmful invasive species have become a more important topic in public, political, and scholarly realms, the legal response to the problem remains sporadic and incomplete. Other than responding to climate change, harmful invasive species may be the aspect of environmental policy most in need of a coherent, comprehensive legal framework—or at least a much more significant legal and policy response.

Many other important environmental issues have, for better or worse, been addressed in statutes of broad scope, such as with the Wilderness Act of 1964, the National Environmental Policy Act of 1969 (NEPA), the Clear Air Act of 1970, the Clean Water Act of 1972, the Endangered Species Act of 1973, the Federal Lands Policy and Management Act of 1976, the National Forest Policy Management Act of 1976, and other similar "framework" environmental legislation.

When legislation sets out the broad general authority and goals for an agency or government, it is sometimes called a "framework" or an "organic" law. An example of an organic law is the brief and elegant National Park Service Organic Act of 1916. Framework laws can be very short, like the National Park Service Organic Act, or very long, such as the Clean Water Act of 1972 (at more than 200 pages of legislation, with subsequent major amendments and implemented through massive regulation). What these broad framework statutes provide is a legal, political, and social framework around which funding, regulation, and policy discourse can occur. These statutes are both the core and the most powerful and successful tools of the modern environmental movement.[2] In turn, they have produced the most visible outcomes: cleaner water and air, protected lands, and the procedural mechanisms to challenge and shape both private and government actions that impact the human environment.

The problem of harmful invasive species calls for a clearer allocation of responsibilities within government and a clearer statement of general principles defining the problem and the proper response. It calls for a framework that can help to promote public and policy discourse, including funding and prioritization. The missing national and state legal identification of the general problem of invasive species is part of the reason that many harmful invasions have gone unnoticed and unaddressed. In short, the problem calls for an organic act.

The first part of this chapter surveys existing federal and state laws addressing nonindigenous species (NIS), focusing on the federal sphere. It suggests a legal paradox. There are a fair number of laws that either explicitly deal with narrow aspects of the harmful invasive species problem, or provide general legal authority that might be applied to harmful invasive species—enough so that this chapter will provide a survey of only the most important authorities. Yet despite the multitude of existing laws that might have some impact on the problem of harmful invasive species, fundamental aspects of the problem have not been addressed. For those familiar with the details of this legal landscape—or not interested in those details—please jump to the second part of the chapter.

The second part of this chapter takes a more theoretical turn and asks the following question: if a compelling case can be made for the need for a more coherent legal framework to address harmful NIS, and indeed for an organic harmful NIS law, why has such a law not been enacted? Thinking about the barriers and avenues to legislation can provide a roadmap with several different policy highways for proponents of a more coherent legal regime for harmful NIS.

I. A Survey of Federal and State NIS Law[3]

Near the end of its 24-year existence,[4] the Congressional Office of Technology Assessment (OTA) produced a highly regarded 1993 report on the problem of harmful nonindigenous species concluded that "the current Federal effort is largely a patchwork of laws, regulations, policies, and programs. Many only peripherally address NIS, while others address the more narrowly drawn problems of the past, not the broader emerging issues." An April 1999 Congressional Research Service (CRS) report titled *Harmful Non-Native Species: Issues for Congress*, concluded that "federal law concerning non-native species is scattered. No laws focus on the broad problems of non-native species, their interception, prevention, and control across a variety of industries and habitats." At another point, the CRS report summarized US federal law this way:

> In the century or so of congressional responses to harmful, non-native species, the usual approach has been an ad hoc attack on the particular problem, from impure seed stocks, to brown tree snakes on Guam. A few attempts have been made to address specific pathways, e.g., contaminated ballast water, but no current law addresses the general concern over non-native species and the variety of paths by which they enter this country.

The general observations of the 1993 OTA and 1999 CRS reports remain true in 2014.

Focusing on federal and state legislative activity since 2005, there appears to be some increasing recognition of the importance of responding to harmful invasive species. This uptick in legislative action and references to invasive species is not distributed across the policy dimensions raised by harmful NIS. Similar to the critique by CRS in 1999, the legislative activity since 2003 focuses on particular invaders, especially those with significant economic impacts, and even more particularly with aquatic invasive species, and the threat from nonnative mussels in particular.

A. Explicit Federal Statutory NIS Authority: "Black List" and "Exclusion" Acts

There is a compensation in the distribution of plants, birds, and animals by the God of nature. Man's attempt to change and interfere often leads to serious results.—Rep. John Lacey (R-Ind.) 33 Cong. Rec. 4871 (1900)

The recognition that NIS might cause harm has been evident in US federal law at least since the Lacey Act, first enacted in 1900 and substantially revised in recent years.[5] The Lacey Act was created to protect native American wildlife, especially birds that were being commercially harvested for their feathers in the Florida Everglades. The sponsors focused on the lack of controls on commerce in wild species among the states as well as between nations, but they also recognized that nonnative species could harm native species and ecosystems (though Representative Lacey and his colleagues wrote the act decades before the term "ecosystems" came into use). The legislative history illustrated these points by depicting harm from sparrows and starlings that had been intentionally introduced in the latter half of the 19th century.

The Lacey Act currently provides the federal government with authority to ban the import, export, or transportation of "any fish or wildlife" or "any plant" that is made illegal by "any law, treaty[,] or regulation" of the United States or of any state. The Act provides for both civil and criminal penalties of a modest financial nature. For example, knowingly or negligently violating the Act may result in a penalty of "not more than $10,000 for each such violation . . . and criminal penalties, up to five years in prison and a $20,000 fine" for each violation. The Lacey Act explicitly leaves US states free to make or enforce laws "not inconsistent" with the federal provisions.

The major limitation on the Lacey Act's general authority to ban animal and plant species is that these powers apply only to animals and plants that are made illegal under federal or state law, whether legislative or administrative. The Lacey Act is built around the idea of a "black list" or forbidden list: it does not authorize the exclusion of animals whose threat is unknown.[6]

While the Lacey Act provides the Secretary of the Interior with the power to exclude several species of particular concern it does not provide for the exclusion of seeds, or plant pests. This gap in the Lacey Act has been partially closed by a host of federal statutes that, together, provide federal officials with power to exclude many kinds of harmful plant pests, seeds, and noxious weeds. These acts include the Plant Pest Act, the Plant Quarantine Act, the Federal Noxious Weed Act of 1974, and the Federal Seed Act.

In May 2000, Congress passed the Plant Protection Act, which consolidated and revised existing laws including the Plant Quarantine Act, the Plant Pest Act, the Federal Noxious Weed Act, and the Department of Agriculture Organic Act. The Plant Protection Act shows signs of Congress' increasing awareness of the importance of NIS issues and the need for more coherent legislative responses. The Plant Protection Act provides a unitary framework for dealing with plant pests and noxious weeds. The statement of findings recognizes that plant pests and noxious weeds threaten "the agriculture, environment, and economy of the United States" and noxious weeds are defined to include "any plant or plant product that can directly injure or cause damage to . . . the natural resources of the United States . . . or the environment."

The 2000 Plant Protection Act suggests the importance of reorganization and simplification of laws that relate to harmful invasive species. The Plant Protection Act expands the regulatory and enforcement powers over plant pests and noxious weeds, including new civil penalty structures. It encourages a steady use of science in assessing risk, the wide involvement of experts and stakeholders in policy making, consideration of "systems approaches," the development of integrated management plans on the basis of geographic and ecological regions, and the authorization of new types of classification systems. While some of these systematic concepts were evident in prior law, the new act joins them to regulatory and enforcement mechanisms and thus offers some hope for more effective, efficient, and informed federal plant pest and noxious weed policies.

Several provisions of the 2000 Plant Protection Act, however, plant their own substantial seeds for mischief. For example, the Act encourages the use of biological pest controls, finding that "biological control is often a desirable, low-risk means of ridding crops and other plants of plant pests and noxious weeds." Biological controls are themselves invasive species—additional biological pollution.[7] Congress does not seem to have considered this aspect of biological controls, or the mixed record of biological controls over the past century.

The Plant Pest Act also preempts state efforts to regulate plant pests and noxious weeds. States and political subdivisions are forbidden to regulate "any article, means of conveyance, plant, biological control organism, plant pest, noxious weed, or plant product" in an effort to control, eradicate, or prevent the introduction of plant pests, noxious weeds, or biological control organisms. States and local subdivisions are also barred from regulating the interstate commerce of these kinds of organisms when there are already federal regulations regarding these organisms.

Given the varied needs of different states, most notably those with highly unique and fragile ecosystems, such as Hawaii, these preemption provisions go well beyond prior law, which, as in the Lacey Act, tended to allow state regulation in these areas that was not inconsistent with federal law. Whether or not these preemption provisions prove to be harmful will depend on how courts and agencies interpret the provision that allows regulation of interstate commerce when regulations "are consistent with" federal regulations, and how sympathetic and wise the Secretary of Commerce will be in response to state requests for waivers based on "special need."

B. National Invasive Species Act: Big Name, Narrow Scope

If awards were given for legislative titles, then anyone concerned with the threat from harmful NIS would give the grand prize to two federal statutes: the National Invasive Species Act (NISA) of 1996 and the Alien Species Prevention and Enforcement Act of 1992 (ASPEA). The major (and useful) purpose of ASPEA, despite its grand title, was simply to confirm the authority to make illegal the shipment through the mail of otherwise illegal organisms, including those species identified under the Lacey Act, the Plant Pest Act, and the Plant Quarantine Act. ASPEA does not itself create any new categories of organisms that are illegal to ship, nor does it create any presumptions or institutions to help in responding to harmful NIS.

NISA reauthorized a 1990 federal statute with a less encompassing but more accurate title: the Non-Indigenous Aquatic Nuisance Prevention and Control Act (NANPCA) of 1990. NANPCA focused on one place (the Great Lakes) and on one pathway for invasive species (ballast water), and was driven by concerns about one harmful NIS (the zebra mussel). NANPCA created a federal interagency Aquatic Nuisance Species Task Force to reduce risk from harmful NIS. The task force was charged with assessing aquatic nuisance species threats to "the ecological characteristics and economic uses of US waters other than the Great Lakes."

In 1996, NISA expanded the focus of the NANPCA to mandate regulations to prevent introduction and spread of aquatic nuisance species. In NISA, Congress encouraged the federal government to negotiate with foreign governments to develop an international program for preventing NIS introductions through ballast water. The geographical scope of the Act was expanded as well, to include funding authorization for research on aquatic NIS in the Chesapeake Bay, San Francisco Bay, Honolulu Harbor, and the Columbia River system.

A series of bills to reauthorize NISA have been introduced in both houses of Congress over the past decade. For example in the Senate, the NISA of 2003 proposed to require mandatory ballast water regulations, and to encourage both further development of ballast water treatment and the use of best available technologies by the shipping industry, though with substantial lag time for adoption. Sen. Susan Collins (R-Me.) introduced the bill with the following observations:

> As with national security, protecting the integrity of our lakes, streams, and coastlines from invading species cannot be accomplished by individual States alone. We need a uniform, nationwide approach to deal effectively with invasive species.

In the House, for example, the Ballast Water Treatment Act of 2008[8] proposed to strengthen NISA, and provide greater authority to the Coast Guard to enforce ballast water standards.

A full review of legal authority addressing invasive species would include regulations and policies promulgated by executive branch agencies. So, for example, under NANPCA and NISA, the United States Coast Guard was given authority to first require voluntary ballast water transfer reports, and then, if compliance was insufficient, to mandate that most ships entering US waters file a mandatory ballast water transfer report.[9] After a 2001 study found that only about 30% of incoming ships voluntarily filed ballast water reports,[10] the Coast Guard finalized regulations in 2004 that mandated the filing of reports showing the pre-coastal transfer of ballast water and imposed penalties on most vessels entering American waters that did not do so.[11] The Coast Guard promulgated revised ballast water regulations in 2012, and the Environmental Protection Agency (EPA) has proposed new rules for permitting of vessels that will further regulate discharges.[12] Relatively recent rule-making includes new "Q37" rules issued by the US Department of Agriculture (USDA) Animal, Plant and Health Inspection Service (APHIS) under the authority of the Plant Protection Act. The revised rules provide new limits on importing some types of nursery stock.[13]

Since 2003, members of Congress have shown increasing awareness of threats specific to their home jurisdiction and especially agricultural, commercial, and industrial interests with strong concerns about invasive species. Several new laws have been enacted—and other laws have been proposed—which, for the most part, focus on specific threats from harmful

NIS in specific regions of the country.[14] These laws—and a longer list of proposed legislation—suggest that harmful invasive species, or at least specific harmful invasive species, are on the Congressional radar. But it is also safe to say that Congress has not yet addressed the general problem of harmful invasive species.

C. General Environmental Policy Acts

Harmful NIS have played a surprisingly small role in the courts, whether through the use by parties harmed by NIS of traditional common law claims (such as nuisance or negligence) or through litigation based on claims under general environmental laws.[15]

There are several major federal environmental policy statutes and a set of public lands statutes that in theory could apply to harmful NIS. The National Environmental Policy Act (NEPA) requires federal government agencies to assess the environmental impact of their actions through the promulgation of environmental assessments (EA) and environmental impact statements (EIS). Yet many actions of the federal government that appear as if they could trigger the requirement of an EIS do not owing to both statutory and regulatory interpretations that limit NEPA to "major" government actions that "significantly" affect the quality of "the human environment."

Claimants have argued that the federal government has failed to take account of the impact of invasive species under NEPA. But even a significantly broader interpretation of NEPA would cover only a modest portion of the full range of harmful NIS issues. NEPA is primarily directed at the actions of federal agencies, and therefore would not apply to the myriad actions of individuals relevant to harmful NIS, or to the actions of state and local authorities. Moreover, NEPA assumes the possibility of expertise in recognizing and assessing future environmental harms from present actions. In the case of potentially harmful NIS, this kind of information and expertise may not be present.[16] Environmental policy decisions in many areas are particularly ill-suited to deal with choices made in the face of limited present information and great future uncertainty. Even when NEPA applies, it requires only analysis of environmental impacts, but does not itself impose substantive barriers, preferences, or limits on government action.

A second major federal environmental statute with some possible application to harmful NIS issues is the Endangered Species Act (ESA). The ESA might apply to NIS whenever a government or private action threatens an endangered species. The ESA might also lead to direct actions against harmful NIS in the development of recovery plans for listed species. Since harmful NIS have been identified as a significant source of ecosystem change (which may lead to pressures on rare or endangered species), and in some contexts as a direct extinction threat through predation, competition, or displacement, the ESA might bar some introductions or lead to some efforts at removal.

The situations where the powerful tools of the ESA apply, however, are likely to be few. If the ESA applies at all in terms of introductions, it will most likely apply only to intentional introductions of NIS, and only to those introductions where a nexus can be found between the NIS and a listed species, a causal link that may be hard to establish. In one case the federal courts have several times upheld an order to the Hawaiian Department of Land and Natural Resources to remove nonindigenous goats and sheep that threatened the endangered palila bird. The rare legal status of the litigation over efforts to protect the palila bird confirms the small likelihood of the ESA and recovery plans becoming a major mechanism for control of harmful NIS.

Other general environmental policy acts have played a role in mandating regulation of some harmful NIS or particular pathways. In 2005 in the case of *Northwest Environmental Advocates v. EPA* a federal district court in California determined that the Clean Water Act (CWA) applies to ballast water. Ballast water discharges from boats have historically been exempt from CWA permitting requirements. The suit was brought against the EPA by Northwest Environmental Advocates, an environmental nongovernmental organization. The lawsuit was joined by the Attorneys General from six Great Lakes states. The trial court concluded that the EPA had exceeded its statutory authority in exempting regular discharges from boats including waste, marine engine, and ballast water discharges, and the trial court decision was affirmed on appeal to the United States Court of Appeals for the Ninth Circuit. At exactly the same time as the Ninth Circuit Decision, Congress passed the Clean Boating Act of 2008 that created a new exemption from the permitting requirement for *recreational vessels* while also requiring EPA to develop best management practices for some pollution from those vessels. The Act had no effect on commercial vessels, which are the vessels whose ballast water poses the primary invasive species risk. Though it has taken another five years, the district court's decision in *Northwest Environmental Advocates* helped to spur EPA to regulate invasive species in ballast water under the Clean Water Act.

Another broad class of federal laws that provide authority to federal agencies that have at times been used for regulation and policy with respect to harmful invasive species are the federal public lands laws, especially the Multiple Use Sustained Yield Act of 1960, the National Forest Management Act of 1976, and the Federal Land Policy and Management Act of 1976. These acts, and related historical and contemporary legislation governing grazing, timber, and other uses of federal lands, provide a broad array of authorities and responsibilities with respect to public lands. Similar legislation aimed at the governance of smaller federal land units includes the National Wildlife Refuge System Administration Act.

In addition to these major federal environmental statutes, a host of more focused environmental and nonenvironmental laws also have some relevance to harmful NIS. Additional sources of law for the many dozens of federal agencies (under many different cabinet-level departments) that respond to harmful NIS include the organic (founding) acts for major agencies, particular legislation directing agencies to carry out a particular task (such as the statutes summarized in this essay), and appropriations bills. The most important federal agencies for dealing with NIS, including the Animal and Plant Health Inspection Service (APHIS), the Agricultural Research Service (ARS), the US Forest Service (USFS), the US Fish and Wildlife Service (FWS), the Bureau of Land Management (BLM), and the National Park Service (NPS), all fall under the authority of two departments—the US Department of Agriculture (USDA) and the US Department of the Interior (DOI). Another federal agency with a significant role in both public lands and the movement of people and goods is the Department of Defense.

Government authority to respond to harmful NIS arises from international law reflected in treaties signed by the United States, including treaties generally enabling and regulating trade.[17] One example is the Convention on International Trade in Endangered Species of Wild Fauna and Flora (CITES), which provides additional authority for border inspections and creates an independent basis (indeed, an independent obligation), even in the absence of listing a species under one of the "blacklist" acts, for exclusion. The 1993 Office of Technology Assessment report lists seven treaties with direct effects on harmful NIS and seven treaties with indirect effects on harmful NIS, including CITES.

D. Executive Orders Addressing Harmful NIS

A judge, like an executive adviser, may be surprised at the poverty of really useful and unambiguous authority applicable to concrete problems of executive power as they actually present themselves. Just what our forefathers did envision, or would have envisioned had they foreseen modern conditions, must be divined from materials almost as enigmatic as the dreams Joseph was called upon to interpret for Pharaoh.—Justice Robert H. Jackson, Youngstown Sheet & Tube Co. v. Sawyer, 343 U.S. 579,634 (1952) (concurring)

Two Executive Orders, one issued by President Carter in 1977 and the other issued by President Clinton in 1999, directly address the problem of harmful NIS. Executive Orders are an odd species of law, issued on occasion by the president. They direct one or more federal agencies to act in a particular policy direction specified by the president. Executive Orders do not themselves create new government powers: general legislative power is vested in the legislative branch (the Congress). The president can, however, rely on powers already vested in the executive branch by Congress, and those limited powers constitutionally committed to the president.

While Executive Orders cannot create new legal authority, the Executive Orders on invasive species can and do draw on the full range of available federal legal authority. This assertion of maximum current federal legal authority highlights the necessity of understanding the greatest possible reach of current laws. Often the issue with regard to a problem with harmful NIS is not an absence of authority but of responsibility to act, or a reflection of insufficient budgetary allocations, and in a unitary executive branch Executive Orders are the policy command of the president (at least in theory).

President Carter issued Executive Order No. 11987 on May 24, 1977. Executive Order No. 11987 has been entirely supplanted by Executive Order No. 13112, issued by President Clinton in 1999.

While now part of federal executive legal history and although it had little lasting impact, President Carter's E.O. 11987 is worth a brief remembrance. Executive Order No. 11987 is an astounding document, as striking and unexpected, though not nearly as profound, as Charles Elton's classic 1958 book *The Ecology of Invasions by Plants and Animals*. Some aspects of harmful NIS were part of public policy and debate by 1977, but NIS as a general environmental issue had yet to strike public and political consciousness.

Executive Order No. 11987 is not only unexpected because of its topic, but also because of its brevity—it is one page long—its clarity (which does not always go hand in hand with brevity), and its political timing. Discussions about this order began within the White House only weeks after Carter took office in January 1977, and the order itself was issued as part of the first public policy statement on the environment by the Carter Administration. The heart of the order provides the following policy directives:

(a) Executive agencies shall, to the extent permitted by law, restrict the introduction of exotic species into natural ecosystems on lands and waters which they own, lease, or hold for purposes of administration; and, shall encourage the States, local governments, and private citizens to prevent the introduction of exotic species into natural ecosystems of the United States.

(b) Executive agencies, to the extent they have been authorized by statute to restrict the importation of exotic species, shall restrict the introduction of exotic species into any natural ecosystem of the United States.

The short Executive Order included at least one other visionary aspect: it directed executive agencies to prevent the export of native (US) species "for the purpose of introducing such species into ecosystems outside the United States where they do not naturally occur." President Carter was not concerned with just US ecosystems; he was concerned with the threat of NIS to the naturalness of all ecosystems.

Executive Order No. 11987 had several flaws that ultimately proved fatal to its virtues. The most significant flaw was that the Executive Order included no complete procedure for implementing its policy directive. The order directed the Secretary of the Interior, in consultation with the Secretary of Agriculture and other agencies, to "develop and implement, by rule or regulation, a system to standardize and simplify the requirements, procedures, and other activities appropriate for implementing" the order. The lack of specificity in this procedural language contrasts with the strong substantive principles of the order.

Executive Order No. 11987 disappeared from federal policy as dramatically as it first appeared. A September 15, 1977, memorandum written by the Council on Environmental Quality (CEQ) summarized the response of all federal agencies to the various aspects of the May 23, 1977, environmental message. Tucked away in this memorandum were a few lines on the question of the DOI's response to the directive to "develop legislation to restrict the impact of exotic plants and animals into the [United States]." The memorandum stated that "legislation is being developed with Agriculture," that agency progress was "adequate," and in what appears to be the final White House file entry on the subject, the "CEQ Progress Evaluation," that there were "delays in interagency meetings and in focus on problems."

Twenty-two years later, on February 3, 1999—in the last two years of his two-term Presidency—President Clinton issued Executive Order No. 13112, a longer and more complex document than Executive Order No. 11987, which it replaced. Executive Order No. 13112 states its goal as to prevent "the introduction of invasive species and provide for their control and to minimize the economic, ecological, and human health impacts that invasive species cause."

In some ways the policy goals are more sweeping than Executive Order No. 11987. Executive Order No. 13112 includes control of existing invasive species as one of its primary goals. "Alien species" is defined in ecological not political terms. An alien species is "with respect to a particular ecosystem, any species, including its seeds, eggs, spores, or other biological material capable of propagating that species, that is not native to that ecosystem." "Introduction" is defined to include "intentional and unintentional escape, release, dissemination, or placement of a species into an ecosystem as a result of human activity." Section 2 of the new Executive Order directs "each federal agency . . . to the extent practicable and permitted by law" to address a range of prevention, monitoring, response, restoration, research, and education goals.

The policy directive to all federal agencies whose actions may affect NIS is sweeping. Unfortunately, saying "everyone" has responsibility is a little like saying no one has responsibility. If the order stopped here, it would be only a more sophisticated, complete, and current version of the Carter effort 22 years earlier.

However, Executive Order No. 13112 also created an Invasive Species Council, made up of all cabinet officers with significant responsibility for NIS. The council was required to issue an Invasive Species Management Plan within 18 months. The council is guided by an Advisory Committee whose responsibility is to "recommend plans and actions at local, tribal, State, regional, and ecosystem-based levels to achieve the goals and objectives" of the management plan.

Executive Order No. 13112 included structural elements that made it more likely to have an impact than E.O. 11987. The interagency council made up of cabinet officers places responsibility as high as it can go. Involving a wide range of cabinet-level officers increases the likelihood of a full airing of views and the revelation of conflicts, and in theory increases the chance of reasonable policy consistency. Requiring a plan provides a device for action and commentary. Creating an advisory committee increases the chance of expert input and invests a number of people and organizations outside the government in the details of the Invasive Species Council's work.

The first National Invasive Species Management Plan was approved by the National Invasive Species Council on January 18, 2001, two days before the inauguration of President George W. Bush. The 80-page plan was full of bureaucratic but relatively nonspecific steps; it was even less clear about its systematic aims.[18] Despite their generality, most and perhaps all of these goals have not been met. It would have been optimistic to think that even a majority of these goals could be met if the plan had appeared at the start or in the middle of a new administration. But the shift to an administration less interested in environmental policy, and the dramatic alteration of priorities occasioned by September 11, 2001, perhaps doomed any plan to fail.

A revised National Invasive Species Management Plan was issued on August 2008, and by its title claims to govern the period 2008–2012.[19] (No plan has been issued to apply to any year after 2012.) The revised National Invasive Species Management Plan is relatively short and is organized around five basic strategic goals—prevention; early detection and rapid response; control and management; restoration; and organizational collaboration. To its detriment, the plan offers no substantial assessment of the success and failure of the 2001 plan ("The [2001] National Invasive Species Management Plan, Meeting the Invasive Species Challenge [2001 Plan] . . . called for about 170 specific actions within nine categories of activity, about 100 of which have been established or completed."). The structure of the 2008–2012 plan—page after page of bullet points and agency assignments—reads like a bad bureaucratic parody. Missing from both the 2001 and 2008–2012 plans is any substantial discussion, definition, or measurement of underlying, substantive (as opposed to bureaucratic, "write more rules") kind of success. How sad.

At the federal level, the effort to organize federal efforts through the National Invasive Species Council does not appear to have achieved what the optimistic Executive Order that led to its creation had sought. The documents and web site produced by the Council suggest that the Council has been at best an irrelevancy, and more likely a failure, at least if viewed in terms of lost opportunity of real and substantial policy change, and a corresponding impact on the problem of new introductions and response to both new and existing invasive species. If a bureaucracy cannot justify its own existence and value in compelling terms, who else can or will?

The story of harmful NIS laws in the states is similar to the federal story—only less so. Generally states with more commerce, and especially foreign commerce, such as California and Florida, have been subject to greater harm from NIS, and therefore have more developed (but still incomplete) laws. The state with the most substantial and aggressive invasive species legal and policy regime is Hawaii. Most state laws respond to particular invaders and pathways; as with federal law there is far greater attention to threats from aquatic invaders, and especially invasive mussels and clams. The special state legislative attention to aquatic invaders appears even in states such as New Mexico that have not yet experienced direct harm.

Hawaii has made its Invasive Species Council a permanent council, has prohibited the importation of portions of invasive plants in addition to a previous prohibition on the importation of whole plants, and has imposed fees on every air or sea freight shipment for invasive species inspection, quarantine, and eradication (even overriding a gubernatorial veto to impose the fee on air freight shipments). Pending legislation would impose severe penalties (at least $100,000 per violation) on "any person or organization who intentionally imports, possesses, harbors, transfers, or transports . . . any prohibited or restricted plant, animal, or microorganism without a permit, with the intent to propagate, sell, or release that plant, animal, or microorganism."

Puerto Rico has what appears on the books to be a relatively strong "white list" invasive species law. In 1999 the Commonwealth of Puerto Rico enacted the "New Wildlife Law of Puerto Rico" that set up both an advisory committee and a technical committee to advise the Secretary of the Department of Environmental and Natural Resources on the importation and ownership of "exotic species." The Technical Committee

> shall assist the Department's biologists specialized in wildlife in the creation of a list of exotic species, of which the importation and ownership as pets shall be allowed. On said list, no species of monkeys or simians shall be included. The importation and ownership of species not included on the list shall be prohibited.[20]

Puerto Rico's experiment with a "white list" approach to importation of species does not appear to have been evaluated by either government agencies, nonprofit entities, or scholars.

There are some critical gaps to both the wide array of federal legal authorities and the mix of state laws. What this mix of laws fails to do is to address what from the standing of invasion biologists, land managers, and agricultural and environmental interests are central issues, such as establishing comprehensive procedures and principles for limiting harmful new NIS introductions, identifying new NIS invasions, tracking the impact of known harmful invasive species, and responding to emerging threats.

At the same time, the summary of federal and state laws in this chapter is far from comprehensive. A comprehensive summary would point to the many dozens of federal statutes that are relevant, or might be relevant, to NIS issues; it would also point to the dozens of federal agencies and hundreds of state agencies that have responded to alien species issues under various kinds of legal authority, including general organic acts for the supervisory agency, annual appropriations bills, and even treaties with foreign governments.

One way to resolve the paradox of no framework law and much piecemeal law is to shift the terrain of the question from: "what laws apply to harmful NIS?" to "what are the costs of not having an organic or framework harmful NIS law?" and "what would the benefits be if an organic harmful NIS law existed?"

Laws can prohibit, regulate, encourage, or require private and government behavior. Private and government actors may see a problem or desire to act in a particular way, and the actor will want to know "can I do this?" and if so "within what legal constraints?" In other words, the proper question may not be whether a lawyer or policy maker might be able to find a basis in current legal authority to defend a specific action, but whether a biologist or policy maker would say that the law adequately guides and mandates appropriate government and private actions, and, more generally, that it responds to the costs and threats imposed by NIS.

No one attentive to the range of invasive species costs and challenges across the United States would consider the current legal regime in the federal system or the states to be coherent or comprehensive. If there is a strong case that the benefits of an organic harmful species law would likely outweigh the costs (and there are always costs to the creation of and administration of law), then a question arises: in the 100 plus years since the Lacey Act, why has no such law been enacted or even proposed?

II. Why No Framework NIS Law Exists

The significance in economic and ecological and aesthetic terms of the problem of harmful NIS provokes the question why, despite the background recognition that the creation of law is both costly (to enact) and risky (to achieve laws where the benefits of the law in practice significantly exceed the costs), that over 100 years since invasive species first emerged on the federal legislative landscape in the Lacey Act there have been no organic (or framework) laws enacted or even proposed in the federal system (or for that matter in any US state). Focusing even on the last several decades as invasive species have emerged as an issue does not present a more optimistic picture: reports going back twenty years to the 1993 OTA report have called for a better legal framework, and more recent scholarship has suggested the need for a comprehensive law.[21]

The answer to this puzzle requires some exploration of why *any* law is proposed and, more importantly, enacted. In formal terms, this question comes under the rubric of a theory of legislation. But it is helpful initially to think in less formal terms.

Law-making is hard, costly and messy—a process sometimes analogized to the making of sausages, with a companion warning that the details are perhaps best left unexplored. But the barriers to the creation of laws are hardly a byproduct of government: in the federal system, and in states that adopt the same general theories of government, the barriers to law creation are a central part of the structural design. The framers of the federal Constitution imposed barriers to legislation through structural constraints, including the limited authority of the federal government (federalism); the separation of power into legislative, executive, and judicial branches; and the rules for the enactment of legislation (including the bicameral legislature, and the requirement of presidential approval and corresponding veto power).

The default position in a political system that imposes high formal barriers and therefore high costs to law-making is against having any law at all. That might seem like an odd observation in a legal system brimming with laws and regulations. But as law-centered as United States government seems to be, there are still vast areas of life and business left entirely to the private sphere, and governed by law primarily through enforcement of private agreements and the protection of private property, and through individual autonomy and social responsibility.

Scholars from a variety of disciplines, with deep roots in philosophy and political science, have produced varying theories of legislation. Scholars of legislation have tended to distinguish between law-making, implementation, and interpretation. It is the first of those topics—law-making—that is most relevant to the question here. Theoretical and empirical work often focus on when legislation is created and how different forces shape legislation, but an equally interesting question is why law is *not* created where there otherwise appears to be facts consistent with theory that would call for law to be made.[22]

Law-making can be the product of a variety of different political, social, structural, and historical forces. Any single theory of law-making will do a poor job explaining different laws and different times. Still, it is useful to recognize three different general theories of legislation that might be applied to the problem of harmful NIS: utility (or cost/benefit); public choice; and, though less familiar, an approach which might be called historical or contingent.

The first theory suggests that legislation should be enacted if its benefits to society, properly considered, exceed its costs. This theory is generally attributed to Jeremy Bentham who, in the beginning of his famous book *Theory of Legislation*, wrote:

> The public good ought to be the object of the legislator; general utility ought to be the foundation of his reasonings. To know the true good of the community is what constitutes the science of legislation; the art consists in finding the means to realize that good.[23]

If you want to know which action should be favored, Bentham says to "calculate their effects in good and evil, and prefer that which promises the greater sum of good."[24] Since Bentham believes utility must be the foundation of a legislator's reasoning, he goes into much detail about what exactly he means by utility. He believes the principle of utility subjects everything to the motives of pain and pleasure.[25] According to Bentham the principle of utility expresses "the property or tendency of a thing to prevent some evil or to procure some good" with evil being pain and good being pleasure. "That which is conformable to the utility, or the interest of a community [or an individual] is what tends to augment the total sum of the happiness of the individuals that compose it."[26] In Bentham's view, once the elements of utility (and inappropriate sentiments) are properly understood, legislation becomes a matter of arithmetic.[27]

Bentham's "utility" has been extended in modern times to a more general approach of cost-benefit analysis, and with much of the scholarly and policy focus on the application of this approach by rule-makers in the executive branch. One more umbrella under which elements of utility analysis have been subsumed in recognition of the complexity of political and administrative decision making is the idea of "civic republicanism."[28]

Utility theory does a poor job of explaining the absence of strong or comprehensive legislation responding to harmful invasive species. The sources, nature, and impacts of harmful invasive species make this problem a poor candidate for resolution entirely by private market forces. Even with a significant discount for the inherent inefficiencies of any legislation and of any bureaucracy, the benefits of more focused and substantial attention to the problem of harmful invasive species are compelling.

A second and particularly influential theory of legislation has emerged from the application of theories from economics to the realm of politics. Economists have suggested that the behavior of legislators can be understood by applying models of rational choice by political actors who want to be reelected and who can sell their influence and the theory of collective action (why it is difficult for groups to coalesce and have their collective preferences adequately reflected in policy). This area is generally known as public choice theory.[29] Public choice theory suggests that legislation will generally favor powerful private interests that are able to invest resources or to organize—to overcome the challenges to collective action—over the kind of public values suggested by Bentham and his many disciples. That legislators will respond to organized and powerful interests can be viewed in a cynical, a neutral or a more sympathetic light. One more sympathetic perspective sometimes identified by the term "neopluralist," suggests that battles between well-organized and powerful interests can produce outcomes favorable to broader public values.[30]

Public choice theory suggests that interest groups are likely to have a larger voice in political decision making than disorganized individuals. Empirical studies have provided a substantial foundation in support of hypotheses about the influence of interest groups on the outcome of particular legislative decisions—who are the winners and losers from legislation—including appropriations decisions. Public choice theory and empirical studies have also been very powerful in assessing decision making by administrative agencies, including extensive discussion of the idea of regulatory "capture" and other failures of reasoning oriented towards public rather than private or narrow values.

But public choice theory has had less success explaining the triggers for legislating in particular areas in the first place. Interest group theory may explain in part, for example, why harmful NIS legislation has focused on particular invasive species with readily identified geographic and economic impacts. Zebra mussels in the Great Lakes have provoked both private and public actors to call for and help to draft federal laws; the lack of commercial or aesthetic advantages from invasive species such as the zebra mussel also help to explain why laws have been enacted in this area. Public choice theory helps to explain why the most immediate interests (the loudest and best organized and financed) receive legislative attention. Thus, Asian carp has become a new federal legislative priority, revealed largely through the appropriations process rather than in new substantive legislation, perhaps to the detriment not only of a more comprehensive invasive species scheme, but to earlier priorities including aquatic nuisance species more generally. Interest groups including environmental organizations and cabinet-level agencies who have expressed substantial concern over the harms from NIS more generally have not seen similar legislative success.

A third approach, and one less identified with a particular theoretical perspective, is to consider legislation as the product of moments—perhaps decades-long moments—in political time and space. Such legislation might reflect the response to single cases or events. Some legislation might reflect the leadership of a single effective legislator (a "policy entrepreneur"), or a group of legislators, or from the executive branch.[31] Policy leadership might come from an individual or organization in the private sector. Some areas seem particularly responsive to crisis-driven legislation, notably in the area of criminal law.[32] Other laws may be the product of longstanding public and political discussion and reflection that builds to a sufficient level of public and political support to produce a law, or reflect periods in history when there was a sustained focus on particular dimensions of public policy. This perspective suggests that legislation, like humor, is ultimately about good timing.

A notable and perhaps the most relevant illustration of the historical contingency of law-production across a general area is that most of the major federal environmental framework legislation was enacted between 1968 and 1976, and no major federal framework environmental legislation has been enacted in the past 35 years.[33]

Applying these three general theories of legislation helps to provide an explanation for the absence of framework harmful invasive species legislation. The substantial economic and other harms from invasive species and the possibility of relatively low-cost regulation that could have a substantial impact on the problem suggest some other barrier to general legislation. Public choice theory helps to explain the relatively focused targets of existing federal and state invasive species legislation.

The utility and public choice theories suggest that the absence of harmful NIS legislation may be a product of proponents of such legislation not yet convincing legislators that the benefits of more general legislation would clearly outweigh the costs. Interest groups may be too focused on the costs of particular invasive species, and less at the general problem. Combining these perspectives suggests the challenge at this point for proponents of a more substantial and coherent legal response to invasive species is a challenge of coordination. In other words, this perspective assumes that if the wide range of private and public actors who appreciate and are concerned about aspects of the invasive species problem worked together, they could convince legislators to act.

But the most powerful explanation for the lack of a framework invasive species may be a historical perspective, which suggests that where there may have been moments where more action could have been taken, strong leaders—policy entrepreneurs—were absent. As demonstrated by the incomplete text of the Carter Executive Order, harmful invasive species were not on the radar of environmentalists and policy makers in the window that opened in the late 1960s through the mid-1970s to revise and introduce framework environmental legislation. If Rachel Carson had focused more on the impact of invasive species in *Silent Spring*, then perhaps such a law would have at least been introduced in the window of dramatic environmental legislation in the late 1960s through the mid-1970s.

The historical perspective is consistent with a story about a lack of a sufficient appreciation of the benefits and costs of potential legislation, and with a lack of coordination among actors who appreciate this point. The historical perspective would highlight the lack of a single invasive species problem of sufficient national or regional standing to generate the kind of national public attention that demands a policy or legislative response. This observation might seem bizarre to scientists, citizens, environmental groups, and regulators who deal with significant harmful invasive species in a particular location. So for example in Southeast Arizona, buffelgrass presents a serious threat to the desert ecosystem, and to economic interests.[34] Nonetheless, buffelgrass is mentioned only in passing on the National Invasive Species Council website, and has yet to be registered in the *New York Times* beyond an occasional passing blog entry and a single word in a travel story.[35]

A federal legislative vessel has appeared in the past few years which might ultimately sail into law. In 2008, the congressional delegate from Guam, Madeleine Bordallo, with the support of eight co-sponsors, introduced H.R. 6311, the Nonnative Wildlife Prevention Act. It was referred to the House Natural Resources Committee, and the Subcommittee on Insular Affairs, Oceans and Wildlife, where it died a natural death. In 2009, in the 111th Congress, Delegate Bordallo introduced H.R. 669, this time with 42 co-sponsors, but the bill met a similar fate in the committee tar pits. In the 112th Congress a similar bill, The Invasive Fish and Wildlife Prevention Act, H.R. 5864, has been introduced in the House by Representative Louise Slaughter, Democrat of New York, along with 28 co-sponsors including Representative Bordallo. It, too, has been sent to committee.

The purpose of the legislation would be to create a process under the control of the Secretary of the Interior "for assessing the risk of all non-native wildlife species proposed for importation into the United States." The bill proposes that the Secretary produce a list of both approved and unapproved species—the "white list" approach. While the bill addressed only one aspect of the multifaceted problem of harmful invasive species, it is an important aspect, and one addressed incompletely in current federal law. Perhaps the bill might serve as the foundation for future bills and hearings.

Lessons from the Climate Change Policy Wars

The question of why there has been only this fragmentary effort (much less success) at creating a framework harmful NIS law can be flipped for scientists, lawyers, policy makers, and individuals concerned with promoting a more substantial response to the problem: under what conditions would federal framework environmental legislation be possible?

It is informative to compare the much more extensive recent literature on the question of why there has not been major federal climate change legislation. As noted, for many scientists and citizens—but also for the military, industry, and environmental groups—climate change is the area where the cost/benefit perspective, and perhaps also the historical perspective, might seem to make the most compelling case.

Multiple environmental scholars have decried the absence of broad federal climate legislation. In a series of articles Richard Lazarus has attacked the modern dysfunction of Congress, and predicted that without structural reforms in Congress, no major environmental legislation will be produced.[36] Lazarus argues that the short-term time horizons for elected officials and the short-term time horizons for most of their constituents disfavor "economic sacrifice for the benefit of other persons (and environmental interests) in distant places and times."[37]

But this argument proves too much: if this is correct it is hard to explain the existence of any environmental legislation—or any legislation, in any area—with impacts distant from a politician's district, and distant in impact from the next electoral cycle (or perhaps several cycles—the politician's expected political lifetime).

Some scholars blame the absence of climate legislation and other major environmental legislation on incoherence in the environmental community between pragmatic (or shallow) and "deep" environmentalism.[38] Scholars often call for more science in legislative decision-making, perhaps imposing on the idea of science a power and policy relevance as incomplete as the view of scientists that problems can be solved with the right law, and that if a law mandates or forbids actions, then that is what in fact will result.[39] Another group of scholars, including Robert Glicksman, Richard Briffault, J. R. DeShazo, and Jody Freeman, debate the potential of state leadership on environmental policy to provoke,[40] or through collaborative federalism to encourage (or to fail to encourage), federal legislation and policy.[41] Other scholars point to the global nature of the climate change problem, and highlight the difficulties of any collective action at a global scale.[42]

The three general perspectives on legislation raise the question whether major climate change legislation—like major harmful invasive species legislation—has not yet been achieved because there is not a sufficient public or political consensus around the need for major legislation. Sufficiently powerful constituencies in support of climate legislation may not yet have formed, and it is not hard to identify powerful economic and political interests opposed to such legislation. Instead of focusing on the difficulty of enacting any legislation, and the intense political partisan conflicts of our times, perhaps the positive case—the disaster (or series of disasters) of sufficient magnitude, for example—has not yet arrived.

Harmful invasive species legislation should be vastly easier to enact than climate change legislation. While the proponents of legislative action are less powerful and clear, so too the opposition to harmful invasive species legislation is less clear. Some of the significant commercial interests that might have concerns about invasive species legislation can also be important advocates in producing wise legislation that takes account of costs and benefits, including the pet industry, agriculture, horticulture, recreational boating, ranching, transportation (shipping, trucking, air cargo), fishing, hunting, and landscaping. But the problem of harmful invasive species has not been seen as one problem, but as many problems, dispersed in time and space, and impacting different interests. The information and knowledge challenges are not trivial; the difficulty of collective action—of organizing what may be sympathetic but disparate groups and individuals—is real. Professor Robert Ahdieh has called this and even more complex problems a challenge not of collective action through the market, but of sufficient coordination in the modern administrative state.[43]

Looking Forward

Thinking through the various paths that lead to actual legislation, proponents of a more coherent and aggressive response to harmful NIS might look to nurture legislative leaders in the Senate to match those who have at least attached their name to the Nonnative Wildlife Prevention Act in the House. More political leadership in the legislative and executive branches, and in the states as well as the federal system, could help to produce the momentum necessary for further reforms. Scientists and advocates might work to address the problem of harmful NIS as a single general problem rather than a series of local problems. An organization with the capacity to work at a national scale could help to organize and link the myriad harmful invasive species challenges across the United States.

The Invasive Fish and Wildlife Prevention Act does not include the range of subjects that a true organic or framework act would include. The creation of a model law that ties together current authorities and fills the gaps could lead to a more likely and better legislative outcome if a particular event or set of stories should raise the level of political or public interest in the issue of harmful invasive species. Scholars could work to address the full range of costs from harmful NIS, and take the time to translate or synthesize technical research to derive the policy implications, and to put them in a form—short and clear—of greatest use to the nontechnical public and nontechnical political leaders. In the absence of a path to wholesale reform, scholars might also focus their efforts on addressing particular gaps and the immediate needs. Piecemeal reform can help to set the stage for broad-scale legislation.

Despite the recent federal legislative proposals, harmful invasive species appears to be declining rather than increasing as a topic of interest, at least among major environmental groups. National organizations such as the Nature Conservancy and the Union of Concerned Scientists that had full-time scientists and policy specialists focused on the issue of invasive species over the last 10 to 15 years have eliminated those positions. Perhaps leadership on the regulation of harmful invasive species will come either from an industry that bears substantial costs from NIS, such as agriculture, or from an industry where regulation of NIS could impose substantial costs, such as the pet trade or horticulture.

In the absence of strong leadership from political actors or an effective campaign to increase attention to harmful NIS, the increase in global trade and many porous borders make it likely that at some point over the next decade more stories will emerge that will bring harmful invasive species to national attention. The wisdom of any response to what will seem like crisis may turn on the degree to which scientists, social scientists, industry leaders, environmental NGOs, and lawyers have planned for that moment.

Notes

1. Miller, M. 2004. The paradox of United States alien species law. In *Harmful Invasive Species: Legal Responses*, edited by M. Miller and R. Fabian,125–184. Washington, DC: Environmental Law Institute (documenting significant increase in awareness of invasive species in legal literature and in new stories starting in the early 1990s).

2. See Lazarus, R. 2006. *The Making of Environmental Law*. Chicago: University of Chicago Press.

3. An earlier version of the section describing federal law appears in Miller, M. 2011. Federal and state laws. In *Encyclopedia of Biological Invasions*, edited by D. Simberloff and M. Rejmánek, 430–437. Berkeley: University of California Press.

4. Congress defunded the OTA in 1995. See Mooney, C. 2005. Requiem for an office. *Bulletin of the Atomic Scientists* 61:41–49.

5. Earlier conceptions of introduced species in the United States have not been studied in depth, though Nineteenth Century Acclimatization Societies sought to introduce birds from England, including sparrows and starlings. Mirsky, S. 2008. Shakespeare to blame for introduction of European starlings to U.S. (originally printed as "Call of the reviled"). *Scientific American*, May 23. http://www.scientific american.com/article.cfm?id=call-of-the-reviled; Osborne, M. 2001. Acclimatizing

the world: a history of the paradigmatic colonial science. *Osiris* 15:135–151; Flannery, T. 2002. *The Future Eaters: An Ecological History of the Australasian Lands and People.* New York: Grove Press (discussing Australian acclimatization societies).

6. See Gorjanc, L.T. 2004. Combating harmful invasive species under the Lacey Act: removing the dormant commerce clause barrier to state and federal cooperation. *Fordham Environmental Law Review* 16:111–140.

7. Miller, M., and G. Aplet. 1993. Biological control: a little knowledge is a dangerous thing. *Rutgers Law Review* 45:285–334; Miller, M. 1994. Virtues of biological control regulation. In *Proceedings, US Department of Agriculture Interagency Gypsy Moth Research Forum 1994*, edited by S. L. C. Fosbroke and K. W. Gottschalk, 55–62. 1994 January 18–21; Annapolis, MD. Gen. Tech. Rep. NE-188. Radnor, PA, US Department of Agriculture, Forest Service, Northeastern Forest Experiment Station. http://www.treesearch.fs.fed.us/pubs/viewpub.jsp?index=4288; Miller, M., and Aplet, G. 2005. Applying legal sunshine to the hidden regulation of biological control. *Biological Control* 35:358–365.

8. H.R. 2830, 110th Cong. (2007)

9. 33 CFR § 151(c) (1999).

10. Ruiz, G.M., A.W. Miller, K. Lion, B. Steves, A. Arnwine, E. Collinetti, and E. Wells. November 16, 2001. Status and trends of ballast water management in the United States: first biennial report of the national ballast information clearinghouse. Smithsonian Environmental Research Center, Edgewater, MD. http://invasions.si.edu/nbic/reports/NBIC_Biennial_Report1_1999–01.pdf.

11. 69 Federal Register 32864–32871 (June 14, 2004) (codified at 33 CFR § 151).

12. 76 Federal Register 76716–76725 (December 8, 2011).

13. 76 Federal Register 31172–31210 (May 27, 2011) (final rule).

14. See, e.g., Nutria eradication and control act of 2003, Public Law 108–16 (2003); Brown tree snake control and eradication act of 2004, Public Law 108–384 (2004); Water resources development act of 2007, Public Law 110–114 (2007) (to control Asian carp on the Mississippi River and to limit the harm from other aquatic nuisance species); Clean boating act of 2008, Public Law 110–288 (2008); Asian carp prevention and control act, Public Law 111–307 (2010). See Miller 2011; see also http://www.invasivespeciesinfo.gov/laws/publiclaws.shtml.

15. See Shannon, M. 2008. From zebra mussels to coqui frogs: public nuisance liability as a method to combat the introduction of invasive species. *Environs Environmental Law and Policy Journal* 32:37–67.

16. See Harper-Lore, B. 2002. Incorporating invasive plant analysis into NEPA. In *Proceedings of the International Conference on Ecology and Transportation, Keystone, CO, September 24–28, 2001*, edited by C. L. Irwin, P. Garrett, and K. P. McDermott, 317–319. Center for Transportation and the Environment, North Carolina State University, Raleigh, NC.

17. See Miller, M. 2003. NIS, WTO, SPS, WIR: Does the WTO substantially

limit the ability of countries to regulate harmful non-indigenous species? *Emory International Law Review* 17:1059–1089; Riley, S. 2005. Invasive alien species and the protection of biodiversity: the role of quarantine laws in resolving inadequacies in the international legal regime. *Journal of Environmental Law* 17:323–359; Stewart, T., and C. Schenewerk. 2004. The conflict between facilitating international trade and protecting US agriculture from invasive species: APHIS, the US plant protection law, and the Argentine citrus dispute. *Journal of Transnational Law and Policy* 13:305–346.

18. The National Invasive Species Council's 2001 invasive species management plan was sharply criticized in a 2002 GAO Report. Invasive species: clearer focus and greater commitment needed to effectively manage the problem, GAO-03-1 (General Accounting Office, Oct. 22, 2002). (In 2004 Congress changed the name of the then 83-year-old agency from the General Accounting Office to the Government Accountability Office.)

19. As of May 29, 2012, the National Invasive Species Council website lists the management plan period alternately as 2008–2012 and 2008–2010, some of the links to the plan do not work; neither does the link to the members of the supposedly current Invasive Species Advisory Council. http://invasivespecies.gov/global /ISAC/ISAC_index.html.

20. See Act No. 241 of Aug. 15, 1999, as amended by Act. No. 176 of Aug. 1, 2004 (adding the prohibition on monkeys).

21. See Miller 2004; Graham, J. C. 2011. Snakes on a plain, or in a wetland: fighting back invasive nonnative animals—proposing a federal comprehensive invasive animal species statute. *Tulane Environmental Law Journal* 25:19–81.

22. See, e.g., McChesney, F. 1997. *Other People's Money: Politicians, Rent Extraction, and Political Extortion*. Cambridge, MA: Harvard University Press.

23. Bentham, J. 1882. *Theory of Legislation*, translated by R. Hildreth. London: Trubner & Co.:1 (original work published 1864). p. 1.

24. Ibid., 87.

25. Ibid., 2.

26. Ibid., 2.

27. Ibid., 31.

28. See generally Sunnstein, C. 1993. *The Partial Constitution*. Cambridge, MA: Harvard University Press; for an effective critique see McGinness, J. 1994. The partial republican. *William & Mary Law Review* 35:1751–1799.

29. See, e.g., two interesting articles assessing theories of public choice in the context of bankruptcy law: Block-Leib, S. 1997. Congress' temptation to defect: a political and economic theory of legislative resolutions to financial common pool problems. *Arizona Law Review* 39:801–871; Walt, S. 2000. Collective inaction and investment: the political economy of delay in bankruptcy reform. *Emory Law Journal* 49:1211–1260 (critiquing public choice theories of legislation). In a similar vein, see Shaviro, D. 1990. Beyond public choice and public interest: a study of the

legislative process as illustrated by tax legislation in the 1980s. *University of Pennsylvania Law Review* 139:1–123.

30. See, e.g., Croley, S. P. 1998. Theories of regulation: incorporating the administrative process. *Columbia Law Review* 98:1–168.

31. See, e.g., Kingdon, J. W. 1995. Agendas, alternatives and public policies. 2nd ed. New York: Harper Collins; Mintrom, M., and S. Vergari. 1996. Advocacy coalitions, policy entrepreneurs, and policy change. *Policy Studies Journal* 24:420–434.

32. See, e.g., Stuntz, W. 2001. The pathological politics of criminal law. *Michigan Law Review* 100:505–600.

33. See Elliot, E.D., B. Ackerman, and J. Millian. 1985. Toward a theory of statutory evolution: the federalization of environmental law. *Journal of Law, Economics and Organization* 1(2):313–340.

34. See Southern Arizona Buffelgrass Coalitioner Center. http://www.buffel grass.org/.

35. See Nelson, B. 2011. A coast-to-coast guide to endangered species. *New York Times*, May 13. http://travel.nytimes.com/2011/05/15/travel/endangered-species -travel-guide.html.

The whoosh of a surfacing orca and the glower of a mother grizzly still have the power to raise goose bumps; a soaring California condor can yet astonish. But chances to admire many of our wildlife neighbors are becoming increasingly uncommon. Invasive buffelgrass is crowding out saguaros and other native cactuses throughout the Southwest, while melting sea ice is threatening the Pacific walrus and polar bear in Alaska. Mosquito-borne diseases are threatening Hawaii's songbirds, and white-nose syndrome is wiping out bats in the East.

36. Lazarus, R. J. 2006. Congressional descent: the demise of deliberative democracy in environmental law. *Georgetown Law Journal* 94:619–681, at 679; Lazarus, R. J. 2007. Environmental law after Katrina: reforming environmental law by reforming environmental lawmaking. *Tulane Law Review* 81:1019–1058, at 1043–44.

37. Lazarus 2007:1043–1044.

38. Schroeder, C. H. 2008. Legislating to address climate change: some lessons from the field. *Environmental & Energy Law & Policy Journal* 3:236–260, at 243.

39. Tarlock, A. D. 2004. Is there a there there in environmental law? *Journal of Land Use and Environmental Law* 19:213–254, at 220–221.

40. DeShazo, R., and J. Freeman. 2007. Timing and form of federal regulation: the case of climate change. *University of Pennsylvania Law Review* 155:1499–1561, at 1559.

41. Briffault, R. A. 2004. Beyond congress: the study of state and local legislatures. *NYU Journal of Legislation and Public Policy* 7:23–30, at 27; Glicksman,

R. L. 2006. From cooperative to inoperative federalism: the perverse mutation of environmental law and policy. *Wake Forest Law Review* 41:719–803, at 802.

42. See, e.g., Schenck, L. 2008. Climate change "crisis": struggling for worldwide collective action. *Colorado Journal of International Environmental Law and Policy* 19:319–379.

43. Ahdieh, R. 2010. The visible hand: coordination functions of the regulatory state. *Minnesota Law Review* 95:578–649.

CHAPTER SIXTEEN

Pathways toward a Policy of Preventing New Great Lakes Invasions

Joel Brammeier and Thom Cmar

Introduction

Freshwater ecosystems across the globe are experiencing tremendous stress caused by aquatic invasive species. In North America, regional freshwater systems like the Great Lakes are experiencing extinction rates five times higher than terrestrial and marine ecosystems.[1] At least 123 freshwater animal species, including fish, mussels, crayfish, and amphibians, have gone extinct across North America since 1900. This disproportionate and disturbing trend is expected to increase owing to anthropogenic factors led and compounded by the impact of climate change.[2] The Great Lakes Regional Collaboration (GLRC) said in its 2005 report that due to an inundation of invasive species the world's greatest freshwater lakes are "succumbing to an irreversible 'invasional meltdown' that may be more severe than chemical pollution."[3]

As an internally connected freshwater system that spans eight US states and two Canadian provinces, supports a multibillion dollar fishery, and provides drinking water for 40 million people, the Great Lakes region is uniquely vulnerable to invasive species impacts. Over the past 200 years, more than 186 aquatic invasive species (AIS), including plants, fish, algae, and mollusks, have become established in the Lakes.[4] The problem was compounded by the opening of the Saint Lawrence Seaway in 1959, which opened the Lakes to traffic from oceangoing vessels. Along with foreign cargo, these vessels bring into the Great Lakes ballast water picked up in foreign ports that is contaminated with a range of aquatic species, including potential invaders, and then dump that contaminated ballast water as they pick up additional cargo. A 2006 study of successful Great Lakes invasions since 1959 attributed approximately 65% of those invasions to vessels' ballast water discharges.[5]

Until the mid-1990s, another significant vector for transfer across regional watershed boundaries received scant attention. Artificial connections between the Great Lakes and other freshwater ecosystems have now been implicated in multiple invasions: the spread of zebra and quagga mussels[6] and the round goby[7] into the Mississippi River basin, the entry of these same species[8,9] and fishhook water flea[10] into New York's Finger Lakes, the invasion of the upper Great Lakes by sea lamprey,[11] and the potential establishment of blueback herring in Lake Ontario.[12] Most recently, the Chicago Area Waterway System (Chicago Waterway) is broadly accepted as the highest risk vector[13,14] for the movement of bighead and silver carp into the Great Lakes watershed.

The environmental and economic impacts of invasive species are well known. Research suggests that the annual cost to the Great Lakes region from invasive species introduced by shipping may be upwards of $200 million annually.[15] The extirpation or major declines in important native species due to effects from aquatic invasive species (such as sea lamprey predation on lake trout, and competition with deepwater ciscoes by introduced alewives and rainbow smelt) are well documented.[16] Invasions limit the ability of the natural ecosystem to support fisheries and wildlife, and to provide for raw water uses such as drinking water. Many invasive species are aggressive and highly adaptable and reproduce quickly, making them very difficult to manage, let alone eradicate, once they are introduced.

Despite the fact that vessels for decades have dumped throughout the Great Lakes ballast water that has introduced and spread harmful invasive species, to date the US Government has yet to create a regulatory regime that effectively addresses this significant cause of environmental degradation and economic harm. Preventive standards should have been required long ago under federal statutes already on the books, but these laws have been underimplemented and underenforced by the responsible federal agencies, in particular the US Environmental Protection Agency (EPA) and the US Coast Guard.

The invasion threat created by artificial canals to and from the Great

Lakes has been known at least as long as the threat from contaminated ballast water.[17] However, the first Congressional authorization to address this vector did not become law until 1996,[18] and even then the effort was experimental and species-specific.[19] Not until 2005 did the federal government prioritize in official policy a collaborative effort to combat the canal vectors.[20] To date, there is no finalized federal implementation plan to fully prevent the movement of invasive species to and from the Great Lakes via artificial canals.

The inability to deploy timely, integrated and effective prevention measures suggests bottlenecks at the federal level that constrain agency ability to convert clear and straightforward data on the sources of invasions into on-the-ground action. While federal agencies are deploying short-term and interim measures to reduce the risk of these vectors, such measures are unlikely to prevent new invasions in the long run. Given that invasive species establishment is generally irreversible and the economic costs and ecosystem impacts of invasion are well-established, the federal government should be pursuing a policy of prevention for species that represent the highest risk to the Great Lakes ecosystem.

Ballast Water and the Great Lakes

Clean Water Act Exemption and Zebra Mussel Invasion

The federal Clean Water Act is the nation's principal law for regulating water pollution and water quality. The Act was enacted by the US Congress in 1972, at a time of high public concern over the impact that America's twentieth-century industrial activity was having on human health and the environment. Congress intended the statute "to restore and maintain the chemical, physical, and biological integrity of the Nation's waters," by adopting a "national goal that the discharge of pollutants into navigable waters be eliminated" and "that wherever attainable, an interim goal of water quality which provides for the protection and propagation of fish, shellfish, and wildlife and provides for recreation in and on the water be achieved."[21] The Clean Water Act conveys significant authority to EPA to oversee a nationwide program, but the Act also embodies "the policy of the Congress to recognize, preserve, and protect the primary responsibilities and rights of States to prevent, reduce, and eliminate pollution."[22] Thus, the statute envisions that EPA's primary role will be to set minimum national standards for pollution control technology and then oversee state implementation, while states are further charged with setting local water quality standards based on the ecological needs and functions of particular ecosystems.[23] In permitting individual discharges under the Act, EPA and the states are required to ensure compliance with both pollution control technology standards and state water quality standards, and compliance with water quality standards must be required without regard to technological feasibility or cost.[24]

Importantly, the Clean Water Act defines "pollutants" to include "biological materials," such as invasive species.[25] Yet the Act's two-step process for requiring reductions in pollution and protection of water quality has never been fully applied to vessels' ballast water discharges. In 1973, as part of a package of regulations to implement the Clean Water Act for the first time, EPA made the fateful decision to exempt vessels' ballast water discharges. Specifically, EPA exempted from the Clean Water Act "any discharge of sewage from vessels, effluent from properly functioning marine engines, laundry, shower, and galley sink wastes, or any other discharge incidental to the normal operation of a vessel" on the ground that "this type of discharge generally causes little pollution."[26] The agency's early Clean Water Act regulations dealt with other priorities and excluded vessel discharges because, in EPA's view, they "were not important to the overall scheme of things at that time."[27] EPA's exemption of vessel discharges from the Clean Water Act remained in effect for over 30 years before it was struck down as unlawful by a federal court in California.[28]

We are still living with the consequences of this exemption today and can make a reasonable assumption that, if EPA had fully applied the CWA's requirements to vessels' ballast water discharges, the zebra and quagga mussel invasions of the Great Lakes may well have been averted. Federal and state authorities did not appreciate the economic and environmental costs of ballast water as a vector for invasive species until the zebra mussels invaded Lake St. Clair in the late 1980s, then spread rapidly throughout the Great Lakes region and beyond. Since their initial introduction, zebra mussels (followed by their cousins the quagga) have invaded large swathes of the country, consuming much of the base of aquatic food webs, degrading water quality, and encouraging the growth of toxic blue-green algae.[29] Zebra mussels have costly impacts on infrastructure in Lake Michigan, clogging water intake pipes and screens.[30] And, recent scientific studies found that zebra mussels have devastated native fish populations: fish biomass has decreased by about 95% in Lake Michigan and Lake Huron since the mid-1990s.[31]

The invasion of zebra and quagga mussels is a dramatic example of how ballast water-mediated invasions do not stop at the shores of the Great Lakes, but spread inland to freshwater ecosystems across North America. Twenty years after the initial zebra and quagga mussel invasion, the invasive mussels had spread to all five Great Lakes and into the St. Lawrence, Mississippi, Tennessee, Hudson, and Ohio River Basins, and have recently established west of the continental divide in California and Nevada, despite an extensive binational effort to prevent a westward expansion.[32]

Coast Guard Regulations

In response to the invasions by mussels and a growing number of other species, Congress chose to enact legislation giving additional authority to the US Coast Guard, rather than to EPA. In 1990, Congress enacted the Nonindigenous Aquatic Nuisance Prevention Act of 1990 (NANPCA), which charged the Coast Guard to "issue regulations to prevent the introduction and spread of aquatic nuisance species into the Great Lakes through the ballast water of vessels" by requiring oceangoing vessels to conduct a mid-ocean saltwater exchange of their ballast water to presumably remove potential invasive species or to "use environmentally sound alternative ballast water management methods" approved by the Coast Guard.[33] In 1996, Congress amended and enhanced NANPCA through the National Invasive Species Act (NISA), which gave the Coast Guard authority to establish national invasive species regulations and required them to review and update them every three years in order to "ensure to the maximum extent practicable that aquatic nuisance species are not discharged into waters of the United States from vessels."[34]

With respect to vessels entering the Great Lakes through the Saint Lawrence Seaway, the Coast Guard's NISA regulations contained a significant loophole: they did not require a vessel to have exchanged or treated its ballast if it declared "No Ballast On Board" (NOBOB) status.[35] A NOBOB ship is one that has pumped out its ballast tanks as much as possible before entering regulated waters. However, NOBOBs are rarely completely dry or free of residual sediment.[36] As a NOBOB moves from port to port, it may take on new water as ballast to maintain trim and stability during operations. This water mixes with residual ballast water, sediment, and any associated nonindigenous organisms, and is later discharged elsewhere. In 2005, after years of scientists and environmentalists calling for closure of the "NOBOB loophole," the Coast Guard established saltwater flushing as a voluntary best management practice for NOBOB vessels, which was later made mandatory for Great Lakes NOBOBs through regulations promulgated in 2008 by the Saint Lawrence Seaway Development Corporation.[37]

Even now that the NOBOB loophole has been closed, the Coast Guard's ballast water regulations still contain significant loopholes for safety and other considerations, do not apply to discharges of ballast water from vessels that operate solely within the Great Lakes (so-called "laker" vessels), and do not fully prevent the introduction or spread of invasive species through ballast water. An EPA study of ballast water exchange found that "a 95 percent exchange of the original water resulted in flushing of only 25 to 90 percent of the organisms studied."[38] Further, EPA found that "where ballast water is taken up and discharged in saltwater ports, it can be expected that mid-ocean ballast water exchange will be even less successful."[39] The EPA study identified a number of "drawbacks to the mid-ocean exchange method of ballast water management," including that many ships are not structurally designed to safely allow ballast water exchange at sea; exchange is sometimes impossible in rough weather due to safety concerns; some organisms can survive under a very wide range of salinity conditions; and despite flushing of the ballast tanks with open ocean water, live invasive organisms can still remain in "pockets" of unexchanged water or in layers of sediment at the bottom of ballast tanks.[40] A 2007 study by the National Oceanic and Atmospheric Association found that saltwater flushing also has highly variable effectiveness in NOBOBs, depending on the age and salinity tolerance of the organisms in the ballast tank.[41]

As of 2007, a new aquatic invasive species was being discovered in the Great Lakes on average every 28 weeks.[42] After Saint Lawrence Seaway regulations closed the NOBOB loophole, that pattern seems to have been reversed; as of this writing, no new aquatic invasive species have been identified in the Great Lakes in over three years. Nevertheless, with thousands of vessels dumping ballast water into the Great Lakes every year while using ballast water management techniques that do not consistently eliminate invasive organisms, there is still a significant risk of ballast water–mediated invasive species establishment.[43]

Ballast Water Treatment Systems

Parallel to these US domestic processes, the international community has slowly reached its own consensus. The International Maritime Organization (IMO) began work on ballast water guidelines in 1992, at the urging of the United Nations Conference on Environment and Development.[44]

The IMO published initial guidelines in 1993, then a more comprehensive set in 1997.[45] The 1997 guidelines recommended a variety of management practices, including ballast water exchange, discharging ballast water to onshore treatment facilities, and minimizing uptake of sediments into tanks.[46] Because the IMO's guidelines were only voluntary, the recommended practices were implemented sporadically across the shipping industry.[47]

After finalizing guidelines in 1997, the IMO began work on a binding international legal regime.[48] The US Coast Guard led the US delegation to these negotiations, which concluded after seven years with the adoption by consensus of a Ballast Water Convention in 2004.[49] The Ballast Water Convention will enter into force twelve months after ratification by thirty states, representing 35% of the world merchant shipping tonnage, which has not yet occurred but is expected in the near future.

The negotiations over the IMO Ballast Water Convention represented a paradigm shift in how regulatory authorities conceptualized ballast water management policy.[50] Recognizing the limitations of saltwater exchange and flushing as management strategies, the IMO negotiations focused on development of a uniform discharge standard that would set concentration-based limits on the number of invasive species that a vessel would be permitted to discharge.[51] To meet this standard reliably, vessels would likely have to install new on-board ballast water treatment technologies modeled after technologies used to clean wastewater in other contexts.[52] Such technologies could include filtration, heat, ultraviolet light, or addition of chlorine or other chemical additives.[53]

Although the IMO negotiators successfully reached consensus on a set of numeric standards for live organism concentration in ballast water discharge, the standards agreed upon were not based purely on a scientific assessment of the level of protection needed to prevent significant risk of ballast-mediated species invasions, but also incorporated "inputs from experts in other disciplines, such as shipping and engineering, risk managers, as well as state, national, non-governmental organization[s] (NGO[s]), and industry representatives."[54] During the negotiations, the US delegation (led by the Coast Guard) proposed a concentration-based discharge standard 1,000 times more stringent than that which was ultimately agreed upon in the negotiations, "based on the large number of organisms that would be discharged even at these low concentrations and the additive densities from multiple ship discharges."[55] Other countries urged that substantially weaker standards be adopted, however, and "the numbers ultimately adopted reflect a negotiated outcome among the many countries with differing views," which "makes it extremely difficult, or impossible, to parse exactly how a decision was made" or to determine "whether the IMO standards are sufficiently protective."[56] Nevertheless, a further analysis of the IMO's own data by the California Advisory Panel on Ballast Water Performance Standards found that nearly half of all commercial vessels that discharge ballast water would meet the IMO standards without conducting any treatment or ballast water exchange whatsoever, indicating that the IMO standards do not represent a substantial improvement in protection over the status quo.[57]

State and EPA Efforts to Compel Regulation under the CWA

In the absence of effective regulation from national or international bodies, a number of States, most notably Michigan and California, stepped up to fill the void by acting under their own authority In 2005, Michigan enacted legislation requiring that all oceangoing vessels obtain a state ballast water control permit requiring use of ballast water treatment technology to engage in port operations in Michigan.[58] To implement this statute, the Michigan Department of Environmental Quality issued a general permit that pre-approves use of four types of ballast water treatment technology and allows vessel operators to apply for an individual permit to use different technology.[59] The constitutionality of Michigan's statute was unsuccessfully challenged by shipping interests.[60]

In 2006, California passed legislation adopting ballast water performance standards recommended by its scientific advisory panel that are based on the standards proposed by the US delegation to the IMO.[61] The California panel found that its recommended standards "are significantly better than ballast water exchange, . . . are in-line [sic] with the best professional judgment from the scientific experts participating in the IMO Convention, and . . . approach a protective zero discharge standard."[62] The California State Lands Commission is still in the process of developing regulations to implement the standards, but it has concluded that "multiple systems have demonstrated that they have the potential to meet California's performance standards" and the technology continues to develop rapidly.[63]

In 1999 a group of West Coast environmental organizations, led by Northwest Environmental Advocates, petitioned EPA in January 1999 to repeal the regulation exempting vessel discharges from Clean Water Act requirements on the ground that the exemption was beyond its authority (ultra vires). The agency's final action denied the petition in its entirety and declined to regulate vessel discharges on September 2, 2003.[64] Three months later, a subset of the original petitioners filed suit in federal district court challenging that denial.[65] The environmentalists' lawsuit was later joined by a coalition of Great Lakes States, including New York, Illinois, Michigan, Minnesota, Wisconsin, and Pennsylvania.[66]

In 2005, the district court ruled in plaintiffs' favor, finding that under the "clear language" of the Clean Water Act, Congress required nonmilitary vessels to obtain a permit "before discharging pollutants into the nation's navigable waters," just like other "point sources" of pollution.[67] After further proceedings, the court ordered EPA to begin regulating vessel discharges within two years, at which time the regulatory exemption would be vacated.[68] In July 2008, a federal appeals court upheld the district court's orders in their entirety.[69]

In response to the *Northwest Environmental Advocates* court's order, EPA issued a Vessel General Permit or VGP to regulate ballast water discharges, along with 25 other "vessel discharge streams."[70] Unfortunately, EPA failed to live up to its obligations under the Clean Water Act in issuing the VGP. The VGP did not require vessels to install treatment technology to clean their ballast water before dumping it, nor did the VGP include requirements that would ensure that invasive species introduced and spread through ballast water would not continue to harm the water quality and biological integrity of aquatic ecosystems.[71] Instead, the VGP merely incorporates requirements from existing Coast Guard and St. Lawrence Seaway Development Authority regulations that oceangoing vessels conduct a mid-ocean ballast water exchange or, if they declare No Ballast On Board status, saltwater flushing of their ballast tanks.[72] Rather than fully implementing the Clean Water Act and using it as a tool to fully prevent the ballast water vector from causing new invasions, EPA essentially abdicated its responsibility by writing status quo regulations into its permit.

After EPA issued the VGP, 12 environmental organizations and the State of Michigan filed litigation that was consolidated in the US Court of Appeals for the District of Columbia, challenging the VGP's failure to live up to Clean Water Act requirements. On March 8, 2011, EPA and these petitioners announced that they had settled these lawsuits, with EPA agreeing to complete scientific reviews and issue a new VGP by November 2012 that would set numeric limits on the concentration of invasive species that vessels would be permitted to discharge in their ballast water.[73] The settlement effectively gave EPA the opportunity to go back to the drawing board and write a new VGP that could, for the first time, treat the "living pollution" in ballast water as aggressively as EPA already regulates conventional toxic pollution, such as oil spills.[74] In November 2012, EPA and the petitioners agreed to extend the deadline for EPA to issue a new VGP to March 2013, and as of this writing EPA continues to work on the permit.

As required by the settlement agreement, EPA issued a new draft VGP on November 30, 2011, for public review and comment.[75] EPA's initial proposal for the next VGP, like a proposal for new ballast water discharge regulations issued by the Coast Guard in 2009[76] and finalized in March 2012,[77] adopts the IMO standards and implementation schedule with virtually no changes or improvements. Despite the fact that, as noted above, it is well established that IMO standards were not designed to achieve complete prevention of new invasive species introductions and thus fully protect water quality standards, EPA asserted that a "profound lack of data" left it unable to design a single concentration-based discharge standard short of zero discharge that would reliably assure complete prevention.[78] On that basis, EPA concluded that ballast water discharge standards that would fully protect water quality from invasive species were "infeasible to calculate."[79] EPA also deferred heavily to the costs to industry in adopting the IMO standards, stating that it "did consider accelerating" the requirements for vessels to install ballast water treatment systems in the draft VGP but "decided against doing so, based on concerns that altering the anticipated [IMO compliance] schedule at this late a date could disrupt the industry's prior planning and perhaps even result in further delays."[80] As a result, although the EPA's proposed new VGP and the Coast Guard's new ballast water regulations represent a significant step forward in terms of incorporating into US federal law the IMO paradigm shift from vessels exchanging their ballast water to treating it to a concentration-based discharge standard, the standard itself that EPA and the Coast Guard propose to adopt, like the one adopted internationally, is not based on a purely scientific assessment of invasive species risk but rather has been weakened to accommodate a broad range of shipping industry concerns and other political factors. Additional litigation will likely occur if EPA finalizes its proposed new Clean Water Act permit, which could spur positive pressure from the courts toward moving the US Government further toward a policy of prevention instead of requiring only partial measures that reduce the risk of ballast-mediated invasions but still allow them to occur.

Artificial Connections to the Great Lakes

Demonstration and Failure

Federal agencies did not prioritize the management of rivers for prevention of invasive species transfer into or out of the Great Lakes until the mid-1990s. Unlike the effort to prevent ballast water pollution, this type of regional watershed-level prevention is a response to the spread of species already introduced to the United States by other means. Of the 39 species determined in 2011 to be "high risk" and potentially able to move between the Great Lakes and Mississippi River, nearly all originated outside of North America.[81] Most of today's domestic prevention efforts now focus on reducing establishment at the watershed and sub-watershed, rather than national, scale.

In light of the understanding of risk and impacts described in the introduction, the federal government's initial moves to address rivers as invasion vectors were modest. The NANPCA of 1990, in addition to establishing a national ballast water program, created the federal Aquatic Nuisance Species Task Force and the Great Lakes Aquatic Nuisance Species Panel. The Panel's 1996 policy position promotes the "examination" of dispersal barrier options and cites the Chicago River.[82]

The policy found its grounding in the NISA authorization of 1996. Congress authorized a "dispersal barrier demonstration" to prevent and reduce the dispersal of ANS between the Great Lakes and Mississippi River through the Chicago Sanitary and Ship Canal.[83] The species selected for the demonstration was the round goby, which was well established in Lake Michigan. Almost from the start, the objective of prevention was undermined by constraints imposed by Congress and the US Army Corps of Engineers (Corps). Having convened a Dispersal Barrier Advisory Panel of experts and stakeholders to advise how to implement this demonstration, by 1998 the Corps determined that the demonstration project could not be allowed to interfere with navigation and flow of water.[84,85] This decision removed a variety of the most effective technologies, such as "canal redesign" and "reverse flow,"[86] from consideration and steered the Corps to choose an electrical barrier. By 1999, or three years prior to the activation of the demonstration electrical barrier, the goby had been discovered below the barrier site before construction had even begun.[87]

Risk and the Requirements of Prevention

Meanwhile, others were beginning to grasp the complete scope of risks posed to the Great Lakes by canals. In 2002, on the cusp of the activation of the demonstration electrical barrier, the US Fish and Wildlife Service (FWS) issued a white paper that set the stage for the critical debate facing the Great Lakes today. This "Perspective on the Spread and Control of Selected Aquatic Nuisance Fish Species"[88] neatly described the invasion vector for Asian carp and other species created by the connection of the Great Lakes to the Mississippi River via the Chicago Sanitary and Ship Canal.

The species of greatest concern to the Great Lakes region today—bighead and silver carp—were first cultivated in the United States in Arkansas aquaculture facilities in the early 1970s. Both species were documented in natural waters in Arkansas outside of these facilities by the early 1980s.[89,90] As filter feeders that are highly efficient at the removal of plankton from water, bighead and silver carp compete directly and successfully with the larval and juvenile forms of a broad spectrum of native freshwater fish. Leaving aside the recent explosion of interest in these fish, knowledge of this invasive behavior and dominant establishment in the Mississippi River basin was common among fishery biologists in 2002.[91] Subsequent research demonstrated the likelihood that bighead and silver carp could establish in the Great Lakes region's bays and tributaries, and that climatic conditions in the Great Lakes region compared favorably to the species' native ranges.[92,93,94,95]

As the first paper discussing the ecosystem impacts of bighead and silver carp to be widely disseminated among policy makers and environmental advocates, the FWS report clarified that the threat to the Great Lakes had entered a more urgent phase, noting presciently that

> Biologists would not be too concerned about this connection if the opportunity for organism exchange was limited to native species as it was in eons past . . . But unlike (the prior 14,000 years), the current organism exchange features species not native to North America. These alien invaders came to the US as a result of international shipping (i.e. transfer and dumping of organism-infested ballast water between continents); and trade in the aquaculture, aquarium and baitfish industries.[96]

It was not the century of connection between these two great watersheds alone that generated worry, but the connection combined with the invasion of the United States by species from around the globe such as bighead and silver carp. The canal had become a liability to the Great Lakes rather than an asset.

In recognition, the City of Chicago and a network of local, state, and federal entities hosted a 2003 expert summit. Three of the four workgroups formed from the nearly 70 attendees indicated that separation of the Great Lakes from the Mississippi River was the option most likely to result in long-term prevention of invasion into the lakes by all aquatic species.[97] While this conference was a discussion of options rather than a statement of official policy, it served as the first broad indicator of the scientific community's opinion that separation, rather than reliance on technology barriers, would be critical to preventing invasion.

The finding was reinforced by numerous reviews of barrier technologies occurring around the same time. In 2004 the Minnesota Department of Natural Resources published an assessment[98] of river barrier technologies indicating that all options besides physical barriers had a less than 100% deterrent value and at least a moderate risk of failure. A 2005 Sea Grant study of canal invasion vectors to Lake Champlain reinforced that "[A physical barrier] is seen as 'preferred' from an ecological perspective. It clearly would offer the most protection against further ANS invasion of Lake Champlain."[99] A comprehensive assessment modeled after the Minnesota Department of Natural Resources (DNR) report prepared for Brammeier et al. in 2008 showed similar findings.[100] Finally, a 2012 US Army Corps of Engineers report concluded that only physical or chemical barriers can target all types of invasive organisms moving in both directions between the Great Lakes and Mississippi River via the Chicago Waterway.[101]

A Renewed Federal Effort

In recognition that the demonstration electrical barrier was inadequate to deter the Asian carp threat, as early as 2002 members of the Dispersal Barrier Advisory Panel turned to motivating the construction of a more sophisticated and powerful option.[102] From 2002 to 2009, the panel worked with the Corps and the Great Lakes states to establish funding streams, Congressional authorizations,[103] and stakeholder buy-in to support the construction of what would ultimately be called "Barrier II." Again under the constraint of not interfering with navigation, the design of Barrier II is believed to deliver the most power output of any other electric fish barrier system in the world.[104,105] Yet the barriers encountered operational and safety difficulties that required operation at parameters less than 100% effective against its targeted fish species and did nothing to prevent the movement of aquatic plants, larvae, eggs, and microorganisms.

Parallel to barrier design and construction was an effort to get Congress to recognize the inadequacy of current technology approaches to prevention. In 2005 a group of experts for the first time described a detailed federal agenda for achieving prevention via the river and canal invasion vector in the 2005 Great Lakes Regional Collaboration.[106] In authorizing what is now known as the Great Lakes Mississippi River Interbasin Study, or GLMRIS, in 2007, Congress removed the constraints imposed during the technology barrier planning processes by stating

> The Secretary (of the Army), in consultation with appropriate Federal, State, local, and nongovernmental entities, shall conduct, at Federal expense, a feasibility study of the range of options and technologies available to prevent the spread of aquatic nuisance species between the Great Lakes and Mississippi River Basins through the Chicago Sanitary and Ship Canal and other aquatic pathways.[107]

Here Congress made the objective of GLMRIS clear: prevention. No mention is made of current uses of the Chicago Waterway, unlike the original authorization of the demonstration electrical barrier that referenced implementation in conjunction with the operation of a lock system and "ongoing operations" of the Corps in the Chicago Waterway.[108]

The Carp Advance

In August 2009, the Corps released results of "environmental DNA" (eDNA) sampling showing the presence of bighead and silver carp DNA above the electrical barrier site. The discovery of evidence of live Asian carp beyond the only barrier deterrent between the Great Lakes and Mississippi River kindled a political and legal firestorm that has reshaped the policy debate over invasive species prevention via canals. Federal agencies duplicated and expanded these findings to multiple locations in the Chicago Waterway and discovered a live carp near Lake Michigan over the next year, fueling the fire even more.

The state of Michigan, joined by five other Great Lakes states, petitioned the Supreme Court to address the threat.[109] The states filed a request for a preliminary injunction against the state of Illinois, the Metropolitan Water Reclamation District (MWRD), and the Corps to close the navigational locks on the Chicago Waterway to stop the flow of water and fish. They also requested that the court reopen the consent decree governing Illinois' use of Lake Michigan water—a practice made possible by the hydraulic connection between the two watersheds. After the Supreme Court, without explanation, declined to exercise jurisdiction over their case, the states filed a similar action against the Army Corps and MWRD in federal district court in Chicago.[110]

Public outcry and official calls for action occurred outside of the courtroom as well. Within 4 months of the initial eDNA findings, members of Congress transmitted a letter to the Corps calling for rapid evaluation of hydrologic separation of the Great Lakes from the Mississippi River.[111] Public outcry outside of Illinois focused almost exclusively on a push to close the locks, as did various legislation introduced in Congress.[112]

Virtually overnight, the discussion of carp prevention that had been restricted to experts in agencies and academia,[113,114] a few interested stakeholders,[115,116,117] and committed members of Congress had erupted into the public discourse. Most telling was that this exponentially larger group of new stakeholders immediately homed in on the prevention measure that had dominated discussions among experts—hydrologic separation. Although at the time closure of the canal locks was erroneously viewed as synonymous with permanently stopping the flow of water, the focus on hydrologic changes to the system was a sea change in the region's approach to invasive species policy. Faced with a critical threat, decision makers intuited the field- and lab-demonstrated reality that technology barriers were less than 100% effective in stopping all organism movements. Having borne the brunt of the damage from prior significant invasions, states like Michigan were understandably uninterested in solutions that promised less than 100% preventive capability.

Getting to Prevention

Independent efforts to pursue the recommendations of the 2003 Chicago summit had begun well prior to the passage of the GLMRIS authorization. Building on the conclusion that physical barriers were the only option likely to result in 100% prevention, the Alliance for the Great Lakes evaluated transportation and hydrology on the Chicago Waterway and described six options[118] for where such a hydrologic separation could occur in 2008. Before the Corps even began the public portion of the GLMRIS study in late 2010, the Natural Resources Defense Council issued a report[119] demonstrating the hydrologic reality behind two of the most likely options and making the case for feasibility.

The most substantive discussion of separation to date is found not with the federal government, but in a privately funded study led by the Great Lakes Commission and Great Lakes–St. Lawrence Cities Initiative representing the states and mayors, respectively. Spurred by regional public opinion and the commitment of both the state of Illinois and city of Chicago to participate, the groups raised $2 million in nongovernment funding in a few months and completed a full engineering analysis of separation options in less than 18 months. On release, the report was praised[120] by elected officials and others across the region for both its technical quality and its demonstration that separation was an achievable and desirable outcome with multiple ancillary benefits.[121]

These independent efforts, however, cannot supplant the federal GLMRIS study. The Chicago Waterway is a federally authorized navigation channel. Prior to speaking to hydrologic separation and any resulting permanent changes to navigation operations, the Corps must complete a feasibility study. While WRDA 2007 authorizes the Corps to review options and technologies, it does not mandate a focus on separation. However, the Corps heard a specific theme during its initial round of comment required under the National Environmental Policy Act. As part of federal Environmental Impact Statement requirements, the Corps hosted a series of public meetings to scope the range of comments on the GLMRIS project management plan (PMP). The majority of commenters providing an opinion registered favor for the preventive hydrologic separation option[122] as well as frustration with the pace of the study, which the Corps estimated it would complete by the end of 2015.[123] In contrast to the WRDA 2007 authorization and the comments of many at the hearings, the PMP creates equivalence between the terms "prevent" and "reduce the risk" in describing the scope of work for GLMRIS,[124] thereby undermining the intent of Congress. The PMP is unclear whether the controls proposed for evaluation will be analyzed for their ability to "prevent" the spread of species. The plan also places mitigation of impacts on current users of the Great Lakes–Mississippi River connections on equal footing with achieving prevention or risk reduction through management measures.[125]

A bipartisan group of Congressional representatives and senators introduced legislation[126,127] in both 2010 and 2011 that would speed up GLMRIS and focus a full feasibility study on separation. In summer

2012 a similar amendment was added to the successful "MAP-21" federal transportation legislation that mandates that the Corps complete the GLMRIS study by December 2013.[128] In November, a federal court effectively closed the Great Lakes states' legal avenue to force separation by denying the injunction that Michigan had requested.[129] The court made it clear that it felt it cannot order separation as the closure of a federal navigable waterway would violate federal law. The Great Lakes and Mississippi River regions now await an indication whether or not the federal government will pursue the separation option as a result of GLMRIS findings at the end of 2013. Pending some unforeseen legal maneuver by the states, this puts the decision whether to proceed with separation and achieve prevention squarely on Congress's agenda.

Looking Forward

The history of incomplete legal and regulatory responses to both the ballast water and canal vectors for introduction and spread of invasive species in the Great Lakes region illustrate the challenge of converting straightforward data on obvious invasion vectors and the potential for additional invasions into regulatory or infrastructure changes that achieve prevention. In both cases, the costs associated with prior aquatic invasions and the risk of further invasions are well established, and we know that the most cost-effective way to avoid those costs is through strategies that would prevent introduction and spread of aquatic invasives rather than attempt to mitigate the risk of harm from invasions while allowing them to occur. However, establishing a policy of prevention for any invasive species vector requires overcoming legal/regulatory, political, and practical obstacles to change.

Policy responses to both the ballast water and canal vectors have been protracted, but in each scenario there is a common thread: over time, a strategy of complete prevention appears to be becoming increasingly legitimate and ripe for implementation. Following the incorporation of the IMO paradigm shift from ballast water exchange to treatment technology in US law, the task still remains to develop a scientifically based set of ballast water standards that will fully protect Great Lakes water quality. There is also growing recognition that ecological separation of the Great Lakes from the Mississippi River Basin is needed to prevent invasive organisms from moving along aquatic pathways between the two ecosystems.

In both cases, the policy solution is clear; the execution requires only political will.

Notes

1. Ricciardi, A, and B. Rasmussen. 1999. Extinction rates of North American freshwater fauna. *Conservation Biology* 15:1220–1222.

2. United Nations press release, November 11, 2009 http://www.un.org/apps /news/story.asp?NewsID=32916&Cr=climate+change&Cr1=.

3. Great Lakes Regional Collaboration. 2005. Great Lakes Regional Collaboration Strategy to Restore and Protect the Great Lakes, p. 17.

4. National Center for Research on Aquatic Invasive Species, Great Lakes Aquatic Nonindigenous Species List, National Oceanic and Atmospheric Administration, http://www.glerl.noaa.gov/res/Programs/ncrais/great_lakes_list.html.

5. Ricciardi, A. T. 2006. Patterns of invasion in the Laurentian Great Lakes in relation to changes in vector activity. *Diversity and Distributions* 12:425–433.

6. US Geological Survey. Zebra mussel and quagga mussel distributions in North America. http://fl.biology.usgs.gov/Nonindigenous_Species/ZM_Progression /zm_progression.html.

7. Benson, A. J. 2011. Round goby sightings distribution. http://nas.er.usgs.gov /taxgroup/fish/roundgobydistribution.aspx.

8. Ibid.

9. US Geological Survey.

10. Domske, H., and C. R. O'Neill. 2003. Fact sheet: invasive species of Lakes Erie and Ontario. http://www.seagrant.sunysb.edu/ais/pdfs/AIS-LErieOnt.pdf.

11. US Federal Aquatic Nuisance Species Task Force. Species of concern: sea lamprey. http://www.anstaskforce.gov/spoc/sealamprey.php.

12. Fuller, P., G. Jacobs, J. Larson, and A. Fusaro. 2011. *Alosa aestivalis*. USGS Nonindigenous Aquatic Species Database, Gainesville, FL. http://nas.er.usgs.gov /queries/factsheet.aspx?SpeciesID=488.

13. Rassmussen, J., H. A. Regier, R. E. Sparks, and W. W. Taylor. 2011. Dividing the waters: the case for hydrologic separation of the North American Great Lakes and Mississippi River Basins. *Journal of Great Lakes Research* 37:588–592.

14. Cudmore, B, N. E. Mandrak, J. M. Dettmers, D. C. Chapman, and C. S. Kolar. 2012. Binational ecological risk assessment of bigheaded carps for the Great Lakes basin. Fisheries and Oceans Canada. http://www.dfo-mpo.gc.ca/Csas -sccs/publications/resdocs-docrech/2011/2011_114-eng.pdf.

15. Lodge, D., and D. Finnof. 2008. Fact sheet: Annual losses to Great Lakes region by ship-borne invasive species at least $200 million: preliminary results.

16. Bails, J., A. Beeton, J. Bulkley, M. DePhilip, J. Gannon, M. Murray, H. Regier, and D. Scavia. 2005. Prescription for Great Lakes ecosystem protection

and restoration (avoiding the tipping point of irreversible changes). http://www
.snre.umich.edu/scavia/wp-content/uploads/2009/11/prescription_for_great_lakes
_08_30_2006.pdf.

17. Mills, E., J. H. Leach, J. T. Carlton, and C. L. Secor. 1993. Exotic species in the Great Lakes: a history of biotic crises and anthropogenic introductions. *Journal of Great Lakes Research* 19:1–54.

18. NISA. 1996. http://www.gpo.gov/fdsys/pkg/BILLS-104hr4283enr/pdf/BILLS -104hr4283enr.pdf

19. Veraldi, F. 1999. Chicago Sanitary and Ship Canal aquatic nuisance species barrier. http://www.lrc.usace.army.mil/pd-s/.

20. GLRC. 2005, p. 19.

21. 33 U.S.C. § 1251(a)(1)-(2). The Clean Water Act optimistically called for the elimination of all pollutant discharges by 1985 and for waters "wherever attainable" to be "fishable" and "swimmable" by 1983. *Id.*

22. 33 U.S.C. § 1251(b).

23. 33 U.S.C. § 1370; *Keating v. F.E.R.C.*, 927 F.2d 616, 622 (D.C. Cir. 1991) ("The states remain, under the Clean Water Act, the 'prime bulwark in the effort to abate water pollution,' and Congress expressly empowered them to impose and enforce water quality standards that are more stringent than those required by federal law.").

24. *E.g.*, *Defenders of Wildlife v. Browner*, 191 F.3d 1159, 1163 (9th Cir. 1999); *Am. Paper Inst., Inc., v. U.S. EPA*, 996 F.2d 346, 350 (D.C. Cir. 1993).

25. 33 U.S.C. § 1362(6).

26. 38 Fed. Reg. 13,528 (May 22, 1973) (later codified at 40 C.F.R. § 122.3).

27. *Northwest Envtl. Advocates v. U.S. EPA*, 537 F.3d 1006, 1011–12 (9th Cir. 2008) (citing Craig Vogt, EPA, EPA Pub. Meeting # 12227, Ocean Discharge Criteria (Sept. 12, 2000, 1 p.m.)).

28. *Northwest Environmental Advocates v. EPA*, 2006 WL 2669042 (N.D. Cal. Sept. 18, 2006), *aff'd*, 537 F.3d 1006 (9th Cir. 2008).

29. Shiao, Y. W. 1995. A historical study of the reproductive pattern of zebra mussels. Zebra Mussel Research Program, US Army Corps of Engineers. Also, see Jonathan Bossenbroek, Evaluating the potential habitat and dispersal of aquatic invasive species (PowerPoint presented to the Natural Resources Defense Council, Chicago, May 3, 2010).

30. Fields, S. 2005. Great Lakes resource at risk. *Environmental Health Perspectives* 113: A164–173.

31. Pelley, J. 2011. Musseling out algae: water quality: invasive mussels may turn the Great Lakes into a biological desert. *Chemical and Engineering News*, March 30.

32. Great Lakes United. 2008. Zebra mussels: facts and figures. March 18. http://www.glu.org/sites/default/files/zm_facts.pdf.

33. Pub. L. No. 101–646, § 1101, 104 Stat 4761 (1990) (codified at 16 U.S.C.

§ 4711); *see also* 33 C.F.R. §§ 151.1500–18 (Coast Guard Great Lakes regulations).

34. Pub. L. No. 104–332, § 1101(c)(2), 110 Stat. 4073 (1996) (codified at 16 U.S.C. § 4711(c)-(e)); *see also* Mandatory Ballast Water Management Program for US Waters, 69 Fed. Reg. 44,952 (July 28, 2004) (codified at 33 C.F.R. §§ 151.2000–65).

35. See 33 C.F.R. § 151.1502 (Great Lakes regulations apply to "each vessel that carries ballast water"); Mandatory Ballast Water Management Program for US Waters, 69 Fed. Reg. 44,952, 44,955 (July 28, 2004) ("our final rule for mandatory [ballast-water management for US waters] does not address NOBOBs").

36. *Assessment of Transoceanic NOBOB Vessels and Low-Salinity Ballast Water as Vectors for Non-indigenous Species Introductions to the Great Lakes* (2005), at 1–2, 2–8. http://www.glerl.noaa.gov/res/projects/nobob/products/NOBOBFinal Report.pdf.

37. Seaway Regulations and Rules: Periodic Update, Various Categories, 73 Fed. Reg. 9, 950 (Feb. 25, 2008) (codified at 33 C.F.R. § 401.30).

38. EPA, Aquatic Nuisance Species in Ballast Water Discharges 9 (Sept. 10, 2001), *available at* http://www.epa.gov/npdes/pubs/ballast_report_attch5.pdf .

39. Ibid.

40. Ibid.

41. Ruiz, G. M., and D. F. Reid. 2007. Current state of understanding about the effectiveness of ballast water exchange (BWE) in reducing aquatic nonindigenous species (ANS) introductions to the Great Lakes Basin and Chesapeake Bay, USA. http://permanent.access.gpo.gov/lps93114/tm-142.pdf.

42. Ricciardi, A. T. 2006. Patterns of invasion in the Laurentian Great Lakes in relation to changes in vector activity. *Diversity and Distributions* 12:425–433.

43. Ruiz and Reid, see note 40; Ricciardi, A., and H. J. MacIsaac. 2008. Evaluating the effectiveness of ballast water exchange policy in the Great Lakes. *Ecological Applications* 18:1321–1322.

44. Global Ballast Water Mgmt. Programme, IMO, The International Response. http://globallast.imo.org/index.asp?page=internat_ response.htm&menu=true.

45. Ibid.

46. IMO, Guidelines for the Control and Management of Ships' Ballast Water to Minimize the Transfer of Harmful Aquatic Organisms and Pathogens, IMO Assemb. Res. A. 868, IMO Doc. A20/Res.868 (Dec. 1, 1997).

47. Sarah McGee, Proposals for Ballast Water Regulation: Biosecurity in an Insecure World, 13Colo. J. Int'l Envtl. L. & Pol'y 141, 142 (2002).

48. Global Ballast Water Mgmt. Programme, Int'l Mar. Org., The International Response. http://globallast.imo.org/index.asp?page=internat_response.htm &menu=true.

49. IMO, Adoption of the Final Act and Any Instruments, Recommendations and Resolutions Resulting from the Work of the Conference: International Convention for the Control and Management of Ships' Ballast Water and Sediments, 2004, IMO Doc. BWM/CONF/36 (Feb. 16, 2004).

50. US Environmental Protection Agency, Ballast Water Self Monitoring, 3 (Nov. 2011) (US EPA, Office of Wastewater Management, EPA 800-R-11-003). http://www.epa.gov/npdes/pubs/vgp_ballast_water.pdf.

51. Ibid.; *see also* Henry Lee II et al., Density Matters: Review of Approaches to Setting Organism-Based Ballast Water Discharge Standards, 12–16 (2010) (US EPA, Office of Research and Development, National Health and Environmental Effects Research Laboratory, Western Ecology Division, EPA/600/R-10/031). http://www.epa.gov/npdes/pubs/vessels_densitymatters_final.pdf.

52. Balaji, R., and O. B. Yakoob. 2011. Emerging ballast water treatment technologies: a review. *Journal of Sustainability Science and Management* 6:126–138 (citing Lloyd's Register Report: Ballast Water Treatment Technology, Current Status (Feb. 2010)).

53. Ibid.

54. Henry Lee II et al., Density Matters: Review of Approaches to Setting Organism-Based Ballast Water Discharge Standards, ix, 12 (2010) (US EPA, Office of Research and Development, National Health and Environmental Effects Research Laboratory, Western Ecology Division, EPA/600/R-10/031). http://www.epa.gov /npdes/pubs/vessels_densitymatters_final.pdf.

55. Ibid. at 15.

56. Ibid. at 15–16.

57. California State Lands Commission, *Report on Performance Standards for Ballast Water Discharges in California Waters*, 4, app. 2 & 4 (2006). http://www .slc.ca.gov/spec_pub/mfd/ballast_water/Documents/CSLCPerformanceStndRpt _2006.pdf.

58. MCL 324.3112(6).

59. MDEQ Ballast Water Control General Permit, No. MIG140000 at 14, http://www.michigan.gov/documents/deq/wb-npdes-generalpermit-MIG140000 _247256_7.pdf.

60. *Fednav v. Chester*, 547 F.3d 607, 624 (6th Cir. 2008).

61. California State Lands Commission, *Report on Performance Standards for Ballast Water Discharges in California Waters* (2006). http://www.slc.ca.gov/spec _pub/mfd/ballast_water/Documents/CSLCPerformanceStndRpt_2006.pdf.

62. Ibid. at 37.

63. Cal. State Lands Comm'n, 2011 Biennial Report on the California Marine Invasive Species Program iv (Jan. 2011). http://www.slc.ca.gov/Spec_Pub/MFD /Ballast_Water/Documents/2011_BiennialRpt_Final.pdf.

64. *Northwest Envtl. Advocates*, 537 F.3d at 1013; EPA, Decision on Petition for Rulemaking to Repeal 40 C.F.R. 122.3(a). http://www.epa.gov/npdes/pubs/ballast _report_ petition_response.pdf.

65. *Northwest Envtl. Advocates*, 537 F.3d at 1014.

66. Ibid. at 1010.

67. *Northwest Envtl. Advocates v. EPA*, No. 03–05760, 2005 WL 756614, at *9

(N.D. Cal. Mar. 30, 2005) (citing statutory definition of "point source," 33 U.S.C. § 1362(14), as specifically referencing "vessels").

68. *Northwest Envtl. Advocates v. EPA*, No. 03–05760, 2006 WL 2669042 (N.D. Cal. Sept. 18, 2006).

69. *Northwest Envtl. Advocates*, 537 F.3d at 1027.

70. 73 Fed. Reg. 79,473 (Dec. 29, 2008).

71. Letter from 14 Conservation Groups to EPA re: Listening Session Seeking Suggestions for Improving the Next National Pollutant Discharge Elimination System (NPDES) General Permit for Discharges Incidental to the Normal Operation of Vessels (Dec. 31, 2010).

72. EPA, Vessel General Permit for Discharges Incidental to the Normal Operation of Vessels (VGP), at 16. http://www.epa.gov/npdes/pubs/vessel_vgp_permit _november%202010.pdf .

73. Press Release, Natural Resources Defense Council, Invasive species settlement: new ballast water permit should help protect American coasts, lakes, and rivers. March 8, 2011. http://www.nrdc.org/media/2011/110308.asp.

74. Ibid.

75. Draft National Pollutant Discharge Elimination System (NPDES) General Permits for Discharges Incidental to the Normal Operation of a Vessel, 76 Fed. Reg. 76,716 (Dec. 8, 2011).

76. Standards for Living Organisms Discharged in Ships' Ballast Water in US Waters, 74 Fed. Reg. 44,632 (Aug. 28, 2009).

77. Standards for Living Organisms Discharged in Ships' Ballast Water in US Waters, 77 Fed. Reg. 17,254 (Mar. 23, 2012).

78. US EPA, *2011 Proposed Issuance of National Pollutant Discharge Elimination System (NPDES) Vessel General Permit (VGP) for Discharges Incidental to the Normal Operation of Vessels: Draft Fact Sheet*, at 124–125.

79. Ibid.

80. Ibid. at 106.

81. Veraldi, F. M., K. Baerwaldt, B. Herman, S. Herleth-King, M. Shanks, L. Kring, and A. Hannes. 2011. Non-native species of concern and dispersal risk for the Great Lakes and Mississippi River Interbasin Study. http://glmris.anl.gov /documents/docs/Non-Native_Species.pdf.

82. Great Lakes Commission. 1996. Policy position of the Great Lakes Panel on Aquatic Nuisance Species. www.glc.org/ans/pdf/RSCHPOLY.pdf.

83. Cite to NISA C.3.A.

84. University of Wisconsin Sea Grant Institute. 2004. Dispersal barrier project history. http://seagrant.wisc.edu/home/Default.aspx?tabid=554, link to "Barrier I."

85. Veraldi, F. 1999.

86. University of Wisconsin Sea Grant Institute. 2004. Dispersal barrier project history. http://seagrant.wisc.edu/home/Default.aspx?tabid=554 link to "Barrier Options."

87. Univ. Wisc. Sea Grant. 2004.

88. Rasmussen, J. 2001. The Cal-Sag and Chicago Sanitary and Ship Canal: a perspective on the spread and control of selected aquatic nuisance fish species. US Fish and Wildlife Service, 4469—48th Avenue Court, Rock Island, IL 61201. 26 pp. http://www.fws.gov/Midwest/LacrosseFisheries/documents/Connecting _Channels_Paper_Final.pdf.

89. Robison, H. W., and T. M. Buchanan. 1988. *Fishes of Arkansas*. Fayetteville: University of Arkansas Press.

90. Kolar, C. S., D. C. Chapman, W. R. Courtenay Jr., C. M. Housel, J. D. Williams, and D. P. Jennings. 2005. Asian carps of the genus *Hypophthalmichthys* (Pisces, Cyprinidae): a biological synopsis and environmental risk assessment. http://www .fws.gov/contaminants/OtherDocuments/ACBSRAFinalReport2005.pdf.

91. See generally Rasmussen, J. 2001 for associated bibliography.

92. Kolar, C., and D. Lodge. 2002. Ecological predictions and risk assessment for alien fishes in North America. *Science* 298:1233–1236.

93. Kolar et al. 2005.

94. Mandrak, N., and B. Cudmore. 2004. Risk assessment for Asian carps in Canada. http://www.dfo-mpo.gc.ca/csas/Csas/DocREC/2004/RES2004_103_E .pdf.

95. Cudmore et al.

96. Rasmussen, J. 2001. p. 2.

97. City of Chicago. 2003. Closing the revolving door: summary of the aquatic invasive species summit proceedings. http://www.cityofchicago.org/content/dam /city/depts/doe/general/NaturalResourcesAndWaterConservation_PDFs/Invasive Species/AquaticInvasiveSpeciesSummitSummary2003.pdf.

98. FishPro. 2004. Feasibility study to limit the invasion of Asian carp into the Upper Mississippi River Basin. http://files.dnr.state.mn.us/natural_resources/invasives /aquaticanimals/asiancarp/umrstudy.pdf.

99. Malchoff, M., J. E. Marsden, and M. Hauser. 2005. Feasibility of Champlain canal aquatic nuisance species barrier options. http://nsgl.gso.uri.edu/lcsg /lcsgto5001.pdf.

100. Brammeier, J., S. Mackey, and I. Polls. 2008. Preliminary feasibility of the ecological separation of the Mississippi River and the Great Lakes to prevent the transfer of aquatic invasive species. http://www.greatlakes.org/Document .Doc?id=473.

101. US Army Corps of Engineers. 2012. Inventory of available controls for aquatic nuisance species of concern. http://glmris.anl.gov/documents/docs/ANS _Control_Paper.pdf.

102. Moy, P. 2002. Notes from the 10 January 2002 meeting of the Dispersal Barrier Advisory Panel. http://seagrant.wisc.edu/home/Default.aspx?tabid=553, link to Jan 2002.pdf.

103. Water Resources Development Act of 2007, Pub. L. No. 110–114, § 3061, 121 Stat. 1041, 1121 [hereinafter "WRDA 2007"].

104. Smith-Root. 2012. Chicago Sanitary and Ship Canal 2B Barrier. http://www.smith-root.com/barriers/sites/chicago-sanitary-and-ship-canal-2b.

105. Chuck Shea, USACE, personal communication with Joel Brammeier, March 5, 2012.

106. GLRC. 2005, p. 19 and appendices.

107. WRDA 2007, § 3061(d), 121 Stat. at 1121.

108. National Invasive Species Act, 16 U.S.C. § 4722(i)(3).

109. *See* State of Michigan's Motion to Re-Open and for a Supplemental Decree, *State of Wisconsin, et al., v. State of Illinois*, Case Nos. 1, 2 & 3 Orig. (U.S. Dec. 21, 2009); State of Michigan's Motion for Preliminary Injunction, *State of Wisconsin, et al., v. State of Illinois*, Case Nos. 1, 2 & 3 Orig. (U.S. Dec. 21, 2009).

110. *See* Complaint, *State of Michigan, et al. v. U.S. Army Corps of Engineers, et al.*, Case No. 10 C 4457 (N.D. Ill. July 19, 2010).

111. 2009. Letter from Great Lakes members of Congress to federal agencies regarding urgent action on Asian carp prevention. http://www.greatlakes.org/document.doc?id=1275.

112. *See, e.g.*, Close All Routes and Prevent Asian Carp Today Act of 2010 (CARP ACT), H.R. 4472, S. 2946, 111th Cong. (2010).

113. Brammeier et al. 2008.

114. City of Chicago. 2003.

115. Hansen, M. 2010. The Asian carp threat to the Great Lakes. Testimony to the House Transportation and Infrastructure Committee Subcommittee on Water Resources and Environment, February 9, 2010. http://republicans.transportation.house.gov/Media/file/TestimonyWater/2010–02–09-Hansen.pdf.

116. Brammeier, J. 2010. Permanent prevention of Asian carp in the Great Lakes. Testimony to the House Transportation and Infrastructure Committee Subcommittee on Water Resources and Environment, February 9, 2010. Available at http://republicans.transportation.house.gov/Media/file/TestimonyWater/2010–02–09-Brammeier.pdf.

117. Buchsbaum, A. 2010. Improving the Asian carp control strategy framework. Testimony to the Senate Energy and Natural Resources Committee Subcommittee on Water and Power, February 25, 2010. http://www.energy.senate.gov/public/index.cfm/files/serve?File_id=06334d70-f5b3–4de2-a340–9b67b7b24174.

118. Brammeier et al. 2008.

119. Henderson, H., T. Cmar, and K. Hobbs. 2010. Re-envisioning the Chicago River: adopting comprehensive regional solutions to the invasive species crisis. http://www.nrdc.org/water/Chicagoriver/files/Chicago%20River.pdf. Technical report at http://docs.nrdc.org/water/files/wat_10102001a.pdf.

120. Great Lakes Commission and Great Lakes-St. Lawrence Cities Initiative. 2012. Quotes to accompany release of Restoring the Natural Divide. http://www.glc.org/ChicagoWaterway/pdf/FINALQuotes.pdf.

121. Great Lakes Commission and Great Lakes-St. Lawrence Cities Initiative. 2012. Restoring the Natural Divide. http://www.glc.org/caws.

122. US Army Corps of Engineers. 2011. GLMRIS environmental impact statement scoping summary report, p 14. http://glmris.anl.gov/documents/docs /GLMRIS_Scoping_Report.pdf.

123. Ibid, p. 10.

124. US Army Corps of Engineers. 2010. GLMRIS project management plan, p. 1. http://glmris.anl.gov/documents/docs/Project_Management_Plan.pdf.

125. Ibid, p. 12.

126. Permanent Prevention of Asian Carp Act of 2010, H.R. 5625, S. 3553, 111th Cong. (2010).

127. Stop Asian Carp Act, H.R. 892, S. 471, 112th Cong. (2011).

128. Moving Ahead for Progress in the 21st Century Act, H.R. 4348, Sec. 1538, 112th Congress.

129. Associated Press. 2012. Judge dismisses Asian carp lawsuit, says states can amend claims against government. December 3. http://bigstory.ap.org/article /judge-toss-asian-carp-suit-if-its-not-changed.

CHAPTER SEVENTEEN

Final Thoughts: Nature and Human Nature

Glenn Sandiford, Reuben P. Keller, and Marc Cadotte

Humans have caused, debated, and studied biological invasions for centuries. Along the way, we have accumulated experience and knowledge about invasive species, and incorporated this ever-expanding body of information into policy choices at all levels, from local to international. That policy history has fluctuated widely, reflecting the ebb-and-flow of attitudes about introduced species (Pauly 1996), but it does show an overall direction since the Columbian-era voyages of discovery (Hall 2003). For eighteenth-century builders of empire, the spread of floral and faunal pests was an incidental and often interesting sidebar to colonial programs of acclimatization. By 1900, measures like the Lacey Act in the United States were being implemented to prevent invasions by foreign biota deemed potentially harmful. The final decades of the twentieth century saw countries like New Zealand take hard-line stances against "biological pollution," whereby native nature has been accorded high levels of *proactive* protection on the basis that risks of harm by exotic species are plausible even if scientific consensus is absent (the so-called precautionary principle). Viewed through the lens of five hundred years, policy for invasive species shows a gradual shift from "cosmopolitan" improvement of nature via selection to "nativist" protection through exclusion.

That historical arc notwithstanding, the issue of invasive species has undergone a paradigm shift in the past quarter-century. Few people in North America (outside science and natural resource agencies) paid attention to biological invasions until the 1988 discovery of zebra mussel in the Great Lakes. The fingernail-sized clam, a native of Eurasia, was thought to have arrived in ballast water of cargo ships. Within three years, zebra mussel had spread into most of the major drainage basins in eastern North America (Mills et al. 1993) and was attracting heavy media coverage because of its costly tendency to clog inlet pipes for water utilities and power plants. That publicity helped propel development of invasion biology as an intellectual discipline. For example, the annual *International Conference on Aquatic Invasive Species* series began in 1991 as the *International Zebra Mussel Research Conference*, and the journal *Biological Invasions* published its first issue in 1999. Zebra mussel also sparked an explosion in public awareness of biological invasions across North America, while the presence and impacts of other invasive species have increased public recognition of the problem in other regions of the world. This includes the cane toad (*Rhinella marina*) in Australia (see chapter 2), a slew of mammals in New Zealand, Nile perch (*Lates niloticus*) in Africa, and golden applesnail (*Pomacea canaliculata*) in Asia. Thus, over the past two decades invasive species have become headline news around the world. An Internet search with the term "invasive species" generates millions of hits. Dozens of mainstream books, with titles like *Alien Invasion*, *Feral Future*, *A Plague of Rats and Rubbervines*, and *Nature out of Place*, chronicle invasions and their impacts. A plethora of movements has mushroomed, variously advocating that invasive species be banned, embraced, controlled, hunted, harvested, even turned into haute cuisine.

Yet this unprecedented public awareness about invasive species has not yielded a comparable transformation in policy (cf. Cangelosi 2002). Granted, significant progress has been made on some policy fronts. Two chapters in our collection, by Joel Brammeier and Thomas Cmar and by Clare Shine, report advances in ballast water regulation and a European Union strategy for invasive alien species, respectively. Change is also permeating international aid agencies, whose well-meaning programs have long contributed to the spread of exotic flora and fauna that subsequently prove invasive and threaten the very development activities they were introduced to support (Msiska and Costa-Pierce 1993; Welcomme 1988). In Southeast Asia, where aquaculture of native fishes has consistently been suppressed in favor of long-used and better understood exotic species like cyprinids and tilapias, some development agencies are establishing indigenous aquaculture programs (Gutiérrez and Reaser 2005). The 193 countries that have ratified the Convention on Biological Diversity (CBD) have committed to "prevent the introduction of, control or eradicate those alien species which threaten ecosystems, habitats or species" (article 8h), and several other international agreements (e.g., the 34 International Standards for Phytosanitary Measures) require nations to make efforts to reduce transport of nonnative species.

Policy advances notwithstanding, however, few international measures include penalties for noncompliance (see chapter 13). The CBD provides nonbinding guidance to nations, but lacks penalties for not following the guidance. Invasive species can still be transported through a myriad of vectors (see chapter 5). The efficacy of border inspections, our last line of defense against bioinvasion, varies widely by nation (Bacon et al. 2012). We still lack an internationally accepted set of criteria for ecological impact assessments of invasive species (Gederaas et al. 2012). Much remains to be done in the realm of invasive species policy—which is the reason for our volume. This chapter about future policy directions is rooted in four principles that are well established and self-evident. These principles, variously affirmed by chapters within our collection, are

- prevention is far more effective and cheaper than control and/or eradication
- the potential for eradicating invasive species decreases the longer they have been established
- science can serve as a guide, but most policy decisions come down to individual and collective judgment
- policy is political

With these four principles as a basis, what are the most pressing needs for invasive species policy?

Knowledge Needs

Basic knowledge about invasive species is still sorely lacking. This has been made clear in chapters throughout the book, including the ballast water example in chapter 1. The move towards international policy for regulating ballast water discharged from ships is proceeding even though the proposed policy's effect on actual invasion rates is currently impossible to determine.

One area particularly requiring more research and understanding is the assessment of risks posed by invasive species, and how these assessments can be wisely incorporated into policy. The process of invasion is enormously complicated. In most instances, broad generalizations are questionable at best. The introduction, establishment, and impacts of any given species depend on an array of biotic and abiotic factors, as well as human influences (cf. Gallardo and Aldridge 2013; Kimbro et al. 2013). By the late 1980s the potential for zebra mussel invasion to the Great Lakes had been predicted for over sixty years (Carlton 1991), and many ships had traveled from the invaded range to the Great Lakes without causing an invasion. Exactly which ship(s) introduced the individual mussels that created the invasion, and why that ship(s) at that time was responsible, will never be known. Much the same is true for species impacts. They are highly context dependent and in many cases difficult to predict and/or explain. Thousands of plant species have been deliberately introduced into North America for agriculture, horticulture, and landscaping. Yet, scientists still lack a comprehensive theory that explains why some flora became problematic, while the majority failed to spread (Jeschke et al 2012).

The probabilistic nature of the invasion process creates difficulties for scientists to understand the processes, and difficulties for policy makers to institute measures that balance the benefits of fewer invasions with the costs of reducing invasion risks. A single discharge of ballast water is highly unlikely to cause an invasion, but many invasions are caused by ballast water discharge. Although some risk factors are known (e.g., Keller et al. 2011), their relative importance, and the ways they interact, remain unclear. This is a challenge for scientists, and leaves policy makers in the unsatisfying situation of creating ballast water regulations that will reduce invasion rates by an unknown amount. Further understanding of the probabilistic nature of invasions, and greater recognition among scientists and policy makers that predictions are inherently uncertain, is needed to push policy forward (Seebens et al 2013).

Judgment and a Biocentric Ethic

No matter the depth and breadth of our knowledge, invasive species policy transcends fact and enters the realm of judgment. It is trans-scientific, as illustrated by Shine, Coates, and Sandiford (chapters 2–4) in their respective chapters about the cane toad in Australia, gray squirrel in Britain, and invasive carps in America. Controversies about exotic species invariably revolve around differences of opinion—about the (un)acceptability of invasion risk, invasion, or impact; (in)appropriateness of biotic introductions; framing of impacts as problem or opportunity; "best" management strategy; and so forth. What is tolerable for some is objectionable to others. Policy responses to such controversies spotlight the difference between knowledge and wisdom. Knowledge is a tool, and wisdom is the craft by which that tool is used. History shows us that knowledgeable, well-meaning people can lack wisdom. This is one reason why, during humanity's centuries-long involvement in biotic transfers around the world, advocates and critics alike have invoked scientific evidence to support their policy position.

We emphasize strongly that the trans-scientific aspect of invasive species policy is *not* a green light for a laissez-faire approach. One problem with the postmodernist critique of nature as a relativist construct (cf. Cronon 1996) is its lack of substantive guidance for decision making. How are policy decisions about nature to be made if, as deconstructionists argue, perceptual and cultural filters distort reality so much that there is no certain knowledge, only constructed "texts"? This quandary was addressed by environmental historian Donald Worster. Yes, agrees Worster, all ideas are historically contingent—but a relativistic equaling of all norms makes Da Vinci no better than graffiti and Disneyland as legitimate as Yellowstone. Historical relativism "frees us from dogma but offers no firm guidance to belief" (Worster 1995:77). Therein lie political implications too. In the ongoing rhetorical contest—it most assuredly *is* a contest—for a dominant definition of nature, a level playing field is desirable, but "a playing field without rules and referees is a free-for-all where bullies win" (Soulé 1995:150). Regrettably, in regard to invasive species, one might add that alarmists, obstructionists, and contrarians can "win" too—at least temporarily, thereby indirectly facilitating further invasions. In particular, a small but vocal group of skeptics, through occasional books and high-profile papers, portray the science of invasion biology as one motivated by nonscientific values (e.g., Theodoropoulos 2003)—and have been partially successful in terms of media coverage. A flurry of popular science articles recently surfaced, presenting scientific evidence that seemingly supports the skeptics' contention that invasive species have been unfairly persecuted (Burdick 2005; also see Sagoff 2007, and responses by Lafrancois and Glase 2007; Russell 2007; Hoshovsky 2007). Yet amidst the seemingly contradictory findings in the scientific literature, some clarity is emerging. For example, we are realizing that studies yield different conclusions about the seriousness of impacts partly because of *scale*—within meter-square quadrants invasive plants cause a large loss in species richness, but this effect diminishes when biodiversity is measured regionwide (Powell et al. 2011). In other words, small-scale effects justify the fight against invasive species, while biodiversity's broad-scale resilience offers hope for restoration of habitats and native species (Powell et al. 2013).

As for constructionists' claim that human activities have so altered the planet that nature is no longer natural but a cultural artifact, the notion of "virgin nature" is an inappropriate metaphor. Virginity knows only absolutes, whereas "nature . . . knows nothing but degrees" (Soulé 1995:155). Even after thousands of years of human impact, including biological invasions, the world's forests and plains and deserts and oceans still retain "the basic taxonomic integrity of biogeographic units: species today still have geographic distributions determined largely by ecological tolerances and geological history and climate, rather than by human activities" (Soulé 1995:157). The skeptics, invoking an argument that blends sweeping generalizations with selective specifics, reject that foundational integrity as outdated and impractical (Davis et al 2011; also see response by Simberloff 2011). But in study after study, including many cited in this chapter and other chapters of this book, science reaffirms species origin as a valid, pragmatic, and relevant basis for invasive species policy in the twenty-first century, especially in the pre- and immediate post-stages of invasion (Shackelford et al. 2013).

Deconstructions of invasive species rhetoric and "belongingness" of species can be informative, but eventually policy makers must make management decisions about such biota (Mooney 2005). On what basis should policy be made? Economics, biological heritage, ecosystem health, ecosystem services, economics, etc.? Currently, most nations have created policy through a case-by-case approach. Plants are typically treated differently to animals. Different modes of introduction are controlled in dissimilar ways and with varying stringency, to the extent that sometimes only individual species are targeted. This piecemeal approach creates a patchwork of contradictions, inconsistencies, and potential for further invasion.

Invasive species policy needs unifying. One idea that has arisen from this book is exploration of an ethic. Is an overarching ethic to guide our management approaches to bioinvasion possible, one that transcends time and place and which can provide a basis for the kind of "organic act" suggested in chapter 15 by Miller?

Bioethics has largely been absent from debate about the "appropriateness" of exotic species in our globalizing world. Most conceptualizations of invasive species have been framed around impacts on economics and ecosystem services. These approaches, preoccupied with what a species *does*, accord little attention to species *identity* and *origin*. Such narrow framing creates policy contradiction, as evident with a river in central North America that has been referred to as the world's "most studied" river (Mills et al. 1966). During the nineteenth century, the Illinois River was renowned for its largemouth black bass (*Micropterus salmoides*), a native fish esteemed by anglers for its feistiness and which also fetched high prices at market. But in 1900, Chicago began channeling its sewage into the Illinois River. The upper river soon had the water chemistry of a sewer, and no fish could survive there (Forbes 1911). Pollution control finally began in the 1930s. Gradual improvement in water quality, combined with other efforts toward restoring the river's biological health, facilitated a rebound in its bass population as well as a reduction in invasive common carp (*Cyprinus carpio*). By the 1980s, the Illinois River was again one of North America's premier black bass fisheries. Unfortunately, that decades-long investment in ecological restoration has now been seriously undermined by the invasion of Asian carp, which constitutes 90% of fish biomass in some stretches of the river (see Sandiford, chapter 4). Eradication of Asian carp at this late stage in their invasion seems impossible, so a campaign is underway in Illinois to commodify them. Although well intentioned, a commitment to commercial carp fishing is not only unlikely to yield any meaningful reduction in abundance of these highly fecund fish, it risks further erosion of the restoration ethic that shaped policy goals for the Illinois River through three-quarters of a century. Boosterism about harvesting of Asian carp is framed around themes of market, nutrition, and cheapness rather than restoration, biological integrity, or protection of economically valuable native species. Utility has superseded heritage, giving rise to policy tension. Priority-wise, are Asian carp being harvested to generate revenue, mitigate their impacts, or restore native species? If a poison pill or other means of carp eradication later becomes feasible (Luciano 2012), will such tools be opposed by commercial fishermen whose livelihoods have become dependent on Asian carp?

The idea of eating invasive species has been embraced by a group of chefs, writers, and other advocates who collectively make it seem hip. This *invasivore* movement has spawned websites, blogs, recipe collections, news stories, and books (Landers 2012). Its followers endorse the likes of kudzu, garlic mustard, crayfish, iguana, armadillo, even the venomous lionfish as edible foods (Olsen 2010). Safety issues aside, the approach has a logic that is persuasive; eating invasive species seems a direct and very personal way to "make a difference." But that difference is invariably negligible. Even if edible, most invasives are too fecund to be impacted by human consumption. A large Asian carp annually spawns more than two million eggs. More significantly, the invasivore approach deflects from the primary policy issue, which is prevention of biological invasions so as to reduce the harm caused by invasive species. Hence, while this approach may have some appeal as "A Diet for an Invaded Earth" (Gorman 2011), it is unlikely to reduce the population size or impacts of invasive species. Furthermore, if the basic biology of invasions are not understood, then the invasivore movement has the potential to mistakenly make people feel that their activities are beneficial, and thus distract them from other, potentially much more beneficial, approaches.

Without an overarching policy ethic for invasive species, agencies and managers will forever be undermined, contradicted, and sidetracked. That ethic must be planetary in perspective, yet must accommodate local context. Its foundational basis must be the growing international commitment to biological preservation and restoration, but it must also acknowledge the economic, cultural, and political complexities that are driving globalization. Some scholars emphasize the *distinction* between policies based on human needs and those based on plant or animal needs—"between an ethics serving *Homo sapiens* and an ethics serving non-human species, be they native or exotic" (Hall 2003). A more fruitful approach would be a visionary ethic that recognizes *overlap* between anthropocentrism and biocentrism (Lowry 1998; Duncan 1998).

In this regard, the issue of climate change offers some precedent. In 2005, after seven years of international wrangling, the Kyoto treaty for reduction of greenhouse gas emissions became effective. Ratified by more than 140 nations, the agreement was a historic breakthrough in negotiations, and as such an important first step in international climate policy (Harvey 2012). Beyond the compelling scientific evidence and recognition of national interests, it was drafted around universal themes of equity and ethics (Garvey 2008; Pinguelli-Rosa and Munasinghe 2002). Even though countries like Canada have withdrawn from the Kyoto protocol, the idea that nations must deal with climate change is still widely accepted, and buttressed by an ever-strengthening body of scientific evidence. A similar breakthrough is needed for invasive species policy. It would include recognition of the value of local and regional biodiversity, the harm that invasive species cause to human economies and agriculture, and the fact that different human populations are impacted differently by invasive species, along with the value of the trade vectors that move nonnative species across the globe. Defining and implementing such a synthetic approach to the global movement of nonnative species is an enormous challenge, but one that must be met in order to rationalize global policies.

Ecology remains the foundation of our knowledge about biological invasions, and thus its scientists merit a leadership role in development of a policy ethic for invasive species (Franz 2001). But ecologists alone cannot complete the task. It requires collaboration that will be interdisciplinary as well as cross-cultural. The conference that gave rise to this book about invasive species policy reflects a promising broader trend of cross-pollination between the environmental sciences and scholars in the social sciences and humanities. Such integration must continue in order for invasive species policy to become anchored in a unifying, holistic ethic.

For example, perhaps the triage model used in medicine could be adapted as a framework for bioinvasion management. As noted by Burgiel (chapter 13), epidemiologists led by the World Health Organization (WHO) are now debating whether outbreaks of infectious diseases affecting humans can be considered invasive species. What can we learn from the international public health community, especially in regard to its early detection/notification and rapid response network, that can quickly mobilize effective cooperative action among nations? Similar strategic planning for invasive species policy is sorely lacking, with only few exceptions (Keller and Perrings 2011). The current status quo is reactive far more than anticipatory.

Policy: Where Science Meets Politics

Natural resource agencies face an ever-increasing array of potential, new, and established invasive species. In allocating resources, should priority be given to preventing potential invaders, eradicating newly arrived species, or alleviating the impact of well-established species? Again, history offers lessons. Eradication can be highly cost-effective—but action needs to be rapid and committed. In 1998, when Asian longhorned beetles (*Anoplophora glabripennis*) were discovered in hundreds of trees in a Chicago neighborhood, city authorities launched a rapid-response eradication effort. Any tree thought to be infected was immediately felled, while an insecticide program targeted healthy trees. In April 2005, seven years after the initial discovery, the longhorned beetle was declared eradicated from Illinois, "a rare success story in the battle against invasive pests" (Hawthorne 2005). Other success stories are highlighted elsewhere in this book, such as the elimination of nutria and muskrat from Great Britain (Coates, chapter 3), and the long-term, albeit expensive, control of sea lamprey in the Great Lakes (Kitchell et al., chapter 10).

A few years earlier, however, Chicago had witnessed failure, this time involving a small fish from central Eurasia. Round goby (*Neogobius melanostomus*) was first discovered in North America in 1990 in a tributary of Lake Michigan (Laird and Page 1996). In 1996, to prevent the aggressive fish spreading through Chicago-area canals, Congress authorized construction of an experimental electric barrier in one of the canals. But because of funding delays, the $1.6-million barrier did not become operational until 2002—three years after round gobies had moved through the canal and entered the Mississippi River basin (Egan 2008).

Such episodes affirm the political realities of invasive species policy. The search for a sustainable path between free-market capitalism and costly bioinvasions entails decisions of potential consequence to multiple constituencies. Especially when the policy process becomes adversarial, the stakes can be high—and communication becomes even more critical. In early 2010, a public relations firm in Illinois was hired by a coalition of business and labor interests opposed to the proposed closure of Chicago-area canals that are a potential conduit for Asian carp to enter Lake Michigan. By the end of 2011, the anti-carp movement was waning, the canals were still operational—and the public relations firm had earned a prestigious communications award for its strategy of framing canal closure as anti-business (Mac Strategies 2011).

Partly because so many perspectives and interests impinge on the issue of invasive species, science has typically had only "weak influence" in shaping policy (Keller et al. 2009:269). *Policy is forged from science and politics.* Scientists thus need to be more engaged in communication aspects of the policy process, and more savvy about it (Lawton 2007; Larson 2007). Restricting publication activity to a format that is policy-relevant but not policy-prescriptive is often appropriate and necessary for scientists to remain objective (Lackey 2001)—but, in some situations, politically naive. Facts are insufficient (Lakoff 2010). No matter how carefully conservation biologists and restoration ecologists word their findings, interpretation thereof among policy stakeholders—including other scientists—is colored by multiple filters (cf. Selge et al. 2011; Bremner and Park 2007; Simberloff 2005). If scientists want to ensure that their message is heard and not corrupted they must communicate in venues beyond academic journals and conferences. Besides countering incomplete or misleading discussion of research findings (Lodge and Shrader-Frechette

2003), they need to make such information more accessible to managers, policy makers, and the public at large, by way of appropriate publications, outreach, gatherings, and other activities beyond the traditionally insular world of academia (cf. Witmer et al. 2009).

At the same time, however, policy makers and government institutions must strive to be consensus builders who bring all available knowledge to the table. At their disposal is a long history of research and understanding, plus a global network of scientists advancing this knowledge on multiple fronts. Ultimately, the onus is on elected officials to transcend political ideology and establish policy that protects biodiversity and economies. Even though the issues are difficult and involve global, long-term consequences, they need to be tackled, especially because bio-invasions rarely are undone.

As globalization intensifies, so too does the phenomenon of bioinvasion. Yet we already have sufficient knowledge and tools to drastically reduce the arrival of new invasive species (Bright 1998; Egan et al. 2013; chapters 6–8, this volume), and to control the spread of existing invaders (Tingley et al. 2013; chapters 9–12, this volume). Our ecological understanding still needs major advances, but there is also great potential for progress through changes in people's perceptions (see Newman, chapter 5). *The principal policy challenge at the moment is not technical, but cultural.* For people and policy to become more proactive and effective, a cultural change needs to occur, one that includes greater appreciation of native nature and cultivation of "a sense of biological history" (Todd 2001:253). As much as we need to learn about nature, so too we must understand human nature.

Literature Cited

Bacon, S., S. Bacher, and A. Aebi. 2012. Gaps in border controls are related to quarantine alien insect invasions in Europe. *PLOS ONE* 7:e47689.

Bremner, A., and K. Park. 2007. Public attitudes to the management of invasive non-native species in Scotland. *Biological Conservation* 139:306–314.

Bright, C. 1998. *Life Out of Bounds: BioInvasion in a Borderless World.* New York: W. W. Norton.

Burdick, A. 2005. The truth about invasive species: how to stop worrying and learn to love ecological intruders. *Discover* 26:34–41.

Cangelosi, A. 2002. Blocking invasive aquatic species: federal law must be updated to stop introductions of nonnative organisms, especially by ships. *Issues in Science and Technology* 19:69–75.

Carlton, J. 1991. Predictions of the arrival of zebra mussel in North America. *Dreissena Polymorpha Information Review (New York Zebra Mussel Clearing House)* 2:1.

Cronon, W. 1996. *Uncommon Ground: Rethinking the Human Place in Nature.* New York: W. W. Norton.

Davis M., M. Chew, R. Hobbs, A. Lugo, J. Ewel, G. Vermeij, J. Brown, et al. 2011. Don't judge species on their origins. *Nature* 474:153–154.

Duncan D. 1998. Natives. *Orion* 17:18–26.

Egan, D. 2008. Shocking trouble for Great Lakes. *Milwaukee Journal Sentinel,* Oct. 5, A1.

Egan, S., M. Barnes, C. Hwang, A. Mahon, J. Feder, S. Ruggiero, C. Tanner, and D. Lodge. 2013. Rapid invasive species detection by combining environmental DNA with Light Transmission Spectroscopy. *Conservation Letters* 6:402–409.

Forbes, S. 1911. Chemical and biological investigations on the Illinois River, midsummer of 1911: a preliminary statement made to the American Fisheries Society, St. Louis, Mo., October 3, 1911. Illinois State Lab. Natural History, Urbana, IL.

Franz, E. 2001. Ecology, values, and policy. *BioScience* 51:469–474.

Gallardo, B., and D. Aldridge. 2013. The "dirty dozen": socio-economic factors amplify the invasion potential of twelve high risk aquatic invasive species in Great Britain and Ireland. *Journal of Applied Ecology* 50:757–766.

Garvey, J. 2008. *The EPZ Ethics of Climate Change: Right and Wrong in a Warming World.* New York: Bloomsbury.

Gederaas, L., T. Moen, S. Skjelseth, and L. Larsen, eds. 2012. *Alien Species in Norway—with the Norwegian Black List 2012.* Norway: Norwegian Biodiversity Information Centre, Norway.

Gorman, J. 2011. A diet for an invaded planet. *New York Times,* January 2, L3.

Gutiérrez, A., and J. Reaser. 2005. *Linkages between Development Assistance and Invasive Alien Species in Freshwater Systems of Southeast Asia.* Washington, DC: SAID Asia and Near East Bureau.

Hall, M. 2003. Editorial: the native, naturalized and exotic—plants and animals in human history. *Landscape Research* 28:5–9.

Harvey, F. 2012. The Kyoto protocol is not quite dead. *Guardian,* November 26.

Hawthorne, M. 2005. Have we finally squashed the beetle? *Chicago Tribune,* April 19.

Hoshovsky, M. 2007. Unnecessary uncertainty [letter to editor]. *Conservation Magazine* 8.

Jeschke, J., L. Aparicio, S. Haider, T. Heger, C. Lortie, P. Pyšek, and D. Strayer. 2012. Support for major hypotheses in invasion biology is uneven and declining. *NeoBiota* 14:1–20.

Keller, R., J. Drake, M. Drew, and D. Lodge. 2011. Linking environmental conditions and ship movements to estimate invasive species transport across the global shipping network. *Diversity and Distributions* 17:93–102.

Keller, R., M. Lewis, D. Lodge, J. Shogren, and M. Krkosek. 2009. Putting bio-economic research into practice. In *Bioeconomics of Invasive Species*, edited by R. Keller, D. Lodge, M. Lewis, and J. Shogren, 266–284. New York: Oxford University Press.

Keller, R., and C. Perrings. 2011. International policy options for reducing the environmental impacts of invasive species. *BioScience* 61:1005–1012.

Kimbro, D., B. Cheng, and E. Grosholz. 2013. Biotic resistance in marine environments. *Ecology Letters* 16:821–833.

Lackey, R. 2001. Values, policy, and ecosystem health. *BioScience* 51:437–443.

Lafrancois, B., and J. Glase. 2007. No two sides to this story [letter to editor]. *Conservation Magazine* 8.

Laird, C., and L. Page. 1996. Non-native fishes inhabiting the streams and lakes of Illinois. *Illinois Natural History Survey Bulletin* 35:1–51.

Lakoff. G. 2010. Why it matters how we frame the environment. *Environmental Communication* 4:70–81.

Landers, J. 2012. *Eating Aliens: One Man's Adventures Hunting Invasive Animal Species*. North Adams, MA: Storey Publishing.

Larson, B. 2007. An alien approach to invasive species: objectivity and society in invasion biology. *Biological Invasions* 9:947–956.

Lawton, J. 2007. Ecology, politics and policy. *Journal of Applied Ecology* 44:465–474.

Lodge, D., and K. Shrader-Frechette. 2003. Nonindigenous species: ecological explanation, environmental ethics, and public policy. *Conservation Biology* 17:31–37.

Lowry, J. 1998. In my town. *Orion* 17:34–43.

Luciano, P. 2012. Taking on Asian carp with pills, lure of sex. *Peoria Journal-Star*, April 13, 1B.

Mac Strategies Group. 2011. Accessed December 20, 2012. http://www.macstrategies group.com/case_studies.html.

Mills, E., J. Leach, J. Carlton, and C. Secor. 1993. Exotic species in the Great Lakes: a history of biotic crises and anthropogenic introductions. *Journal of Great Lakes Research* 19:1–54.

Mills, H., W. Starrett, and F. Bellrose. 1966. Man's effect on the fish and wildlife of the Illinois River. *Illinois Natural History Survey Biological Notes* 57.

Mooney, H. 2005. Battling invaders. *American Scientist* 93:553–555.

Msiska, O., and B. Costa-Pierce, eds. 1993. History, status and future of common carp (*Cyprinus carpio* L.) as an exotic species in Malawi. *International Center for Living Aquatic Resources Management Conference Proceedings*, 40.

Olsen, E. 2010. Florida Keys declare open season on the invasive lionfish. *New York Times*, November 22.

Pauly, P. 1996. The beauty and menace of the Japanese cherry trees: conflicting visions of American independence. *Isis* 87:51–73.

Pinguelli-Rosa, L., and M. Munasinghe, eds. 2002. *Ethics, Equity and International Negotiations on Climate Change*. Northampton, MA: Edward Elgar.

Powell, K., J. Chase, and T. Knight. 2011. A synthesis of plant invasion effects on biodiversity across spatial scales. *American Journal of Botany* 98:539–548.

Powell, K., J. Chase, and T. Knight. 2013. Invasive plants have scale-dependent effects on diversity by altering species-area relationships. *Science* 339:316–318.

Russell, J. 2007. Overlooked damages [letter to editor]. *Conservation Magazine* 8.

Sagoff, M. 2007. Aliens among us. *Conservation Magazine* 8.

Seebens, H., M. Gastner, and B. Blasius. 2013. The risk of marine bioinvasion caused by global shipping. *Ecology Letters* 16:782–790.

Selge, S., A. Fischer, and R. van der Wal. 2011. Public and professional views on invasive non-native species—a qualitative social scientific investigation. *Biological Conservation* 144:3089–3097.

Shackelford, N., R. Hobbs, N. Heller, L. Hallett, and T. Seastedt. 2013. Finding a middle-ground: the native/non-native debate. *Biological Conservation* 158:55–62.

Simberloff, D. 2005. The politics of assessing risk for biological invasions: the USA as a case study. *Trends in Ecology and Evolution* 20:216–222.

Simberloff, D. 2011. Non-natives: 141 scientists object [letter to editor]. *Nature* 475:36.

Soulé, M. 1995. The social siege of nature. In *Reinventing Nature? Responses to Postmodern Deconstruction*, edited by M. Soulé and G. Lease, 137–170. Washington, DC: Island.

Theodoropoulos, D. 2003. *Invasion Biology: Critique of a Pseudoscience.* Blythe, CA: Avvar Books.

Tingley, R., B. Phillips, M. Letnic, G. Brown, R. Shine, and S. Baird. 2013. Identifying optimal barriers to halt the invasion of cane toads *Rhinella marina* in arid Australia. *Journal of Applied Ecology* 50:129–137.

Todd, K. 2001. *Tinkering with Eden: A Natural History of Exotics in America.* New York: W.W. Norton.

Welcomme, R. 1988. *International Introductions of Inland Aquatic Species.* FAO Fisheries Technical Paper No. 294. Rome: FAO.

Witmer, G., G. Keirn, N. Hawley, C. Martin, and J. Reaser. 2009. Human dimensions of invasive vertebrate species management. *Proceedings of Wildlife Damage Management Conferences.* http://digitalcommons.unl.edu/icwdm_wdmconf proc/141.

Worster, D. 1995. Nature and the disorder of history. In *Reinventing Nature? Responses to Postmodern Deconstruction*, edited by M. Soulé and G. Lease, 65–85. Washington, DC: Island.

Index

Page locators in *italics* refer to figures.